2005

GRASSLAND DYNAMICS

LONG-TERM ECOLOGICAL RESEARCH NETWORK SERIES
LTER Publications Commmittee

1. Grassland Dynamics: Long-Term Ecological Research in Tallgrass Prairie
Edited by
Alan K. Knapp, John M. Briggs, David C. Hartnett, and Scott L. Collins

GRASSLAND DYNAMICS

Long-Term Ecological Research in Tallgrass Prairie

Edited by

Alan K. Knapp
John M. Briggs
David C. Hartnett
Scott L. Collins

LTER

New York Oxford • Oxford University Press 1998

Oxford University Press

Oxford New York
Athens Auckland Bangkok Bogota Bombay
Buenos Aires Calcutta Cape Town Dar es Salaam
Delhi Florence Hong Kong Istanbul Karachi
Kuala Lumpur Madras Madrid Melbourne
Mexico City Nairobi Paris Singapore
Taipei Tokyo Toronto Warsaw

and associated companies in
Berlin Ibadan

Copyright © 1998 by Oxford University Press, Inc.

Published by Oxford University Press, Inc.
198 Madison Avenue, New York, New York 10016

Oxford is a registered trademark of Oxford University Press

All rights reserved. No part of this publication may be reproduced,
stored in a retrieval system, or transmitted, in any form or by any means,
electronic, mechanical, photocopying, recording, or otherwise,
without the prior permission of Oxford University Press.

Library of Congress Cataloging-in-Publication Data
Grassland dynamics : long-term ecological research in tallgrass
prairie / edited by Alan K. Knapp . . . [et al.].
p. cm.—(Long-term ecological research network series ; 1)
Includes bibliographical references and index.
ISBN 0-19-511486-8
1. Prairie ecology—Research—Kansas—Konza Prairie Research
Natural Area. 2. Konza Prairie Research Natural Area (Kan.)
I. Knapp, Alan K., 1956- . II. Series.
QH105.K3g73 1998
577.4'4'09781—dc21 978334

9 8 7 6 5 4 3 2 1

Printed in the United States of America
on acid-free paper

577.44
G769

We dedicate this volume to the memory of Dr. Lloyd C. Hulbert, whose tireless efforts led directly to the selection and establishment of the Konza Prairie Research Natural Area, and for his vision in the design and early implementation of the experimental design upon which much of the long-term research at Konza Prairie is based.

Preface

Grasslands are often thought of as simple ecosystems. This has led to generalizations that grasslands lack the complexity of, and hence are less interesting from an ecological perspective than, forest or desert biomes. Such generalizations are incorrect, especially in tallgrass prairies such as those located in the Flint Hills of Kansas. William Least Heat-Moon studied and then wrote about these grasslands in the book *PrairyErth* (1991) and concluded that "the prairies are nothing but grass as the sea is nothing but water" (p. 28). Indeed, instead of being structurally simple, monotonous, or uninteresting, tallgrass prairies include a mosaic of distinct herbaceous-dominated communities that vary from those resembling shortgrass prairie in dry topoedaphic locations to those resembling wetlands in lowland sites. The Konza Prairie and many other tallgrass prairies also are dissected by thin strips of deciduous forests that extend along prairie streams and represent some of the westernmost extensions of the forest biomes to the east. Overlaid on this mosaic of vegetative communities is a continental climate with extreme intra- and interannual variability, a frequent fire regime that is critical for the persistence of tallgrass prairie, and a variety of large and small herbivores. The net result is a fascinating and ecologically complex ecosystem that is dynamic temporally and spatially and in which both biotic and abiotic interactions are critically and equally important. Ironically, the lack of extensive vertical structure throughout most of the tallgrass prairie provides unique opportunities for ecologists to manipulate this complex ecosystem at a scale that is both appropriate and replicable.

The extreme temporal variability and the wide variety of key abiotic and biotic interactions in tallgrass prairie warrant a commitment to long-term ecological research if a predictive understanding of the system is the goal. We believe that the synthesis represented by this volume, which is a product of over 15 years of research focused on a fully replicated, watershed-level, fire and grazing experimental design at the Konza Prairie in northeastern Kansas, has brought us closer to this goal. But an assessment of our progress is needed. This synthesis represents such an assessment.

Our specific goals for this volume include providing the reader with a comprehensive site description of Konza Prairie and the tallgrass prairie region of the Flint Hills of Kansas; an overview and summary of some of the key long-term data sets that form the basis of the Konza Prairie Long-Term Ecological Research (LTER) Program; a synthesis within and across the core areas of research emphasis at Konza Prairie; and a foundation for newly initiated and future ecological studies of tallgrass prairie. Because there are no other tallgrass prairie sites with a similar fire/grazing experimental design or with comparable long-term data sets, this synthesis is biased decidedly toward Konza Prairie research. We offer no apologies for this, but we do recognize the limitations of such a focused synthesis.

The data that constitute the bulk of this volume would not have been possible without the support of the Nature Conservancy, and its past and ongoing role in the Konza Prairie LTER Program is gratefully acknowledged.

Major funding for research was provided by the National Science Foundation (NSF) LTER Program; several other programs within the NSF Division of Environmental Biology; the U.S. Geological Survey; the National Aeronautics and Space Administration; the U.S. Department of Agriculture; the U.S. Department of Energy; the Environmental Protection Agency; the Mellon Foundation; the State of Kansas, Kansas State University; the Kansas Agricultural Experiment Station; and the Division of Biology.

In addition, we acknowledge the support of Tom Callahan for his role in the NSF LTER Program; Jerry Franklin as the LTER network coordinator during the time most of the research was completed; prior senior principal investigators of the Konza Prairie LTER Program: G. Richard Marzolf, Donald W. Kaufman, and Timothy R. Seastedt; past Konza Prairie site directors Lloyd C. Hulbert, Donald W. Kaufman, Ted Barkley, and Jim Reichman; past and present faculty members of the Section of Ecology and Systematics at Kansas State University; and a large group of postdoctoral scholars, research associates/assistants, graduate students, and especially undergraduates, who as a group and under the motivational spell of Rosemary Ramundo have harvested, sorted, dried, and weighed aboveground plant biomass from an estimated 21,446 quadrats since 1981.

The preparation of this volume would not have been possible without the technical and logistical support of many people. In particular, we thank Eva Horne of the Division of Biology for her excellent assistance in all aspects of manuscript preparation; Eileen Schofield of the Kansas Agricultural Experiment Station for careful manuscript editing; Kirk Jensen of Oxford University Press for supervising the project; and the staff of the Sevilleta LTER site and Archbold field station, where editorial retreats for this volume were held. Photographs were contributed by Craig Freeman, Mickey Ransom, and Bob Robel. Finally, we are grateful to several of our colleagues for providing reviews of many of the chapters that constitute this synthesis. These reviewers include Tom Bragg, University of Nebraska at Omaha; Dave Coleman, University of Georgia; Doug Frank, Syracuse University; Marty Gurtz, U.S. Geological Survey; Tom Jurik, Iowa State University; and Osvaldo Sala, University of Buenos Aires.

Data and supporting documentation for the research included in this volume and additional information about the Konza Prairie LTER Program can be found at http://climate.konza.ksu.edu.

Manhattan, Kansas
December 1996

A. K. K.
J. M. B.
D. C. H.
S. L. C.

Contents

V. Toward the Future

Contributors

John M. Blair
Division of Biology
Kansas State University
Manhattan, KS 66506

John M. Briggs
Division of Biology
Kansas State University
Manhattan, KS 66506

Scott L. Collins
Ecological Studies Program
RM 635
National Science Foundation
Arlington, VA 22230

Walter K. Dodds
Division of Biology
Kansas State University
Manhattan, KS 66506

Edward W. Evans
Department of Biology
Utah State University
Logan, UT 84322

Philip A. Fay
Division of Biology
Kansas State University
Manhattan, KS 66506

Craig C. Freeman
R. L. McGregor Herbarium
University of Kansas
2045 Constant Avenue
Lawrence, KS 66047

Lawrence J. Gray
Department of Biology
Ottawa University
Ottawa, KS 66067

David C. Hartnett
Division of Biology
Kansas State University
Manhattan, KS 66506

Bruce P. Hayden
101 Clark Hall
Environmental Science Department
University of Virginia
Charlottesville, VA 22903

Geoffrey M. Henebry
Department of Biological Sciences
Smith Hall 135
Rutgers University
101 Warren Street
Newark, NJ 07102

Donald W. Kaufman
Division of Biology
Kansas State University
Manhattan, KS 66506

Glennis A. Kaufman
Division of Biology
Kansas State University
Manhattan, KS 66506

Alan K. Knapp
Division of Biology
Kansas State University
Manhattan, KS 66506

James K. Koelliker
Department of Civil Engineering
Kansas State University
Manhattan, KS 66506

Gwendolyn L. Macpherson
Department of Geology
University of Kansas
Lawrence, KS 66045

M. Duane Nellis
Eberly College of Arts and Sciences
201 Woodburn Hall
West Virginia University
Morgantown, WV 26506

Charles G. (Jack) Oviatt
Department of Geology
Kansas State University
Manhattan, KS 66506

Clenton E. Owensby
Department of Agronomy
Kansas State University
Manhattan, KS 66506

Rosemary A. Ramundo
Division of Biology
Kansas State University
Manhattan, KS 66506

Michel D. Ransom
Department of Agronomy
Kansas State University
Manhattan, KS 66506

Charles W. Rice
Department of Agronomy
Kansas State University
Manhattan, KS 66506

Timothy R. Seastedt
EPO Biology and Institute of Arctic and Alpine Research
University of Colorado
Boulder, CO 80309

Ernest M. Steinauer
Department of Botany/Microbiology
University of Oklahoma
Norman, OK 73019

Haiping Su
Argonne National Laboratory
EAD-900
9700 South Cass Ave.
Argonne, IL 60439

Timothy C. Todd
Department of Plant Pathology
Kansas State University
Manhattan, KS 66506

Clarence L. Turner
Minnesota Department of Natural Resources Office of Planning
500 Lafayette Road
Box 10
St. Paul, MN 55155

William A. Wehmueller
USDA–Natural Resources Conservation Service
Manhattan, KS 66502

Gail W. T. Wilson
Department of Plant Pathology
Kansas State University
Manhattan, KS 66506

John L. Zimmerman
Division of Biology
Kansas State University
Manhattan, KS 66506

GRASSLAND DYNAMICS

Introduction

Grasslands, Konza Prairie, and Long-Term Ecological Research

Alan K. Knapp
Timothy R. Seastedt

Tallgrass Prairie and Past Ecological Research

Grasslands are complex yet underappreciated ecosystems. Even though they occupy, or occupied, large areas of all continents except Antarctica (Walter 1985), the development of the science of terrestrial and ecosystem ecology was biased toward forests, with grasslands often considered less stable than forested communities (Clements 1936; Odum 1969; Vogl 1974). Grasslands ecosystems are notable for two characteristics: they have properties that readily allow for agricultural exploitation through the management of domesticated plants or herbivores and a climate that is quite variable both spatially and temporally (Borchert 1950; Frank and Inouye 1994; Chapter 2, this volume). Grassland climates can be described as wet or dry, hot or cold (typically in the same season) but on average are intermediate between the climates of deserts and forests. To the frustration of those dependent on (former) grassland sites for agricultural production, interannual climatic variability is extreme.

The tallgrass prairies of North America (Fig. 1.1) are temperate, mesic grasslands that lie between forested ecosystems to the east and north and less productive, mixed-grass and shortgrass prairies to the west and south. Great Plains grasslands occupy over 1.5 million km^2 of land area and are primary resources for livestock production in North America. Much of the grassland region is characterized by a temperature-moisture regime under which plants with different photosynthetic pathways (C_3, C_4), monocots and dicots, and angiosperms and gymnosperms can coexist in a tenuous relationship. A general lack of woody vegetation helps define grasslands, but precipitation and temperature alone cannot be controlling factors because both wetter and drier and warmer and cooler

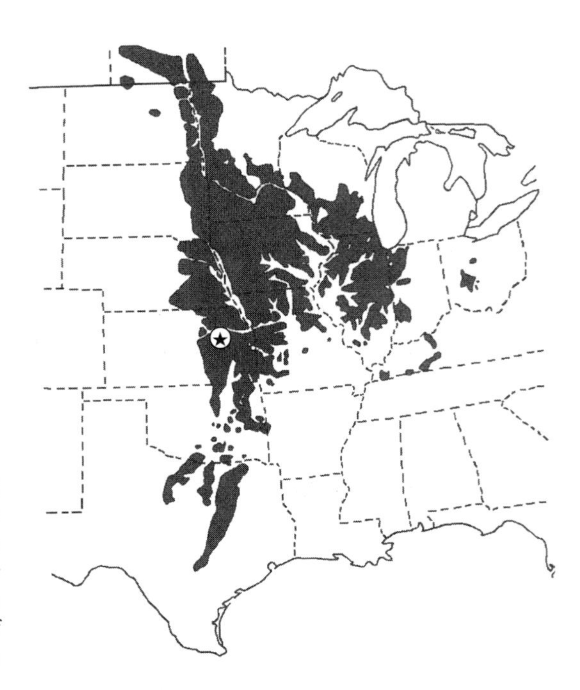

Figure 1.1. Estimated pre-European settlement extent of tallgrass prairie in North America (from Küchler 1974b). Star denotes the approximate location of Konza Prairie.

regions are dominated by woody plants (Fig. 1.2). The Flint Hills region of North America (Fig. 1.3) represents the largest contiguous area of unplowed tallgrass prairie (1.6 million ha) and, relative to other grasslands, is floristically diverse, with over 600 higher plant species recorded (Great Plains Flora Association 1986). In the tallgrass prairie in particular, precipitation is sufficient to allow dominance by trees, but frequent fire and perhaps chronic herbivory have been implicated as factors leading to a grass-dominated system (Wells 1970b; Bragg and Hulbert 1976; Axelrod 1985). Overall, grasslands originated and have persisted due to a complex interplay among a variable continental climate, biotic factors such as herbivory, and positive interactions such as that between the annual production of large quantities of flammable fuel and a relatively uniform physiognomy, both of which support frequent fire (Cowles 1928). The tallgrass prairie is no exception to these generalizations, and its dependence on fire has been well documented (Collins and Wallace 1990).

The factors that led to the establishment of tallgrass prairie and, in particular, the organic-rich soils derived from the dominant biota facilitated the agricultural exploitation of this grassland type. Consequently, much of the historically persistent tallgrass prairie has been converted to cropland or is used for other agricultural purposes (Samson and Knopf 1994). Indeed, this conversion continues today. As a result, the continued existence of the remaining tracts and remnants of tallgrass prairie has become more important as efforts intensify to conserve ecosystems and the species they contain (Samson and Knopf 1994). Coupled with the recent increase in research on prairie ecosystems, this leads to a greater

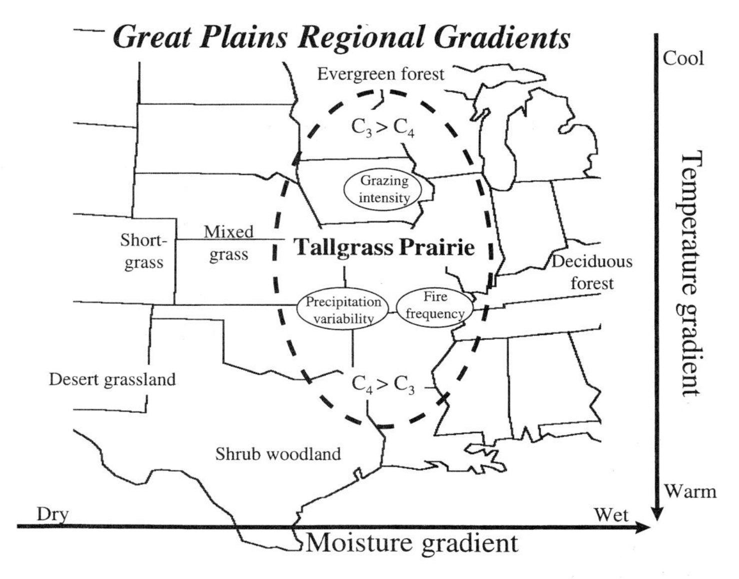

Figure 1.2. Grassland types in North America segregate along gradients of precipitation and temperature and are bordered by biomes dominated by woody vegetation. Tallgrass prairie has the highest grazing intensity and fire frequency of any grassland type, and precipitation variability may be as great as that in shortgrass prairie. West to east, NPP increases several-fold as precipitation increases (see Chapter 12, this volume). Although the temperature gradient is not as extreme as the precipitation gradient, dramatic shifts in photosynthetic physiology occur from north (C_3 dominance) to south (C_4 dominance).

urgency for an updated synthesis of information on the ecology of tallgrass prairie.

A substantial literature exists on individual species, species interactions, and community dynamics in tallgrass prairie, but studies focusing on the tallgrass prairie as a structural and functional unit are less common. Terms such as *forestation*, *deforestation*, and *desertification* are common in the ecological literature, but analogous terms associated with grasslands are lacking, with the exception of the desertification of desert grasslands (Schlesinger et al. 1990). One possible explanation for this is that ecologists have found it easier to define transitions from the perspective of end points—forests or deserts—rather than from the more tenuous (nonequilibrium) midpoints that many grasslands, and the tallgrass prairie in particular, represent. Interpretations of biological phenomena within tallgrass prairie, whether they are population-, community-, or ecosystem-level in scope, and syntheses across these levels require a nonequilibrium perspective that is only now being explored (Seastedt and Knapp 1993; Huston and De Angelis 1994; Lodge et al. 1994; Reice 1994). This perspective, and its utility for grasslands, is discussed more fully later in this chapter.

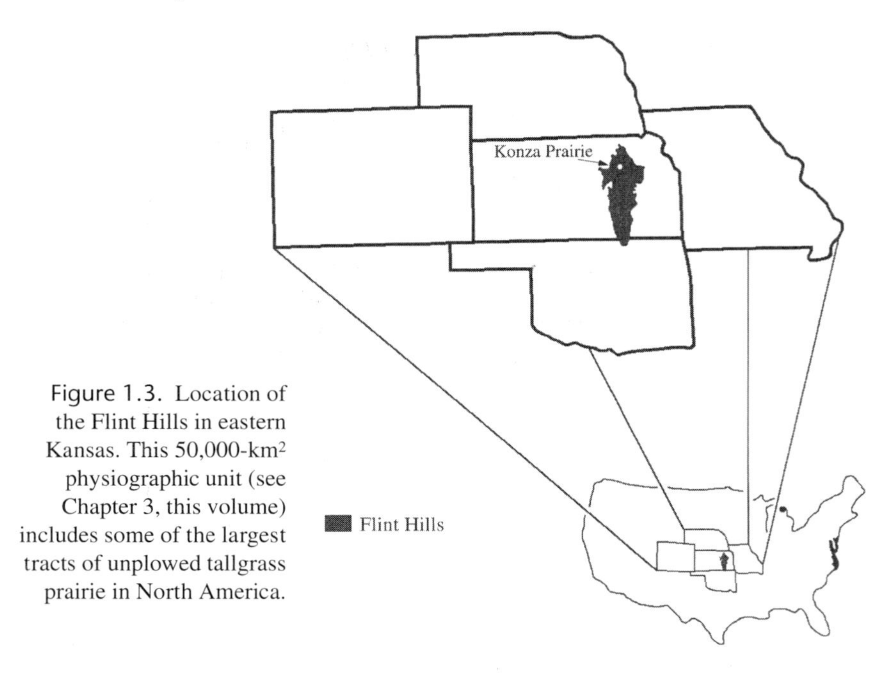

Figure 1.3. Location of the Flint Hills in eastern Kansas. This 50,000-km^2 physiographic unit (see Chapter 3, this volume) includes some of the largest tracts of unplowed tallgrass prairie in North America.

Syntheses of ecological research in tallgrass prairie have been undertaken at least twice: by John Weaver near the end of his career as the preeminent grassland ecologist of North America (Weaver 1954, 1968) and by Risser et al. (1981) at the conclusion of the International Biological Program (IBP). Weaver's synthesis was based on plant community composition and succession studies and was influenced strongly by the predominant land-use practice of grazing and the occurrence of droughts in the 1930s and 1950s. His synthesis encompassed 50 years of research at a wide array of sites. The IBP synthesis incorporated a broader, ecosystem perspective and also included studies from a number of sites. However, IBP studies usually were limited to documentation of seasonal patterns or 2- to 3-year studies. Although both of these efforts have been invaluable to grassland ecologists, the present volume represents a unique and complementary tallgrass prairie synthesis. The research synthesized here spans organismic through ecosystem perspectives designed to capture the mean, and possibly more important, the variance (Sullivan 1996) in key ecological patterns and processes that lead to the structure and function of a tallgrass prairie. This volume summarizes research based at a single site, the Konza Prairie Research Natural Area in northeastern Kansas. It represents the collective efforts of more than 50 scientists and 20 years of research, much of which is still ongoing. Clearly, this synthesis is not an end point but rather a reference junction from which we can gain insight from the past, as well as look to the future.

It is our contention that the research program under way at Konza Prairie is ideally suited for such syntheses because of the long time frame of study, the diversity of ecological patterns and processes studied, and the experimental nature of the research. Elaboration on these characteristics follows.

Figure 1.4. Dr. Lloyd C. Hulbert setting an experimental fire on Konza Prairie. (M. Gurtz)

The Konza Prairie Long-Term Ecological Research Program—Hulbert's Vision

Konza Prairie was conceived and established with a commitment to long-term ecological research. In an early description of the experimental design implemented at the site, Dr. Lloyd C. Hulbert expressed his conviction that the value of the area would increase "proportionately to the length of time the treatments are continued" (Hulbert 1973). Lloyd Hulbert was an ecologist with exceptional vision; he also possessed astonishing energy and determination (Fig. 1.4). Beginning in the 1950s, he led efforts at Kansas State University to secure a large, undisturbed tract of tallgrass prairie dedicated to ecological study. After minimum site criteria were established and a considerable number of obstacles were overcome, a 371-ha portion of what is now Konza Prairie was purchased in 1971 by The Nature Conservancy. This land subsequently was deeded to Kansas State University (Hulbert 1985). Initially envisioned as a site devoted primarily to the study of long-term effects of fire in ungrazed prairie, Konza Prairie grew through acquisitions of adjacent land during the next 8 years until it reached 3487 ha, or nine times its original size. This expansion allowed Hulbert and other scientists at Kansas State University to develop a more comprehensive experimental plan built on watershed-level manipulations of fire frequency and grazing activities by both native and domesticated large ungulates (Fig. 1.5). The original experimental plan (Hulbert 1985), with only minor modifications, remains central to the research efforts at the site today.

The National Science Foundation's Long-Term Ecological Research Program

Coincident with the land acquisitions for Konza Prairie, a series of workshops sponsored by the National Science Foundation (NSF) explored and developed

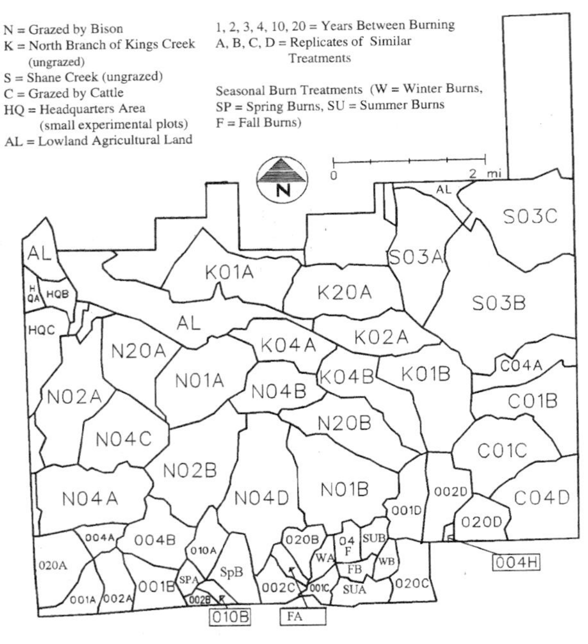

Figure 1.5. Experimental design of Konza Prairie. Watershed units, each designated by a code indicating the fire and grazing treatment, vary in size from approximately 3 to 200 ha.

the concept of a Long-Term Ecological Research (LTER) Program. Key features of this program were that it would be site-based and provide a relatively stable funding source for multiple years. This was in contrast to the more traditional, investigator-based and shorter-term funding mechanisms already in place at the NSF (Callahan 1984). The need for long-term ecological studies had been recognized for many years by ecologists (Likens 1983; Franklin 1989). The rationale for the LTER Program included explicit recognition of the low rate of change of many significant ecological processes, the importance and prevalence in most ecosystems of rare events and episodic phenomena with long return intervals, the tremendous interannual variability of many ecological processes, and the value of long-term databases for providing the context for shorter-term studies.

In 1980, Konza Prairie was selected as one of the initial six LTER sites funded by NSF. At present the LTER network includes 20 sites, all of which have research programs that are focused to various degrees on the following "core areas" of study: (1) pattern and control of primary production; (2) spatial and temporal distribution of populations selected to represent trophic structure; (3) pattern and control of organic matter accumulation in surface layers and sediments; (4) patterns of inorganic input and movements through soils, groundwater, and surface water; and (5) patterns and frequency of disturbance to the ecosystem (Callahan 1984). These core areas are not viewed as constraints to innovative research; rather, they provide an infrastructure and serve as points of common interest from which individual sites diverge to research programs appropriate for their system.

The union of Hulbert's experimental plan for Konza Prairie with the NSF LTER initiative has resulted in a statistically robust, ecological research program designed to elucidate patterns and processes important in tallgrass prairie. In contrast to a monitoring approach to long-term studies, the Konza Prairie LTER Program is an example of an experimental paradigm with a long time frame. As such, it contains what Krebs (1991) has argued are the "essential components of the twenty-first century" research program (p. 3).

Experimental Design and Rationale for the Konza Prairie LTER

The watershed-level experimental design for Konza Prairie (Fig. 1.5) was based on several underlying assumptions about the tallgrass prairie ecosystem. Chief among these is the role of fire. There is no doubt that fire has been an important force in tallgrass prairies, and many other grasslands, for millennia (Sauer 1950; Komarek 1968; Daubenmire 1968; Wright and Bailey 1982; Vogl 1974). Fire alters the energy, water, and nutrient relationships of the tallgrass prairie primarily through the removal of detritus and the subsequent alteration of the microclimate (Knapp and Seastedt 1986). However, the frequency of these fires, their seasonality and origin, and the spatial extent and uniformity of presettlement fires are poorly known. As large as Konza Prairie is, it does not provide sufficient land to experimentally assess all of these variables. Instead, fire frequency was selected as an ecologically important aspect of the historical regime and an aspect of fire easily amenable to manipulation. Thus, Konza Prairie today has replicate watersheds that have been burned annually for 20 years and other watersheds burned at 2-, 4-, 10-, and 20-year frequencies (Fig. 1.5). All watersheds include distinct topographic gradients, with both soil depth and type varying from uplands (maximum elevation of 444 m) to lowlands (320 m).

During the last 20 years, watersheds on Konza Prairie have been burned in the spring (April 10 $\pm$ 20 days depending on weather conditions). The timing of these experimental fires was selected by Hulbert (1973) for two reasons. Although the prairie will burn at any time of year (Bragg 1982; Howe 1994b), the probability of lightning ignition is greatest in autumn and spring, with spring having more frequent thunderstorms (Hulbert 1973). Moreover, most of the Flint Hills is burned in the late spring as an agricultural practice to increase the production of domestic livestock (Owensby and Anderson 1967; Anderson et al. 1970; Bragg 1995). A small number of watersheds are burned at other times of year (summer, fall, winter), but these treatments were not initiated until 1994.

A second major ecological factor influencing grasslands is herbivory (Mc-

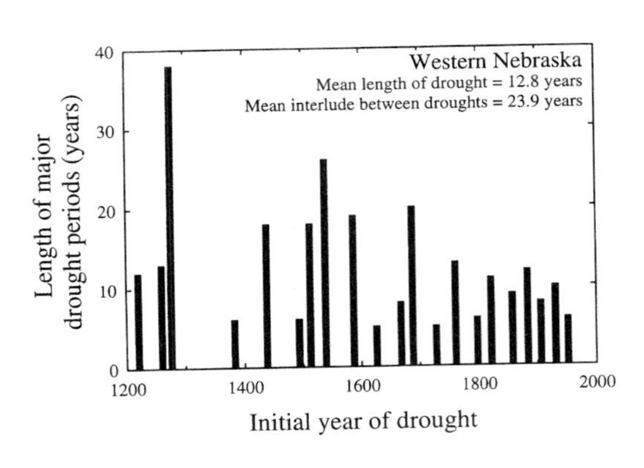

Figure 1.6. Long-term record of major droughts in the Great Plains. Estimates were made by Weakly (1962) using tree ring data. According to this analysis, the 10-year "dust-bowl" drought of the 1930s was slightly below average in duration.

Naughton 1985; Milchunas et al. 1988; Sala et al. 1988), and the tallgrass prairie is no exception. Herbivory is an important ecological factor both aboveground and belowground (Seastedt and Crossley 1984; Seastedt 1985b; Fahnestock and Knapp 1993; Vinton and Hartnett 1992; Evans and Seastedt 1995), with herbivores ranging in size from microscopic nematodes to large ungulates. Herbivory may mimic some of the effects of fire (Knapp and Seastedt 1986) by also influencing energy, water, and nutrient relationships. And strong interactions certainly exist between herbivory and fire (Vinton et al. 1993).

Of the herbivores most easily manipulated, large native ungulates (*Bos bison*) were selected as experimental agents for long-term study on Konza Prairie. Historically, *B. bison* was the prominent large herbivore in the tallgrass prairies until it was replaced by the domestic livestock of European settlers in the middle to late 1800s. *Bos bison* was reintroduced to Konza Prairie in 1987 (Fig. 1.5). An additional benefit of manipulating large native ungulates is the opportunity for comparative studies with domesticated ungulates (cattle). Such studies have been initiated and provide a critical link between the more basic and applied research programs in the region.

The third major factor influencing grasslands is a variable climate. Weaver's work in grasslands was influenced strongly by the great midwestern droughts of the 1930s and the 1950s. Although these droughts were not typical events in the short term, they probably were important ecological forces with long return intervals (Borchert 1950; Weakly 1962). Indeed, Weakly (1962) estimated that over the last 800 years, major droughts in the Midwest persisted for an average of 12 years with a 24-year return interval (Fig. 1.6). Years with flooding rainfall amounts and series of wet years are less well studied, but these also may critically influence aquatic and terrestrial systems by resetting trajectories of ecological patterns and processes.

Although the ecological importance of extreme climatic events should not be understated, recognition of the high degree of "background" variability in climate at Konza Prairie may be equally important for understanding the dynamics of this system. For example, interannual variability in precipitation generally decreases as precipitation increases. At Konza Prairie, interannual variability in annual precipitation is more similar to that of drier sites within the LTER net-

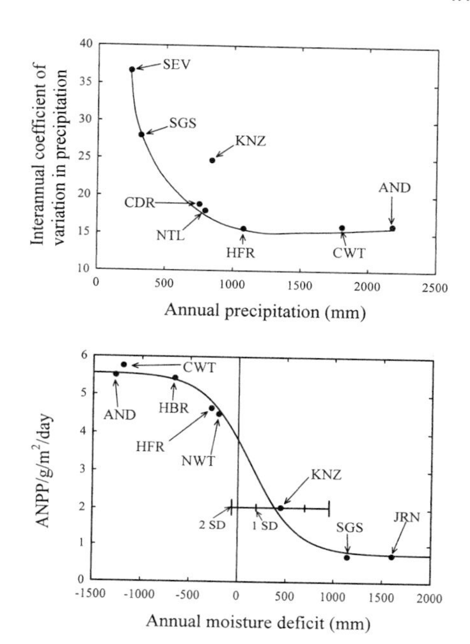

Figure 1.7. Top: The coefficient of variation (CV) of annual precipitation plotted versus mean annual precipitation for eight LTER sites in North America. The CV for Konza Prairie (KNZ) is distinctly greater relative to the trend for the other sites. Bottom: Relationship between aboveground net primary productivity (ANPP in $g/m^2/d$) and the average annual moisture deficit (potential evapotranspiration – precipitation) for eight LTER sites. Also shown for Konza Prairie is the interannual variability in moisture deficit (standard deviation, SD) for a 26-year period (1970–1995). Note that Konza Prairie falls on the steep portion of the curve and that two SDs encompass moisture deficits from shortgrass to forested sites. ANPP data are expressed on a growing-season basis (defined as the average number of days when minimum temperatures are greater than –5° C). Shallow lake evaporation was used as a surrogate for potential evapotranspiration (data from U.S. Army Corps of Engineers). Overall relationship developed by Dr. J. Koelliker, Department of Civil Engineering, Kansas State University. AND = H. J. Andrews Experimental Forest, OR; CDR = Cedar Creek Natural History Area, MN; CWT = Cowetta Hydrologic Laboratory, NC; HBR = Hubbard Brook Experimental Forest, NH; HFR = Harvard Forest, MA; JRN = Jornada Experimental Range, NM; NTL = North Temperate Lakes, WI; NWT = Niwot Ridge, CO; SEV = Sevilleta National Wildlife Refuge, NM; SGS = Shortgrass Steppe, CO.

work (Fig. 1.7). Moreover, when annual moisture deficits (potential evapotranspiration – precipitation) are compared across sites, Konza Prairie is intermediate between shortgrass/desert grasslands and forested sites (Fig. 1.7). However, the high degree of interannual variability in precipitation and temperature at Konza Prairie leads to high interannual variance in estimates of moisture deficits. When this variability is overlaid on the interbiome pattern of net primary productivity versus moisture deficit, productivity at Konza is predicted to vary from the low levels found in shortgrass steppes to the high levels of deciduous forests (Fig. 1.7). Long-term records of primary production support this prediction (Briggs and Knapp 1995; Chapter 12, this volume). Such a wide range in productivity does not occur in forested sites because of low variability in precipitation, whereas in more xeric sites, structural constraints of the dominant vegetation (meristem or growth rate limitations) limit production responses during years with high precipitation (or low moisture deficit). Thus, biotic responses to climate variability appear to be maximized in tallgrass prairie because the dominant vegetation can persist through drought periods in which water stress is comparable to that in desert biomes (Knapp 1984c), yet these grasses also can produce and maintain high leaf area during wet years.

Unlike fire or grazing, manipulation of climate is not a viable option for watershed-level research. Instead, the influence of climatic variability and extremes in climate on the system are evaluated primarily through long-term study. Nonetheless, studies of both water and CO_2 enrichment are under way at selected sites on and adjacent to Konza Prairie.

Hulbert's long-term experimental plan for Konza Prairie incorporates explicit study of all the major factors influencing tallgrass prairie (i.e., climate, fire, grazing). A simple additive view of these factors can be misleading, however, because there are emergent attributes of the prairie that are critical for understanding this complex system. One such attribute, which serves as the unifying theme for this volume, is the extreme spatial and temporal variability inherent in fire, grazing activities, the continental climate, and the interactions among these factors across edaphic gradients (Briggs and Knapp 1995; Schimel et al. 1991; Knapp et al. 1993a). Recognition of this variability and the potential for the system, or portions of it, to respond to the historical, as well as the extant, state of these factors is essential for the interpretation of research in tallgrass prairie. Moreover, the interactions among these factors, in concert with their variability, are key elements of many of the processes that give rise to the patterns and attributes that define this grassland (Collins 1987; Vinton et al. 1993; Seastedt and Knapp 1993). These defining characteristics of tallgrass prairie, derived from research at Konza Prairie, are listed in Table 1.1; each will be explored further in subsequent chapters.

Chronology of Konza Prairie LTER Studies

Implementation of the Konza Prairie experimental design (Fig. 1.5) and the LTER research emphasis occurred in phases corresponding with funding cycles. During LTER I (1981–1986), comparative investigations of biotic responses to fire and climatic variability were initiated. Long-term research sites and sampling protocols were established with an emphasis on studies of the extremes of annually burned versus unburned watersheds at upland versus lowland topographic positions. During LTER II (1986–1991), research efforts were expanded to include a wider range of fire frequencies, and LTER researchers began to address more complex issues of scale and make use of remotely sensed satellite

Table 1.1. Defining biotic and abiotic characteristics of tallgrass prairie as developed from LTER studies at Konza Prairie in northeastern Kansas

1. Tallgrass prairie is adapted to extreme variability in climate, fire, and grazing.
2. Tallgrass prairie is dominated primarily by perennial C_4 grasses. C_3 grasses, forbs, and shrubs are subdominant, particularly when fire is frequent. Plant reproduction is primarily vegetative.
3. Tallgrass prairie has a characteristic, temporally dynamic, and diverse assemblage of animals and annual and perennial plants. However, few species are endemic to this biome.
4. Tallgrass prairie is characterized by high rates of net primary productivity (NPP) both above- and belowground. However, interannual variability in NPP can be extreme, and water, nitrogen, and light each may limit NPP at various points in the landscape or at different times of the growing season.
5. Tallgrass prairie has significant allocation and storage of carbon and nutrients in belowground biomass and nonliving organic pools, especially when compared with forested systems.
6. Tallgrass prairie aquatic systems are characterized by ephemeral streams with permanent pools fed by oligotrophic groundwater. Headwater riparian areas are dominated by grasses, but lower reaches are dominated by woody plants.

data to explore landscape-level questions (Plates I, II). A number of focused plot-level experiments were established, many of which are ongoing (see the description of the Belowground Plot experiment, Chapter 14, this volume). In October 1987, 30 North American *B. bison* were released into a 469-ha portion of Konza Prairie, and the total area grazed by *B. bison* was expanded to 949 ha in 1992 as the size of the herd increased. Within LTER III (1991–1996), grazing by large ungulates was incorporated formally into the research program, and the landscape perspective was expanded to include sampling at a variety of topographic positions. As a result of the long-term research emphasis on fire and climatic variability versus the more recent incorporation of grazing by *B. bison* into the research program, the results presented in this volume will emphasize the longer-term data sets.

A Nonequilibrium Theme for Synthesis

Recently, Wu and Loucks (1995) reviewed the historical development of, and several current issues relative to, equilibrium versus nonequilibrium perspectives in ecology. Their argument, that the classical equilibrium view in ecology has failed to provide a useful predictive model of most natural systems because of the rarity of equilibrium conditions in nature, is certainly true for grasslands. In contrast, elements of the nonequilibrium models they reviewed—such as the focus on transient dynamics, stochastic processes, fluctuations in environmental variables as dominant processes, the important role of historical factors, nonlinear interactions, sensitivity to critical changes that cross threshold boundaries (bifurcation), and inherently low predictability (Chesson and Case 1986; DeAngelis et al. 1985; Wu and Loucks 1995)—are the features that make nonequilibrium models ideal for interpreting grassland dynamics. Interestingly, these elements, which may lead to wide fluctuations in measurable patterns and processes, do not preclude the long-term persistence of ecosystems that display these traits (Pimm 1991). This is consistent with the known 5- to 7-million-year history of grasslands in North America (Anderson 1990).

Patterns and processes in tallgrass prairie are best studied using a nonequilibrium paradigm. Tallgrass prairies are the wettest of the grasslands in the Great Plains and as such are not chronically limited by water availability, in contrast to many other grassland types (Briggs and Knapp 1995). Indeed, the tallgrass prairie is a unique ecosystem because it represents an area where water, energy, or nutrients may each be the primary limiting resource that constrains system response at any point in time or space. Our research indicates that all of these resources strongly interact to determine ecosystem structure and function (Seastedt et al. 1991; Schimel et al. 1991; Knapp et al. 1993a; Seastedt and Knapp 1993). Although other systems may undergo seasonal shifts in resource limitation, the degree of spatial and temporal unpredictability in this grassland (Frank and Inouye 1994; Hubbard 1994) contrasts with that in most other biomes, where single variables predominantly limit energy flow into and through the ecosystems.

Put another way, what truly makes the tallgrass prairie a challenging system to study is the high frequency of switching between limiting factors during and among years. Fire, herbivores, and climatic extremes all influence, and their effects are influenced by, this variation in supply and demand for essential resources. Hence, interactions and feedback among abiotic and biotic factors are strong, and these can either ameliorate or exacerbate the variation in biotic responses at hierarchical levels ranging from the single leaf to the landscape.

Thus, this extreme level of variance leads us to approach our synthesis from a decidedly nonequilibrium viewpoint, as defined by Wu and Loucks (1995). The need to understand and predict biotic responses to climate change has led to recent interest in nonequilibrium systems and biotic responses at a variety of scales (Sprugel 1991; Pastor and Post 1993; Smith and Shugart 1993; Huston 1994). Moreover, several models have been developed that assess the consequences of multiple-resource limitations (Chesson and Case 1986; Tilman 1988; Gleeson and Tilman 1992; Bloom et al. 1985; Chapin 1991), some with explicit recognition that grassland responses are best understood from a nonequilibrium perspective (Westoby et al. 1989; Seastedt and Knapp 1993; Gigon and Leutert 1996; Sullivan 1996). The underlying theme in our synthesis is predicated on recognition that the combination of climate, fire, and grazing in tallgrass prairie creates a stochastic environment of vastly fluctuating resource availability. Our research demonstrates that biotic responses to these fluctuations dominate individual, population, community, and ecosystem characteristics. Indeed, frequent alterations in essential resource availability and demand in tallgrass prairie may generate characteristic transient behaviors unknown from systems dominated by single-resource constraints (Seastedt and Knapp 1993; Huston and DeAngelis 1994). These nonequilibrium behaviors, in turn, may generate additional nonlinear system characteristics, all of which can vary across the landscape (Schimel et al. 1991; Knapp et al. 1993a; Benning and Seastedt 1995).

In addition to the stochastic variation and nonequilibrium dynamics of tallgrass prairie, the complexity of ecological interactions is another theme that emerges from long-term study. The apparent physical and physiognomic simplicity of the tallgrass prairie belies an enormous complexity of biotic and abiotic interactions and linkages. Fire regimes influence patterns of herbivory. Plant responses to fire are a result of both direct effects and indirect effects mediated through fire's influence on plant-insect and plant-fungal interactions. Ecosystem-level processes such as net primary production are mediated at the population level through the tillering dynamics of the dominant species. These examples illustrate the biotic complexity of the tallgrass prairie, where interactions among ecological processes are at least as significant as their independent effects, and the complexity of biotic interactions extends to multiple dimensions beyond trophic structure. Although our focus will be on the nonequilibrium nature of the system, that more traditional "equilibrium" ecological processes such as competition also operate and are important at various temporal and spatial scales within the ecological hierarchy (Bazzaz and Parrish 1982; Collins 1987).

Throughout the chapters that follow, our goal is to provide the reader with examples that reinforce these two themes, while incorporating a systems-level perspective that is essential to understanding the dynamics of this grassland ecosystem and its biota. Of equal importance is the predictive value a nonequilibrium paradigm has for this grassland as land-use pressures and potential climate changes alter the system. Such a perspective not only allows one to conceptualize the basis for variation in pattern and process in this system but also teaches us that ecological complexity is not dependent on topographic relief and, ultimately, engenders appreciation for the persistence of the remaining tracts of the tallgrass prairie ecosystem in North America.

Acknowledgments This is contribution no. 97-181-B from the Kansas Agricultural Experiment Station, Kansas State University, Manhattan.

Part I

Physical Environment

2

Regional Climate and the Distribution of Tallgrass Prairie

Bruce P. Hayden

The Konza Prairie lies near the southeastern margin of the North American prairie (Transeau 1935) at the latitude where grasslands exhibit extensive development (Fig. 2.1). The term *prairie* as defined by Transeau (1935) and used here includes communities that range from steppes to tallgrass prairies. To understand the climatic causes of prairies, a distinction between prairies and grasslands is needed. The term *grasslands* is reserved here as the more generic term, because many grasslands exist that generally are not called prairies. Whittaker (1975) and later McNaughton et al. (1982) characterized the extent of grasslands on a climate nomogram (Fig. 2.2) in which the x-axis is mean annual precipitation and the y-axis is mean annual temperature. Whittaker showed that grasslands occur over almost all mean annual thermal climates, excluding icebound landscapes, but only within a narrow range of mean annual precipitation. Even this delineation for grasslands is narrow. The Virginia Coast Reserve LTER site includes submerged marine grasslands (*Zostera*), emergent *Spartina alterniflora* grasslands, fresh-to-brackish wetland *Spartina patens* grasslands, mesic *Amophila,* and semiarid *Aristida* grasslands on the sandy barrier islands. These grasslands are within tens to hundreds of meters of each other and therefore have not been "established" or "caused" by large-scale climatic processes. None of these coastal grasslands could be called biomes. The prairie of North America meets the test of biome status and has been long associated with a climate suitable for prairie development and unsuitable for forests (Clements 1936; Barnes 1959; Coupland 1992). Konza Prairie, near the southeastern margin or ecotone of the North American prairie, although climatically established, requires fire and perhaps grazing to realize the climatic imperative and limit the encroachment of forests.

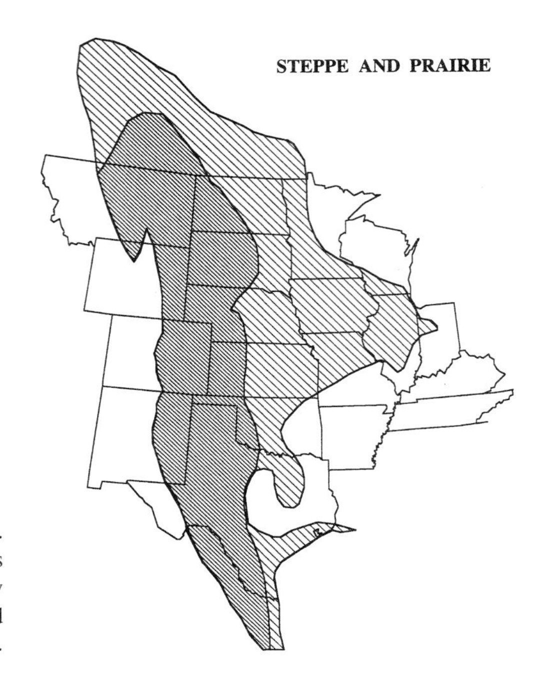

Figure 2.1. The Prairie Peninsula. The lightly shaded area is designated prairie, and the heavily shaded area is steppe. Adapted from Carpenter (1926).

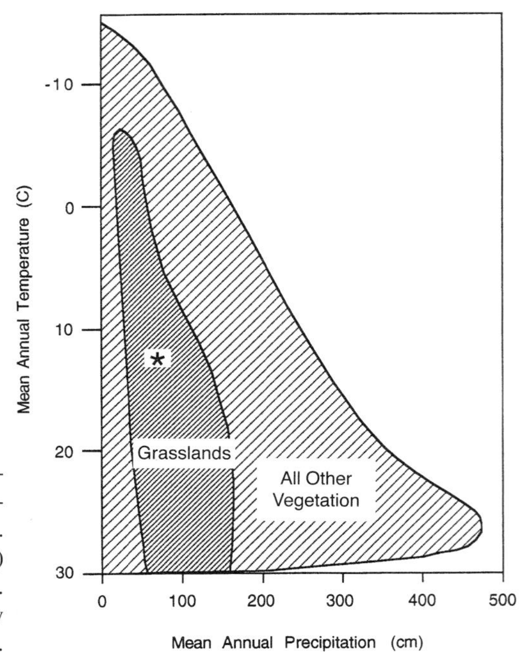

Figure 2.2. A temperature-precipitation nomogram for grasslands and other vegetation. Adapted from Whittaker (1975) and McNaughton et al. (1982). Konza Prairie is indicated by the asterisk.

The notion that grasslands persist in response to the prevailing climatic equilibrium is long-standing and often articulated (Clements 1936; Whittaker 1951, 1953). For example, Barnes (1959, pp. 244–245) noted, "The present distribution of natural grasslands, while not solely the result of the operation of natural forces . . . does nevertheless continue to reflect primarily climate." Hare (1951, p. 953) defined ecological climatology based on the concept that "vegetation and soils are mirrors of the normal climate" and that "soil and vegetation maps divided into regions ought therefore to give us a useful method of defining rational climate regions." Walter and Breckle (1985, p. 16) state the modern paradigm: "Climate is the sole primary factor which influences all others—the soil, the vegetation and, to a lesser extent, the fauna; it is in turn affected by these factors only at the level of the microclimate." Feedbacks from vegetation to climate are now recognized (Anthes 1984; Hayden 1994; Pielke and Avissar 1990), and there is little doubt that climate is a critical determinant of biome scale patterns of vegetation. Although the frequency and magnitude of fire are critical in many grassland and savanna regions and the designation of climax may be in question, this chapter will focus on the role of climate in the genesis of the North American prairies and the consequences of climate changes for these systems.

Rainfall and Temperature

The climate of a region commonly is characterized by its temperature and rainfall regimes (Fig. 2.3). Borchert (1950) contrasted the rainfall regime of the prairie with those of the deciduous forests of the U.S. Southeast and with the coniferous forests of Canada and the U.S. Northeast. He noted that the prairies of North America receive less than 200 mm of winter rainfall (November through March), with few heavy rainfall events. This is in marked contrast to the precipitation regime in the forested southeastern United States. In similar manner, average annual snowfalls on the prairie are generally under 800 mm (under 80 mm liquid water content). Higher snowfalls are found to the east and north of the prairie, i.e., in the coniferous forests of Canada and the mixed forests of the eastern United States. Konza Prairie gets 521 mm of snowfall each year (52 mm of liquid water). The prairies rely on summer precipitation in significant measure. At Konza Prairie, 75% of the annual average precipitation (835 mm) falls during the growing season and is highly variable from year to year (Fig 2.3).

For prairies, the difference in rainfall between the 10 driest and 10 wettest summers in a sample of 40 years (1899–1939) typically was larger than the average rainfall for the entire period (Borchert 1950). Much of the annual rainfall and most of the summer rainfall arises from thunderstorms associated with fronts and squall lines in the region. Across the prairies, thunderstorm frequencies (recorded as days with the sound of thunder) increase eastward from the foothills of the Rocky Mountains. Konza Prairie has about 45 days with the sound of thunder each year. With the advent of weather radar, new thunderstorm data sets have become available to detail the geographic variation in thunderstorm occurrence. The three highest radar echo return classes are associated

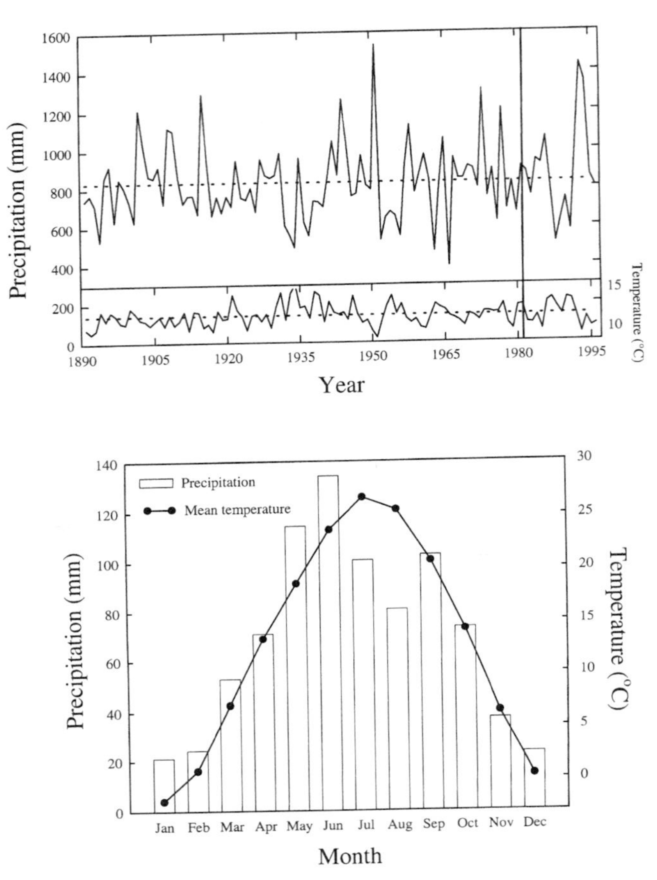

Figure 2.3. Top: Long-term record of annual precipitation and average annual air temperature for Manhattan, Kansas (12 km N of Konza Prairie). The horizontal dashed lines represent long-term mean values. The vertical line at 1981 denotes when the Konza Prairie LTER Program began. Bottom: Long-term average mean monthly precipitation and air temperature for Manhattan, Kansas.

with the intense rainfall rates (over 56 mm/hour) typical of thunderstorms. The annual number of hours that have such high rainfall rates for at least part of the hour recorded is shown for Kansas in Figure 2.4. Unlike thunderstorm days, based on sound of thunder, the radar data indicate substantial geographic pattern across Kansas. Konza Prairie and the Flint Hills area have high annual frequencies of intense-rainfall thunderstorms (250–300 hours per year).

Both Henry (1930) and Borchert (1950) noted the contiguous nature of drought across the midcontinent. This suggests that droughts have a regional cause rather than resulting from the chance occurrence of summer thunderstorms. Borchert (1950) noted that "when the summer rains and clouds fail, the mean temperatures soar." Rose (1936) found a general negative correlation between summer rainfall and high temperatures throughout the prairies. Cline

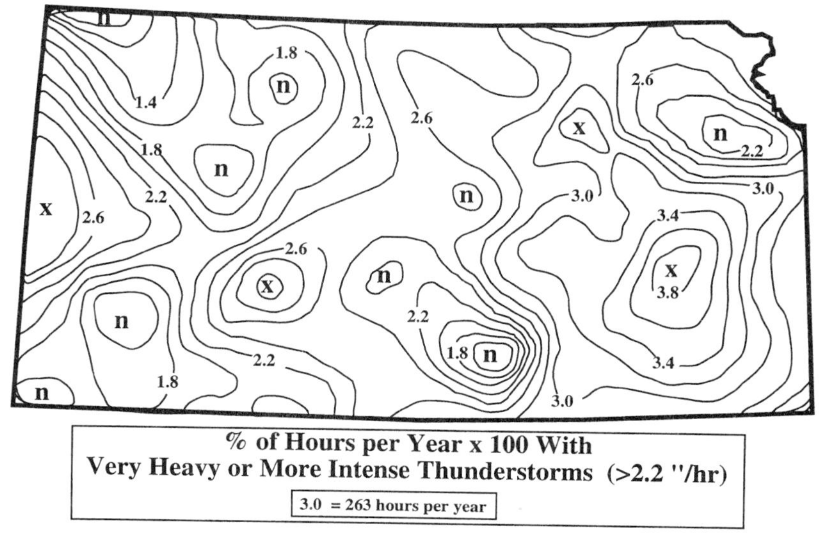

Figure 2.4. Annual Kansas thunderstorm frequency based on radar return echoes. Contours indicate hours within which rainfall rates over 56 mm/hr were recorded. The symbol "n" indicates a local minimum, and "x" indicates a local maximum of thunderstorm frequency.

(1894) attributed the high temperatures to persistent hot, dry winds from the west. Borchert (1950) also noted the coincidence of westerly airstreams during dry, hot summers. However, without advective heat from elsewhere, adequate soil water for evapotranspiration precludes temperatures much above 32°C. With inadequate soil moisture, heat dissipation from the surface occurs mostly through sensible heating of the atmosphere. Model studies of the heat waves that accompany drought in the prairie confirm the role of in situ heat production during periods of drought (Adem and Donn 1981).

Borchert (1950) summarized the common climatic attributes of prairie as (1) low winter snow and rainfall; (2) high probabilities of large rainfall deficits in summer; (3) fewer days of rainfall as compared with the forested regions to the north, south, and east; (4) low summer cloud cover; (5) low summer relative humidity; (6) large positive departures from average temperature; (7) frequent hot, dry winds in summer; and (8) frequent large departures from average climatic conditions. At a site, these weather elements and statistical attributes of climate arise from the occurrence of particular synoptic weather patterns. A characterization of these weather systems in an average or climatic sense then should serve as a proxy for the various weather elements that often are used to derive the actuarial climatic statistics for this region.

Natural Climatic Complexes of North America

In 1950, Borchert published the first genetic analysis and classification of the climate of the North American prairie. His work was stimulated by the observation that, during the great droughts of the 1930s, the frequency of subsiding air increased over the Rocky Mountains and subsequently spread out across the prairies and suppressed rainfall. This led him to begin a systematic analysis of the records of winds at weather stations east of the Rocky Mountains. In his analysis of North American wind fields, he found a single climatic parameter, the duration of air masses of Pacific Ocean origin, whose geographic pattern matched the geography of the North American prairie. He suggested that the essential determinant of the prairie was the degree to which aridity was realized by the dominance of westerly airstreams and the resulting precipitation suppression. He noted that a dominance of maritime tropical air masses occurred to the south and east of the prairie, and a dominance of continental polar air occurred to the north and east.

Borchert's analyses were expanded by Bryson (1966), Mitchell (1969), Bryson and Hare (1974), and Wendland and Bryson (1981). In his 1966 paper, Bryson showed that over much of North America east of the Rocky Mountains, the boundaries between air masses of various origins outlined the boundaries of the major biomes. Mitchell (1969) subsequently analyzed the frontal boundaries in the intermountain West. In the classification of natural climatic complexes of North America presented here, the Rocky Mountains and the Sierra-Cascades are treated as "stationary" fronts because the masses of air on either side are as thermodynamically distinct as fronts and, thus, equivalent to fronts in this classification. The resulting classification is genetic and does not require subjective levels of climatic variables to define regions or use vegetation. Coupland (1992) noted in regard to climate and vegetation classification that "it is hard to find a classification system for one of these that has not used data concerning the others in its formulation." The classification presented here is based on meteorological analyses of wind fields, streamlines, fronts, and the air masses described by Borchert (1950), Bryson (1966), and Mitchell (1969) and is totally independent of the biogeography of North America.

Figure 2.5 shows a composite map of all monthly mean frontal positions published by Bryson (1966) and Mitchell (1969). The Sierra Nevada and Cascade ranges and the Rocky Mountains also are shown. The regions between frontal zones have characteristic seasonal sequences of air masses (Table 2.1). Several types of air masses occur in North America. Air that flows out of Canada and into the United States is of two types: *continental polar air* (cP) and *Arctic air* (A). The cP air is formed within high-pressure systems residing over the high-latitude land masses, whereas Arctic air forms over the Arctic Ocean. Airstreams from the Pacific Ocean are called *maritime polar air* (mP) and are of two types: that from a southerly origin (mPs) and that from a more northerly origin (mPn). These two masses of air are modified thermodynamically as they pass inland and over the Sierra and Cascade ranges and later transit across the Rocky Mountains. On occasion, usually in winter, high-pressure systems reside in the

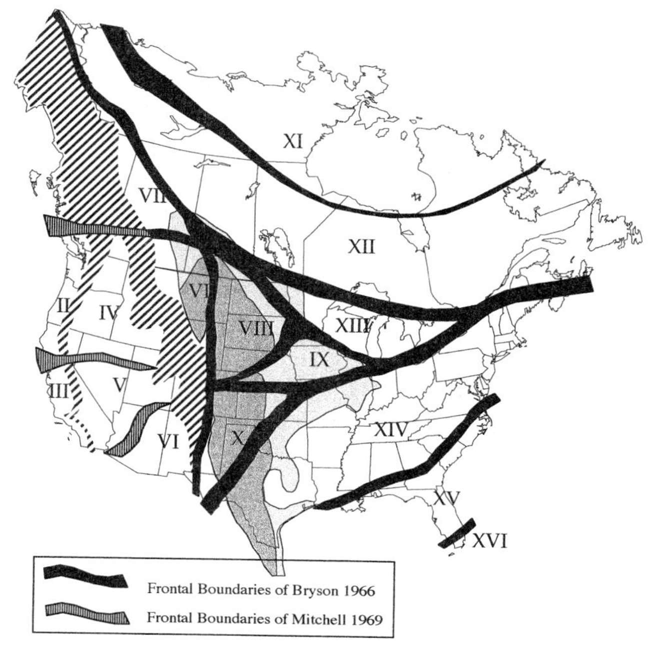

Figure 2.5. Natural climatic complexes of North America. Diagonal hatching indicates mountain ranges. Shaded areas are the same as in Figure 2.1. See Table 2.1 for additional information on the climatic complexes indicated by roman numerals.

Great Basin of the western United States, and a type of air called *Great Basin air* (GB) evolves there. In similar fashion, a high-pressure cell resides over the southeastern United States in fall and winter. Polar air sinks into these high-pressure systems and at the surface is recirculated poleward. This rather mild winter air mass is called *return polar air* (rP). Finally, warm, moist maritime tropical air (mT) from the subtropical oceans is common in the United States during the warmer months of the year. Because water temperatures are around 24°C, the air likewise is about this warm when it comes ashore. On transit across the land, it dries out via rainfall and warms up via sensible heating of the air by the land. The unmodified type of maritime tropical air is restricted largely to the coastal plains of the United States.

Fifteen natural climatic complexes, identified by roman numerals, are defined by the frontal boundaries and the mountain ranges (Fig. 2.5). Each area prescribed is defined here as a natural climatic complex and has a unique climatic signature. The types of air masses, their seasonal occurrence, and the nonmontane vegetation found in each natural climatic complex are summarized in Table 2.1.

Four of the natural climatic complexes defined by mean frontal boundaries and mountain ranges (VII, VIII, IX, and X) are associated with the prairies of

Table 2.1. Genetic classification of climate and vegetation

Region	Vegetation Type	Air Mass Dominance
I	Northern coastal coniferous forest	mPn in all months
II	Southern coastal coniferous forest	mPn in fall + winter + spring mPs in summer
III	Chaparral	mPs in all months
IV	Sagebrush grassland	mP*n in all months
V	Sagebrush-saltbush desert	mP*s in spring + summer + fall Great Basin air in winter
VI	Desert	mP*s in fall + spring mT in summer Great Basin air in winter
VII	Highland steppe	mP**n in all months
VIII	Northern steppe	mP**n in fall + early winter A+cP in later winter/spring mT in summer
IX	Corn belt	mT in summer mP**n in fall cP in winter
X	Southern steppe	mT in spring + summer mP**s in fall + winter
XI	Tundra	A+cP in all months
XII	Boreal forest	A+cP in fall + winter + spring mP*a in summer + early fall
XIII	Transition forest	A+cP in winter A in spring mT in summer + early fall
XIV	Eastern deciduous forest	Return polar air in winter mT in spring + summer + fall
XV	Pine/oak and pine savanna	Return polar air in winter Coastal zone mT in spring + summer + fall
XVI	Everglades	Coastal zone mT in all months

Note: T = tropical; P = polar; A = Arctic air; Rp = return polar air
m = maritime; c = continental
s = southern; n = northern; a = Alaska
* = modified by Sierra and Cascade Ranges
** = modified by Sierra and Cascade Ranges and the Rocky Mountains

North America (Fig. 2.5). Taken as a collective, they are coincident with the prairie. Region VII is a steppe area at higher elevations that intergrades into the foothills of the Rocky Mountains. These grasslands often are interspersed with sagebrush. In all months of the year, they are dominated by airstreams from the northern Pacific Ocean modified first by passage over the Cascade Mountains and then by transit across the Rocky Mountains. Maritime tropical air rarely reaches this region in summer. Although the airstreams within this region on average are from the west, excursions of continental polar and Arctic air do enter the region in winter and spring. Because they occur on fewer than half of the days in the month, they do not appear in mean statistics of wind flow. Seasonal rainfall maxima occur in winter from passing cyclones and largely as snowfall.

All other prairie regions are dominated by maritime tropical air in summer, and thunderstorms provide the dominant water supply for soil water recharge.

Natural climatic complex region VIII is the northern shortgrass prairie or steppe. This region is under the influence of modified air masses from the northern Pacific in fall and early winter (Table 2.1). Continental polar air and Arctic air dominate in winter and spring. In summer (June, July, and August), maritime tropical air from the Gulf of Mexico dominates the monthly statistics. Summer rainfall typically constitutes more than 60% of the annual rainfall.

Natural climatic complex region IX outlines what is commonly called the "corn belt." This region is characterized by two additional months of maritime tropical air, April and May, compared with prairies to the west. The maritime tropical air and thunderstorm season begins in April and lasts through September. This long season of thunderstorm rainfall supports the tallgrass prairie of the region.

Natural climatic complex region X, like the corn belt to the north and east, gets 6 months of maritime tropical air, April through September. Soil water replacement occurs during the growing season. However, unlike the prairies to the north, the prairies of region X receive mild airstreams of southerly Pacific air that pass around the southern end of the Sierra Nevadas and across Arizona and New Mexico (Fig. 2.5). Although excursions of cold polar air occur intermittently in winter, on the average, mild southwesterly airstreams dominate.

Biomes, Species Ranges, and Natural Climatic Complexes

Borchert (1950) and later Bryson (1966) and Mitchell (1969) made the case that the regions between frontal zones tended to have a characteristic vegetation and that the frontal zones themselves were coincident with the ecotones between biomes. The comparison between the natural climatic complexes and the associated frontal boundaries with vegetation cover is very encouraging but nonetheless a sample size of one. Krebs and Barry (1970) found that the frontal zone model worked well in Asia for the boreal forest-tundra ecotone. Testing at the species level will permit a more robust evaluation of the utility of natural climatic complexes as proxies for biomes.

A new test is offered here in the form of a multivariate analysis of floral and faunal ranges, the results of which then are compared with mean frontal boundaries and natural climatic complexes. The hypothesis is that assemblages of North American prairie species with approximately common ranges fit within the natural climatic complexes shown in Figure 2.5. For this analysis, species were selected using the criteria that their ranges must include the state of Kansas. A grid that included 49 cells was laid over a base map of the prairie. A transparent version of this gridded base map was laid over each species range map, and the presence of a species in at least 25% of the grid cells was recorded as a 1. Absence of species occurrence was scored as a 0. By this method, each range of maps was converted to an incidence matrix of 1s and 0s. Two separate data sets were developed from published range maps. Ranges of trees from Little (1971, 1977) and of selected pteridophytes and gymnosperms from the Editorial Committee of the Flora of North America (1993) were used to develop a data set of plant species. The resulting incidence matrix of range maps of plant species consisted of 49 grid cells by 101 species. An incidence matrix of fish, reptile, amphibian, and mammal range maps also was developed, using the same 49 grid cells with 155 species. The data used in this analysis were taken from the species range maps of Collins (1959). Both the fauna and flora incidence matrices were analyzed using varimax rotated principal components with grid cells as variables and species as cases. Incidence matrices have been studied in this way previously (Swan 1970; Williams 1978), and the most common problem, the "horseshoe effect," arises from the difficulty of dealing with the assemblages of absent species. This problem is most serious with incidence data along long, linear transects and usually is not serious where two-dimensional data are used. In addition, the use of the orthogonal varimax rotation further reduces this problem.

Flora and Fauna Species Assemblages

An analysis of the flora incidence matrix of 49 grid cells by 101 species analysis resulted in four principal components that passed the N-rule for statistical significance (Overland and Prisendorfer 1982). In total, 78.1% of the total variance in the data set was explained by the first four principal components (PC1, 33.9%; PC2, 16.1%; PC3, 12.5%; and PC4, 15.6%). Figure 2.6 shows component loadings derived from fields of the 49 grid cell component loadings for each of the first four components. Also shown is a schematic map of the locations of the major frontal boundaries in the same region taken from Figure 2.5. The species with factor scores with absolute values over 1.0 are listed for each of the four components in Table 2.2. This table provides a list of the member species for each of the species assemblages. Some species belong to more than one assemblage.

The first principal component (PC1, 33.9%) for flora is a northern grasslands assemblage of plants (Table 2.2). These species are found largely to the north of the winter position of the front that separates return polar air from the U.S. Southeast and maritime polar air from the west (see front B in Fig. 2.6). The contours of loadings are largely parallel to this front, with the highest loadings in the Dakotas. The contours also parallel the front that represents the northern extent of maritime tropical air in spring (April and May). This northern flora assemblage is centered in natural climatic complex VIII (Fig. 2.5 and Table 2.2). This region is under the influence of Arctic and continental polar air in late spring and has a reduced growing season in comparison with the corn belt immediately to the east. A steep gradient in the loadings of this component is present in Kansas and runs from southwest to northeast.

The second principal component (PC2, 16.1%) for flora is centered in Iowa, the corn belt, and represents a flora associated with the tallgrass prairies (Table 2.2). This assemblage of species occurs in natural climatic complex IX, the region between fronts B and C (Fig. 2.6). Front B is the northern extent of maritime tropical air in April and May. Spring comes early to this region. To the

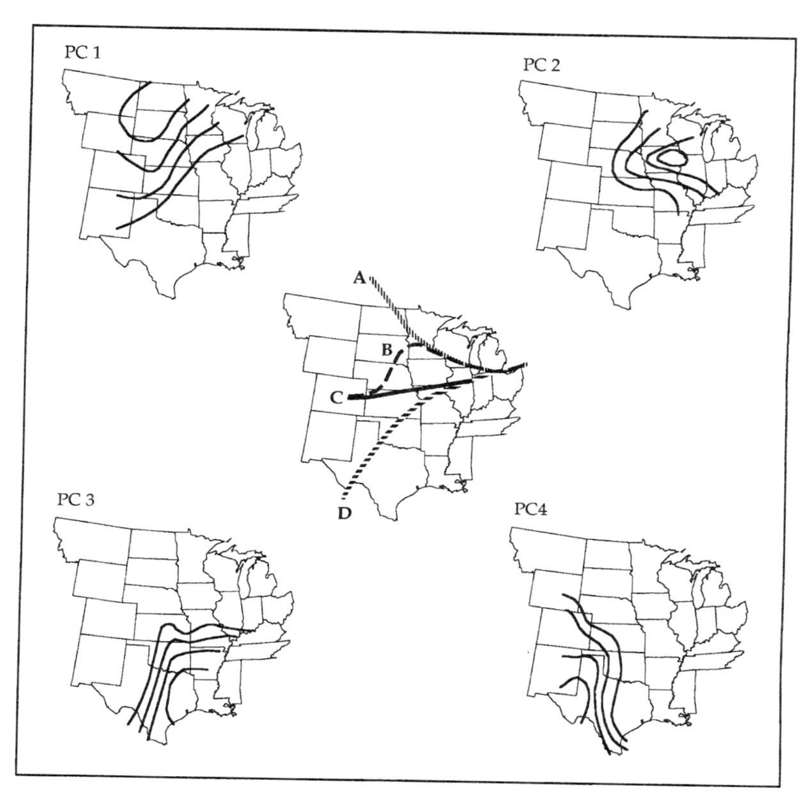

Figure 2.6. Maps of the component loadings for the first four principal components (PCs) of the incidence matrix of plant species ranges. Component loadings of 0.8, 0.6, 0.4, and 0.2 are included, with the highest loading at the center of each region indicated. In the center of the illustration is a schematic representation of the fronts shown in Figure 2.5. A is the early winter position of the polar front. B is the front at the northern limit of maritime tropical air in April and May. C is the front that separates maritime polar air north and maritime polar air south in winter. Front D is the northern limit of return polar air in fall and winter.

south of front C is either the winter maritime polar air of southern North Pacific origin or return polar air from the U.S. Southeast. Northeastern Kansas and Konza Prairie are within the contours of loading of PC2. Again, a steep southwest-to-northeast gradient across northeastern Kansas occurs in the loadings of this component. Because this gradient represents a zone in species composition, it might well be considered an ecotone.

The third principal component (PC3, 12.5%) for flora represents the western extension of the flora typical of the southeastern United States (Fig. 2.6) and is typical of natural climatic complex XIV. This region is bounded by front D in winter, with return polar air to the east and south and maritime polar air to the west and north. The contours of this component cross the southeastern portion of Kansas, indicating a gradient in species composition for this flora.

Table 2.2. Plants associated with the first four principal components of species ranges

Component 1	Component 2	Component 3	Component 4
Cystopteris fragilis	*Azolla caroliniana*	*Azolla caroliniana*	*Pellaea atropurpurea*
Phegopteris hexagonoptera	*Polystichum acrostichoides*	*Botrychium virginianum*	*Cheilanthis alabamensis*
Botrychium virginianum	*Woodsia obtusa*	*Isoetes melanopoda*	*Cheilanthis feei*
Equisetum hyemale	*Thelypteris palustris*	*Lycopodiella appressa*	*Ceilanthes tomentosa*
Equisetum ferrissii	*Adiantum pedatum*	*Ulmus rubra*	*Ophioglossum engelmannii*
Equisetum arvense	*Ulmus rubra*	*Ulmus americana*	*Equisetum hyemali*
Equisetum laevigatum	*Salix nigra*	*Quercus marilandica*	*Equisetum ferrissii*
Ulmus americana	*Quercus velutina*	*Platanus occidentalis*	*Equisetum laevigatum*
Salix amygdaloides	*Quercus rubra*	*Morus rubra*	*Selaginella underwoodii*
Quercus marocarpa	*Prunus serotina*	*Juglans nigra*	*Salix nigra*
Populus deltoides	*Ostrya virginiana*	*Gleditsia triacanthos*	*Morus rubra*
Fraxinus pennsylvanica	*Morus rubra*	*Carya tomentosa*	*Ptelea trifoliata*
Celtis occidentalis	*Juglans nigra*	*Acer negundo*	*Bumelia lanuginosa*
Acer negundo	*Gleditsia triacanthos*	*Sambucus canadensis*	
Salix exigua	*Fraxinus americana*	*Rhus glabra*	
Prunus virginiana	*Carya cordiformis*	*Rhus copallina*	
Prunus americana	*Acer saccharinum*	*Ptelea trifoliata*	
	Sambucus canadensis	*Prunus mexicana*	
		Cornus drummondii	
		Cephalanthus occidentalis	
		Cercis canadensis	

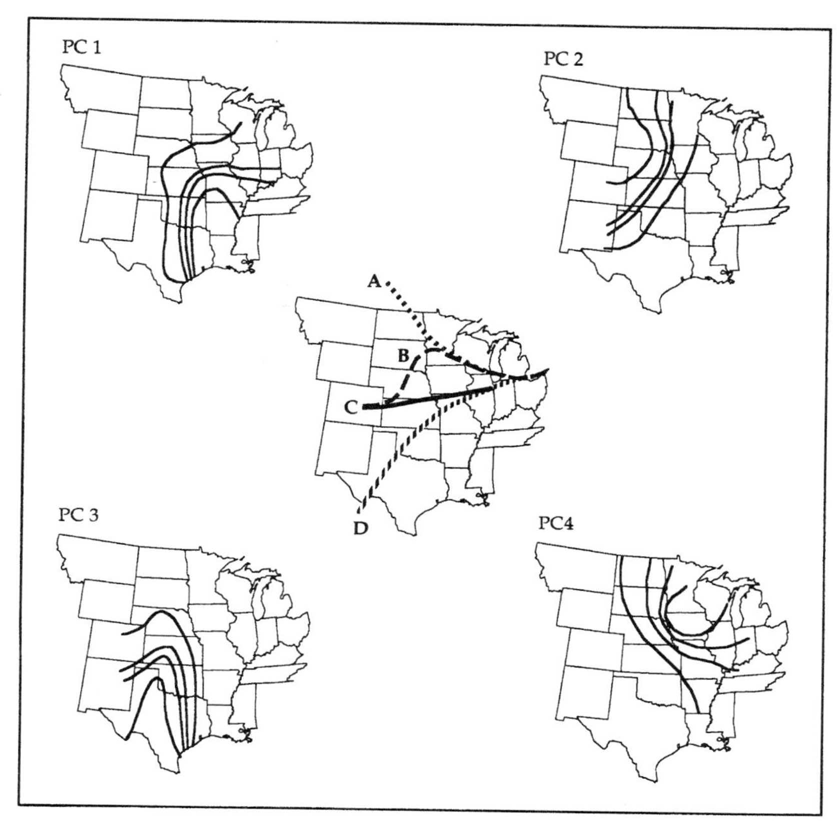

Figure 2.7. Maps of the component loadings for the first four principal components (PCs) of the incidence matrix of animal species ranges. Component loadings of 0.8, 0.6, 0.4, and 0.2 are included, with the highest loading at the center of each region indicated. In the center of the illustration is a schematic representation of the fronts shown in Figure 2.5. A is the early winter position of the polar front. B is the front at the northern limit of maritime tropical air in April and May. C is the front that separates maritime polar air north and maritime polar air south in winter. Front D is the northern limit of return polar air in fall and winter.

The fourth principal component (PC4, 15.6%) for flora is an assemblage of species with origins in the U.S. Southwest. The highest loadings on this component are in west Texas, southern New Mexico, and undoubtedly Mexico, i.e., between fronts C and D (Fig. 2.6). The contours of this component extend into western Kansas with an east-to-west gradient and are associated with natural climatic complex X.

The fauna incidence-matrix of 49 grid cells by 155 species analysis also resulted in four principal components that passed the N-rule for statistical significance (Overland and Prisendorfer 1982). In total, 73.2% of the total variance in the data set was explained by the first four principal components (PC1, 15.3%; PC2, 21.8%; PC3, 20.0%; and PC4, 16.0%). Figure 2.7 shows component load-

Table 2.3. Fauna associated with the first four principal components of species ranges

Component 1	Component 2	Component 3	Component 4
Tadarida molossa	*Urocyon cinereoargenteus*	*Phrynosoma douglassi*	*Phrynosoma douglassi*
Bufo compactilis	*Leptotyphlops dulcis*	*Rana areolata*	*Agkistrodon piscivorus*
Leptotyphlops dulcis	*Bufo compactilis*	*Macroclemys temmincki*	*Macroclemys temmincki*
Bufo punctatus	*Marmota monax*	*Lepus townsendii*	*Holbrookia maculata*
Citellus franklinii	*Carpiodes cyprinus*	*Zapus hudsonius*	*Eumeces laticeps*
Rhinocheilus lecontei	*Tamias striatus*	*Eumeces laticeps*	*Crotalus viridis*
Hypsiglena torquata	*Kinosternon subrubrum*	*Noturus flavus*	*Vulpes velox*
Citellus spilosoma	*Rhinichthys atratulus*	*Moxostoma carinatum*	*Sceloporus undulatus*
Hiodon terigisus	*Thamnophis elegans*	*Hyla crucifer*	*Dipodomys ordii*
Lepus townsendii		*Rana palustris*	*Reithrodontomys montanus*
Bufo debilis		*Diemictylus viridescens*	*Lygosoma laterale*
Taxidea taxus		*Rana pipiens*	*Terrapene carolina*
Perognathus flavus		*Chelydra serpentina*	*Perognathus flavus*
Bufo cognatus		*Ambystoma tigrinum*	*Gambusia affinis*
Perognathus flavescens		*Mephitis mephitis*	*Rana areolata*
Arizona elegans		*Mustela frenata*	*Perognathus hispidus*
Rana pipiens		*Rattus norvegicus*	*Masticophis flagellum*
Chelydra serpentina		*Canis latrans*	*Micropterus punctulatus*
Ambystoma tigrinum		*Procyon lotor*	*Natrix rhombifera*
Mephitis mephitis		*Castor canadensis*	*Microhyla carolinensis*
Mustela frenata		*Lasionycteris noctivagans*	*Trionyx muticus*
Rattus norvegicus		*Lasiurus cinereus*	*Scaphiopus bombifrons*
Canis latrans		*Eptesicus fuscus*	*Agkistrodon contortrix*
Procyon lotor		*Moxostoma aureolum*	
Castor canadensis		*Alosa chrysochloris*	
Lasionycteris noctivagans		*Agkistrodon piscivorus*	
Lasiurus cinereus		*Chrysemys picta*	

Eptesicus fuscus	*Synaptmys cooperi*
Sylvilagus audubonii	*Myotis lucifugus*
Onychomys leucogaster	*Ambystoma opacum*
Carpiodes cyprinus	*Ambystoma texanum*
Rhinichthys atratulus	
Stizostedion vitreum	
Vulpes velox	
Tantilla nigriceps	
Marmota monax	
Thamnophis marcianus	
Cyonomys ludovicianus	

ings derived from fields of the 49 grid cell component loadings for each of the first four components. Also shown is a schematic map of the locations of the major frontal boundaries in the same region taken from Figure 2.5. The species with factor scores over 1.0 are listed for each of the four components in Table 2.3, which also identifies assemblage membership by species.

Each of the four flora assemblage maps (Fig. 2.6) has a corresponding fauna assemblage map (Fig. 2.7) with essentially the same pattern of component loadings: flora PC1, flora PC2, flora PC3, and flora PC4 are equivalent to fauna PC2, fauna PC4, fauna PC1, and fauna PC3, respectively. As was the case with the flora assemblages, each fauna assemblage is represented by the maps of component loadings, and each component has steep gradients through Kansas.

In conclusion, the natural climatic complexes of the Prairie Peninsula are associated with specific floral and faunal assemblages of species. The congruence of this biogeography with independently defined climatic regions, and the connection between biota and climata as Bryson and Hare (1974) expressed it, have direct implications for the connection between climate change and changes in regional biodiversity. The geographic boundaries of these species assemblages all cross Kansas and indicate the ecotonal nature of the flora and fauna in the Konza Prairie region. Therefore, long-term study of the fauna and flora at Konza Prairie may serve as a "bellwether" for environmental change (Chapter 16, this volume).

3

Geomorphology of Konza Prairie

Charles G. (Jack) Oviatt

Konza Prairie lies in the Flint Hills of eastern Kansas, a region of stream-dissected hills eroded from flat-lying to gently dipping, chert-bearing limestones and shales of Permian age. Konza Prairie is part of an erosional landscape produced during millions of years of exposure to weathering and to stripping by streams tributary to the Kansas River. The major drainage basins on Konza are Kings Creek, Shane Creek, Pressee Branch, Swede Creek (tributaries of McDowell Creek), and Deep Creek (Fig. 3.1). Also prominent is the bench and slope topography in the uplands created by the erosion of contrasting bedrock units—resistant limestone layers alternate with less-resistant mudstone layers in a layer-cake pattern (Fig. 3.2). Because the landscape is undergoing long-term erosion, surficial deposits are generally thin and relatively young, and they accumulate in temporary storage sites, such as ridgetops and valley bottoms, before being transported out of the system to the Kansas River.

Previous work on the geology and geomorphology of Konza Prairie area has been mostly of a reconnaissance nature. The first systematic study of the geology of Riley and Geary Counties was by Jewett (1941), who mapped the geology at a scale of 1:126,720. Beck (1949) mapped the Quaternary deposits of Riley County, and Byrne et al. (1949) mapped the bedrock and Quaternary deposits at a scale of 1:31,680. Frye (1955) reviewed the sketchy history of erosion of the Flint Hills, and Mudge (1955) outlined drainage changes in northeastern Kansas related to glaciation. The history of glaciation in Kansas was summarized by Aber (1991). On Konza Prairie itself, Smith (1991) mapped the bedrock and Quaternary deposits at a scale of 1:10,150 and studied the geomorphology and geomorphic history of Kings and Shane Creeks. Ross (1995) mapped the geology and geomorphology of the N04D watershed (Chapter 1, this volume) at a

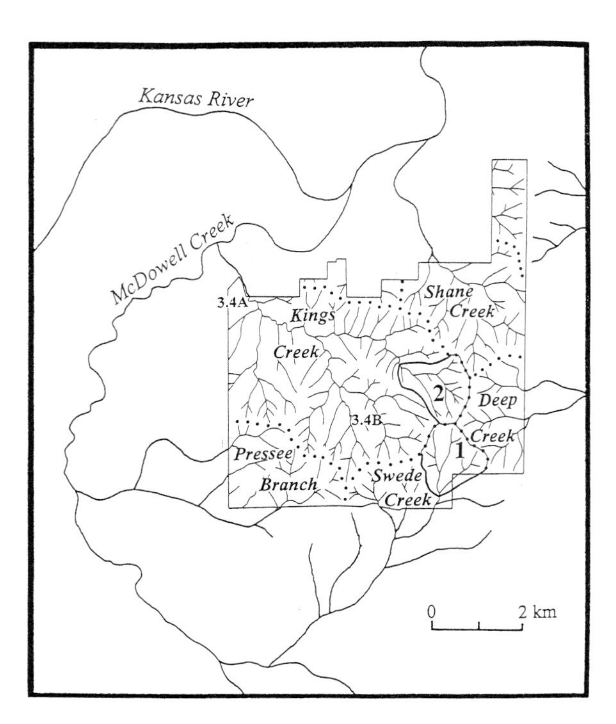

Figure 3.1. Drainage networks on Konza Prairie. Dotted lines represent major drainage divides. Drainage basins 1 and 2 are referred to in the text and in Table 3.1. Numbers 3.4A and 3.4B indicate approximate locations of composite cross sections show in Figure 3.4.

scale of 1:2,000. Much of the material in this chapter is summarized from these previous studies.

Bedrock Geology

The bedrock has a direct influence on the landforms and surficial processes on Konza Prairie. Limestone and mudstone (or shale) of Permian age (Jewett 1941; Miller and West 1993) form the backbone of the Flint Hills on Konza Prairie (Fig. 3.2), although the rocks generally are poorly exposed except in roadcuts or stream cutbanks. The Permian rocks comprise the Council Grove Group and the lower part of the Chase Group (Fig. 3.2). Limestone units form benches, and shale (mudstone) units form slopes, thus giving rise to the "terraced" topography of the Konza and the Flint Hills. Each limestone unit has a distinctive character that, in combination with its position in the stratigraphic sequence with the others, allows it to be recognized in the field (Jewett 1941; Smith 1991). No faults or folds occur in the bedrock, and the only structures are a slight dip (0.19°) to the west-northwest and a system of vertical joints or fractures in two orthogonal sets, N 25–35°W and N 55–65°E. The slight dip and the joints are important in considerations in the groundwater flow on Konza Prairie.

The Florence Limestone Member of the Barneston Limestone underlies the highest hills on the Konza Prairie and contains a large volume of chert that

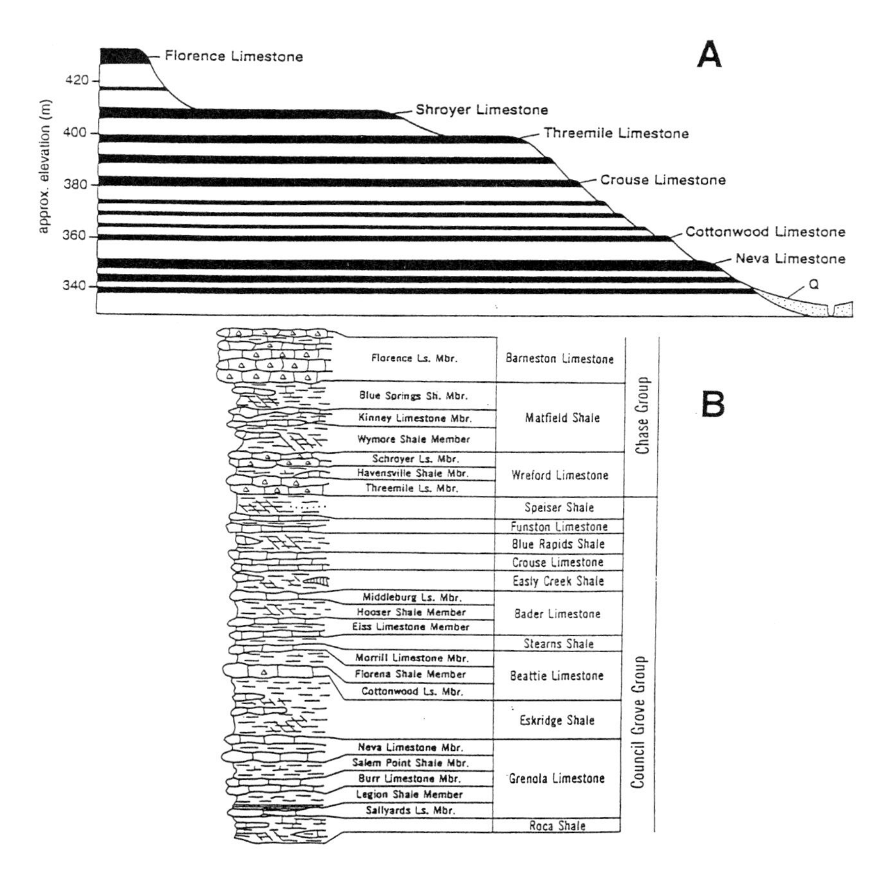

Figure 3.2. Bedrock units on Konza Prairie. (A) Schematic cross section through a hill on Konza Prairie showing bedrock units and their typical topographic expression (modified from Figure 4 of Smith 1991). Q marks Quaternary valley-fill alluvium and colluvium. (B) Stratigraphic column showing Permian rocks on Konza Prairie (from Zeller 1968).

weathers out to mantle ridgetops and slopes. It is the youngest bedrock unit on the Konza and forms the drainage divide between Kings Creek and streams in the southern and southeastern parts of Konza. Two older limestone units, the Shroyer Limestone and Threemile Limestone Members of the Wreford Limestone, also contain significant amounts of chert. The Shroyer Limestone forms a broad bench below the Florence Limestone, including the drainage divide between Kings Creek and Shane Creek.

The stratigraphically and topographically lowest bedrock unit is the Grenola Limestone, although only the upper part of this unit, the Neva Limestone Member, is well exposed on the Konza. The Neva Limestone forms a prominent shrub- and tree-covered shoulder at the base of the hills in the middle and lower parts of Kings Creek and Shane Creek valleys. A number of springs issue from the Neva in the valley bottoms.

Table 3.1. Morphometry of two drainage basins on Konza Prairie

	Basin 1 (Low-Relief)	Basin 2 (High-Relief)
Area	1.66 km^2	1.65 km^2
Maximum elevation	444 m	444 m
Minimum elevation	402 m	373 m
Relief	42 m	71 m
Bifurcation ratio[a]	3	4
Total stream length	9.6 km	11.8 km
Drainage density	5.8 km/km^2	7.2 km/km^2

Note: See Fig. 3.1 for locations. Basin 1 is small subbasin of the Swede Creek drainage basin (including Konza watersheds 001D, 002D, and 020D; Fig. 1.5). Basin 2 is a small subbasin of the Kings Creek drainage basin (watersheds K01B and K04B; Fig. 1.5).

[a]Bifurcation ratio is the ratio of the number of streams of a given order to the number of streams of the next higher order.

Prominent bedrock units between the Neva and Threemile include the Cottonwood Limestone Member of the Beattie Limestone and the Crouse Limestone. The Cottonwood forms a conspicuous narrow bench colonized by numerous shrubs that take advantage of the groundwater that discharges from the jointed limestone. Although the topographic bench formed by the Crouse Limestone is subtle in many places, its limestone and dolomite weather to distinctive flat plates, and a number of springs issue from the Crouse in valley bottoms.

Geomorphology

Landforms

The dendritic drainage pattern of streams on Konza Prairie illustrates the strong fluvial influence in the shaping of the landscape (Fig. 3.1). The valley morphometry of streams that flow more or less directly to the Kansas River (Kings Creek and Shane Creek) contrasts with those that follow much longer paths and, therefore, gentler gradients to the Kansas River (Pressee Branch, Swede Creek, and Deep Creek) (Fig. 3.1; Table 3.1). In the Kings and Shane Creek basins (referred to in the following as high-relief drainage basins), first-order streams have steep gradients (Table 3.1), especially where they head on the Florence Limestone, and first-order valley cross sections are V-shaped to rounded or bowl-shaped. Second-order stream valleys are generally V-shaped and have narrow floors, but the valleys of third- and fourth-order streams open up considerably and have flat floors. Kings Creek and Shane Creek are both fifth-order streams where they exit Konza.

In contrast to the steep drainage basins of Kings and Shane Creeks, the headward parts of Pressee Branch, Swede Creek, and Deep Creek have lower gradients, and even the first- and second-order valleys are more open and flat-floored. In addition, valley-fill deposits are thicker and more widespread in these drainage basins. However, in both the southern low-relief drainages and the northern high-relief drainages, the first- and second-order streams are entrenched in the Quaternary valley-fill deposits, suggesting a common extrinsic cause for the entrenchment (possibly cattle grazing during historic time or climate change).

In an attempt to test the hypothesis that bedrock jointing determines the orientation of stream valleys, Smith (1991) measured the orientations of 353 first-order streams in the low-relief basins and 486 first-order streams in the high-relief basins. His reasoning was that because first-order streams flow across and are eroding directly into the bedrock, they are the most likely parts of the drainage system to show a connection between erosional weaknesses in the bedrock (joints) and the streams. His data suggest that no preferred orientation of first-order streams exists in either the high-relief or low-relief basins on Konza Prairie, and that the net effects of joints on stream orientation are insufficient to cause significant deviation from the randomly oriented dendritic stream network.

Drainage densities (stream length/drainage-basin area) are 5.8 km/km^2 in a typical low-relief basin on the Konza and 7.2 km/km^2 in a typical high relief basin (Table 3.1), values that are within the expected range of dendritic stream systems that drain nonresistant, flat-lying sedimentary rocks in humid-temperate regions of the central and eastern United States (Strahler 1964). Drainage density is a measure of the relationship between climate and the geology of a watershed; it can range from as low as about 2 km/km^2 in a humid climate with permeable bedrock and dense vegetation to greater than 800 km/km^2 where soils are impermeable and vegetation is sparse. The relatively low values measured on the Konza are typical of terrains in which a significant proportion of the precipitation that might otherwise run off quickly as overland flow and in channels infiltrates into the soil and bedrock or is evaporated or transpired quickly after a storm.

Morphometric analyses of drainage basins on the Konza (Fig. 3.3) indicate that the stream network is generally in equilibrium with the underlying geology. Bifurcation ratios of 3 and 4 in basins 1 and 2, respectively (Table 3.1), are within the normal range of values of drainage basins where the geologic structures do not have a strong influence on the drainage pattern (Strahler 1964). A hypsometric analysis of the N04D watershed (a tributary of Kings Creek) yielded a hypsometric integral of 60% (Ross 1995), suggesting that the basin is transitional between an inequilibrium (youthful) stage and an equilibrium (mature) stage of development (Strahler 1952). The morphometric results indicate that the Konza Prairie landscape was formed dominantly by fluvial processes over a significant interval of geologic time, and that inputs and outputs to the geomorphic system are nearly in equilibrium. Geomorphic processes other than stream erosion must have had a relatively minor role in landscape evolution over the long term (millions of years). Shorter-term cycles of deposition and erosion, which are discussed in the following, are superimposed on the long-term trend of erosion and represent perturbations in the system caused by changes in some extrinsic factor, such as climate.

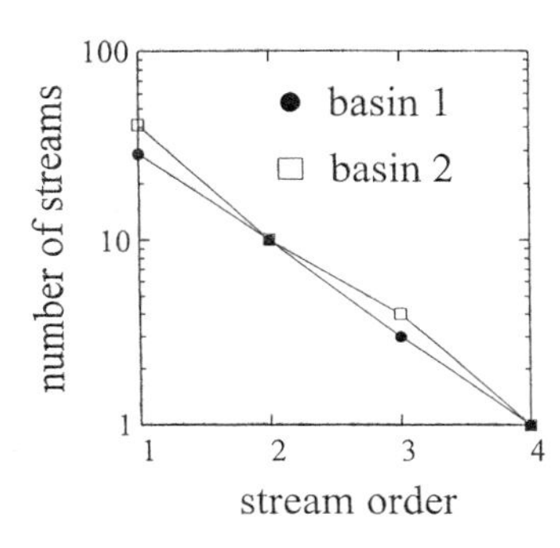

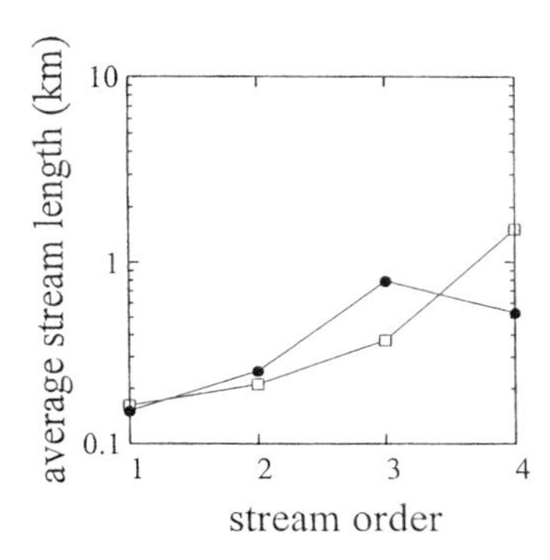

Figure 3.3. Plots of two morphometric parameters (number of streams and average stream length) versus stream order for basins 1 and 2 on Konza Prairie (Fig. 3.1). The approximate log-linear relationships indicate that the fluvial systems are in equilibrium; that is, long-term dynamic balance exists among all of the variables that constitute or affect the fluvial system, such as geology, climate, vegetation, tectonics, and land use on a long time scale. These plots illustrate two of Horton's (1945) laws of drainage-basin morphometry.

Geomorphic History

The details of the erosional history of the Flint Hills are poorly known because deposits of Tertiary or early Pleistocene age are scarce in the region, and few of them have been dated numerically. The following summary of Tertiary through early Pleistocene history is based on the work of Frye and Leonard (1952), Frye (1955), Aber (1988), and other sources. By the late Tertiary, streams flowing eastward from the Rocky Mountains had transported sand and gravel to a number of local basins and over a broad area in the western Great Plains region to deposit the stratigraphic unit referred to as the Ogallala Formation. The upper surface of the Ogallala, which is now dissected, was probably continuous with an erosional surface that extended across central and eastern Kansas and beveled the gently dipping Permian rocks, from which the Flint Hills later would be carved. As the climate became drier and as glacial cycles began in the late Pliocene and early Pleistocene (between 3 and 2 million years ago), major stream systems, such as the Kansas River, probably began downcutting below the level of the erosional surface, and their tributaries dissected the underlying rocks into hills and valleys. Regional uplift on a small scale also may have encouraged downcutting. By approximately 700,000 yr B.P., when the "Kansan" ice sheet reached its maximum southern limit near the Kansas River (Aber 1991), the Flint Hills region had attained a degree of topographic relief (i.e., dissection into hills and valleys) similar to the modern landscape (Dort 1987).

The southwestern limit of the "Kansan" ice sheet was 20 to 30 km northeast of the Konza Prairie (Aber 1991), and no alluvial or other deposits older than late Pleistocene in age have been identified on Konza. Therefore, glacial processes had no direct influence on the evolution of the Konza landscape. During "Kansan" glaciation, and possibly during the many other episodes of global glaciation when ice sheets advanced as far south as Nebraska and Iowa, northeastern Kansas probably was subjected to periglacial climates. Geomorphic processes associated with permafrost and extremely cold-dry conditions probably were repeatedly active on the Konza during the Pleistocene.

The oldest known surficial deposits on the Konza Prairie are windblown silt deposits (loess) on upland ridges, although they have not been dated directly. Soil genesis studies by Wehmueller and Ransom (Chapter 4, this volume) indicate that at least two loess stratigraphic units are present on the flat-topped Florence Limestone and Shroyer Limestone benches. The total thickness of loess is less than 1 m. The younger of the two loess units may be equivalent to the Peoria Loess, which has been dated elsewhere in the central United States between 25,000 and 12,000 yr B.P. (Forman et al. 1992). The older loess unit may be equivalent to the Loveland Loess, which probably represents more than one period of deposition, and has been dated at other localities between 135,000 and 70,000 yr B.P. (Forman et al. 1992). Both loess units on the Konza were deposited during periods of late Pleistocene glaciation, when the climate in Kansas was much drier and colder than it is today and eolian-dust deposition rates were greater. Smith et al. (1970) measured dust influx rates of about 50 kg/ha/month near Manhattan, Kansas during the 1960s. Dust deposition is a significant input of mass to the Konza geomorphic system.

The next younger surficial deposits on the Konza consist of colluvium (slope deposits) and alluvium (stream deposits) less than 10,000 years old. Colluvium ranges in thickness from a few centimeters on steep hillslopes to several meters in toe-slope positions. Fine-grained light brown alluvium and colluvium are widespread in valley bottoms of third- and higher-order streams in the high-relief drainages and throughout the basins of low-relief basins on the Konza. The light brown sediment is similar in grain size, color, and composition to loess but differs in its bedding and distribution in the valley bottoms. Therefore, it probably is reworked loess that has been washed off the hillsides, mixed with weathered bedrock debris, and accumulated in valley bottoms. The deposits form sloping terraces (T4) whose surfaces stand 1.5 to 6 m above modern stream channels (Fig. 3.4). Charcoal collected from this material in the lower part of Kings Creek has yielded a radiocarbon age of 8,920 ± 120 yr B.P. (Smith 1991; Table 3.2). Although it is not reasonable to assume that the light-brown alluvium is exactly the same age everywhere on the Konza, similar deposits are clearly the oldest valley-bottom deposits throughout the area and probably date from the early Holocene (the last 10,000 years of Earth history).

At least two alluvial terrace-fill deposits postdate the light-brown alluvium in the Kings Creek and Shane Creek valleys, although the deposits have been dated in only a few localities. The terrace-fill deposits generally are dark brown and relatively coarse-grained; in many places the alluvial fill contains lenses of cross-bedded chert gravel. Charcoal from the older of the two alluvial units, which underlies terrace T3 in lower Kings Creek valley, yielded a radiocarbon age of 1,770 ± 80 yr B.P. (Smith 1991; Table 3.2; Fig. 3.4). In the N04D drainage basin (a tributary of Kings Creek) a paleosol that developed on a terrace surface

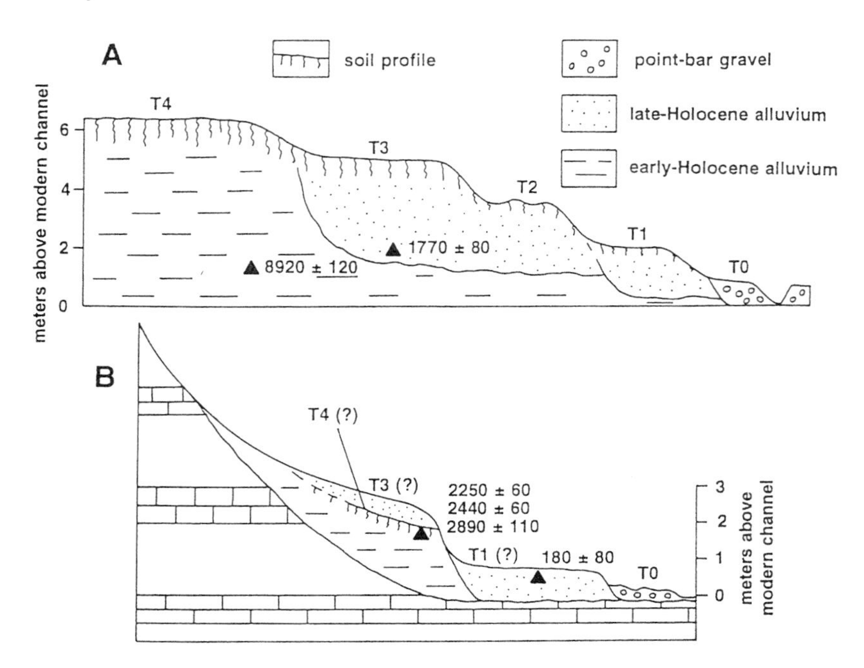

Figure 3.4. Schematic cross sections showing stratigraphic and geomorphic relationships between fluvial deposits and terraces in the Kings Creek drainage basin on Konza Prairie. No horizontal scale. (A) Composite from lower Kings Creek (modified from Fig. 18A of Smith 1991). (B) Composite from the fourth-order reach of watershed N04D. Approximate locations are shown in Figure 3.1. Terrace T0 consists of modern point-bar deposits; terraces T1, T3, and T4 are fill terraces; and terrace T2 is a cut terrace. Refer to Table 3.2 for information on the radiocarbon ages.

(T4 ?) on the older light-brown alluvium has been dated at between 2,900 yr B.P. (soil carbonate nodules) and 2,300 yr B.P. (soil humates), and is overlain by gravelly alluvium that forms the T3 (?) terrace (Fig. 3.4; Table 3.2). Although more data are needed to make reliable conclusions, the streams in the Kings Creek basin apparently began depositing alluvium after about 2,000 yr B.P. following a period of downcutting.

Fluvial landforms and deposits younger than the post-2,000-year T3 terrace include a cut terrace in lower Kings Creek (T2) and one or two fill-terrace deposits (T1) throughout the Kings Creek basin (Fig. 3.4). These features are poorly dated but suggest that the landscape is dynamic and that streams continue to cyclically downcut and aggrade their valley floors. Charcoal from alluvium beneath the T1 terrace in the N04D drainage basin has yielded a radiocarbon age of 180 ± 80 yr B.P. (Table 3.2; Fig. 3.4). The youngest alluvial deposits consist of point bars and narrow, thin terrace fills (T0) adjacent to the modern ephemeral channels. Almost all stream channels on the Konza Prairie are now entrenched, although short reaches of some second- and third-order streams are presently aggrading.

Table 3.2. Radiocarbon ages of samples from Konza Prairie

Age	Material Dated	$\delta^{13}C$ ‰ (PDB)	Adjusted Age[a]	Geologic Significance	Lab #	Reference
180 ± 80	Charcoal	—	—	Charcoal[b] from alluvium beneath 1-m terrace in N04D drainage basin	β-49073	Ross 1995
1,770 ± 80	Charcoal	—	—	Charcoal from alluvium beneath 5-m terrace in lower Kings Creek	β-29963	Smith 1991
2,120 ± 60	Paleosol humate	−16.9[c]	2,250 ± 60	A-horizon of a buried soil[d] developed in early Holocene alluvium in N04D drainage basin; buried by alluvium	β-47895	Ross 1995
2,320 ± 60	Paleosol humate	−17.4[c]	2,440 ± 60	A-horizon of a buried soil[d] developed in early Holocene alluvium in N04D drainage basin; buried by alluvium	β-50737	Ross 1995
2,890 ± 110	Paleosol carbonate	—	—	Calcium-carbonate nodules from the Bk horizon of a buried soil[d] developed in early Holocene alluvium in N04D drainage basin; buried by alluvium	I-17172	Ross 1995
8,920 ± 120	Charcoal	—	—	Charcoal from alluvium beneath 6-m terrace in lower Kings Creek	β-23396	Smith 1991

[a]Radiocarbon age adjusted using $\delta^{13}C$ content of sample relative to −25 ‰ average for wood.

[b]The charcoal in this sample may have been transported to the deposition site by the stream, and therefore its age would be greater than or equal to the age of the alluvium, or the charcoal could be the remains of a tree root that burned underground, which would make its age less than or equal to the age of the alluvium.

[c]Determined on organic matter. These values are similar to those reported by Cerling et al. (1989) for organic matter in modern soils from the Konza, and suggest that the mix of C_3 and C_4 plants at the site when the paleosol formed was similar to that in the prairie at the sites sampled by those authors.

[d]Samples β-47895, β-50737, and I-17172 were collected from the same buried soil. Samples β-47895 and I-17172 were collected from a backhoe soil pit by W. A. Wehmueller, and β-50737 from a stream cutbank less than 100 m from the pit.

From the available limited data, the Holocene fluvial geomorphic history of the Konza Prairie can be summarized as follows. Valley floors had been carved from bedrock to approximately their present levels by the early Holocene (Fig. 3.5). Although no alluvial deposits older than early Holocene have been identified on Konza Prairie, late Pleistocene deposits may be present in the lower parts of Kings Creek or other large stream valleys. Deposition of light-brown, fine-grained alluvium was under way in the Kings Creek valley by 9,000 yr B.P. and may have continued for several thousand years (Fig. 3.5), although it is unknown when deposition ended or whether more than one depositional episode occurred during the early Holocene. Limited evidence indicates that soils were forming in valley bottoms between 3,000 and 2,000 yr B.P. (Fig. 3.5), suggesting that streams were not actively depositing sediment on their floodplains, and channels may have been entrenched and actively eroding. In lower Kings Creek, the chan-

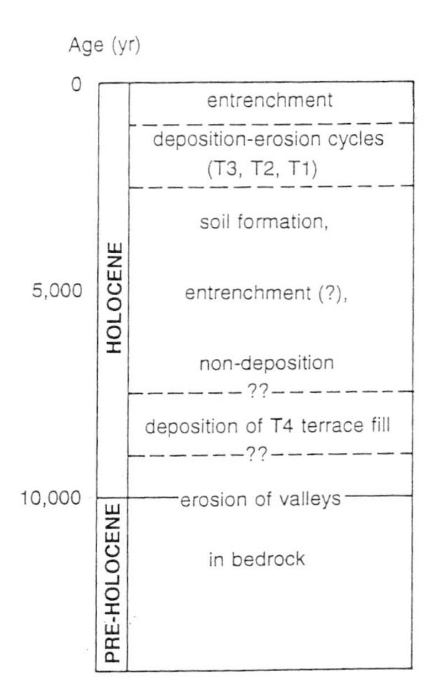

Figure 3.5. Summary of the Holocene fluvial geomorphic history of Konza Prairie.

nel was entrenched about 5 m below its former floodplain level prior to 1,800 yr B.P., by which time aggradation had begun again in the valley bottom. It is unknown when this second period of valley aggradation ended, but it was followed by another cycle of downcutting and aggradation prior to the last downcutting event. Most streams presently are entrenched.

Although more data are needed to refine and test the fluvial geomorphic history presented in Figure 3.5, the Konza history, as it is presently known, bears some similarity to the history of other stream basins in the central Great Plains, as noted in reviews of published alluvial chronologies (Johnson and Martin 1987; Mandel 1994). In particular, periods of floodplain stability and soil formation in valley bottoms during the late Holocene have been documented in a number of drainage basins and include a period between 3,000 and 2,000 yr B.P. In addition, during the early to middle Holocene, many small stream valleys in the Midwest experienced erosion in the uplands and removal of sediment from valley floors. The sediment removed from the small valleys was deposited downstream in larger valleys (Mandel 1994). Sediment eroded from the Konza Prairie during the middle Holocene would have been transported to the McDowell Creek and Kansas River valleys.

Changes in the behavior of Konza Prairie streams during the Holocene most likely were caused by climate changes, but changes in vegetation, fire frequency, or grazing patterns, which are linked to climate, also may have played a role. The climate in the central Great Plains during the later phases of the last glacial maximum and the deposition of the Peoria loess (~18,000–12,000 yr B.P.) was probably cold and dry. With the melting of the Laurentide ice sheet after 14,000 yr B.P. and a peak in summer insolation between 12,000 and 9,000 yr B.P., zonal circulation increased in strength over the Great Plains as the glacial anticyclone weakened, temperatures increased, and the region became effectively drier (Bryson and Hare 1974; COHMAP Members 1988). Precipitation may have increased slightly at 12,000 yr B.P. but declined to full-glacial levels during the middle Holocene and then increased slightly again by 3,000 yr B.P. (Kutzbach 1987). The strong westerly (zonal) air flow over the Great Plains reached its peak during the middle Holocene, and grasslands expanded to the east (Wright 1970). A shift in atmospheric circulation after about 4,000 yr B.P. to a more meridional pattern allowed the incursion of warm, moist air masses into the Great Plains from the Gulf of Mexico (Bryson and Hare 1974), thus causing an increase in precipitation and in the frequency of large storms and floods (Knox 1983). The hydrologic effects of these climate changes included flushing of sediment from small stream valleys during the early and middle Holocene and aggradation in the same valleys during the late Holocene. Smaller-scale, cut-and-fill cycles, which are not necessarily correlatable from one drainage basin to another, are superimposed on this general pattern.

No paleovegetation records are available from the Konza Prairie itself, but evidence from a number of other sites in the region (as reviewed by Fredlund and Jaumann 1987) can be used to infer the vegetation changes during the late Pleistocene and Holocene that probably occurred on the Konza Prairie (Chapter 5, this volume). Pollen data suggest that prior to about 24,000 yr B.P., grasslands constituted the dominant vegetation in the central Great Plains region. Grassland vegetation persisted during the last full-glacial period from about 24,000 until 12,000 yr B.P., when spruce, mixed spruce and deciduous trees, and aspen increased in abundance (Fredlund and Jaumann 1987). Macrofossils of spruce charcoal, needles, twigs, and cones have been recovered from a number of sites in the central Great Plains (Wells and Stewart 1987), but fossil mollusk assemblages in loess suggest that the vegetation consisted of grasslands with scattered shrubs or isolated tree stands rather than a closed forest (Rousseau and Kukla 1994). After 12,000 yr B.P., spruce abundance rapidly decreased, accompanied by at least a local rise in oak, hickory, elm, and related deciduous trees. By about 9,000 yr B.P., arboreal species had declined in abundance, and grasslands had attained the dominance they have held through most of the Holocene (Fredlund and Jaumann 1987).

No data are currently available on the Holocene history of fire or grazing on the Konza Prairie, both of which would affect sediment yield from headward parts of stream systems. Changes in these variables probably would cause changes in the stream environment at a smaller spatial scale and shorter temporal scale than would climate change. Such changes might be locally important at certain times, but they would be more difficult to decipher from the geologic record than are regional climate changes. However, because of the importance of fire and grazing in the Konza ecosystem, more work on the long-term history of these factors is needed.

Current Geomorphic Processes

Few quantitative data are yet available from observations of current geomorphic processes on the Konza Prairie. In simulated rainfall experiments on a low-gradient (6%), upland site on the Konza Prairie, Koelliker and Duell (1990) found that burning initially decreased the infiltration capacity of the soil by 5 mm/hr, but that overland flow did not attain high enough velocity or depth to significantly erode the soil. They concluded that higher overland flow velocities are unlikely to develop in the tallgrass prairie and that overland flow is unlikely to significantly erode even steeper slopes in the Flint Hills. Field observations (Ross 1995) indicated that in the N04D drainage basin, overland flow must be concentrated from an area of 0.13 ± 0.10 ha above the head of a typical first-order stream to produce sufficient velocity and depth to create a channel. The average length of overland flow (distance from drainage divide to first-order channel head) is 80 ± 40 m. Under the present climate and levels of grazing and fire frequency, overland flow probably is not significantly eroding hillslopes, except where it cascades over amphitheater headcuts (discussed in the following). Erosion is concentrated primarily in stream channels. Creep of hillslope deposits is undoubtedly a locally important, but as yet unmeasured, process.

Erosional features referred to as amphitheater headcuts by Smith (1991) are prominent at the heads of first-order streams in the low-relief basins. Amphitheater headcuts are curved scarps, 0.5 to 1.5 m in height, formed in loess and fine-grained residuum at the heads of first-order streams near drainage divides. Undisturbed prairie above the headcuts drops off in vertical or near-vertical slopes to bare, rapidly eroding soil below the headcuts. Small slumps are common on the steep headcut slopes and provide abundant sediment that is transported downstream through entrenched reaches of the first-order streams. Although amphitheater headcuts are present in the high-relief drainage basins, they are not as well developed there in the low-relief drainage basins, possibly because the loess and other fine-grained deposits are thinner in the high-relief basins.

Stream entrenchment and bank erosion are active processes throughout the Konza. Most first-order streams flow directly across bedrock, and higher-order streams flow across either bedrock or alluvium of variable thickness. A typical second- or third-order stream has entrenched reaches in which the channel bottom consists of bare rock or is mantled by boulders and cobbles and ends downstream in flat-floored vegetated reaches, where the channel is poorly defined and the sediments are fine-grained and appear to be accumulating. These aggrading reaches are terminated at their downstream ends by abrupt knickpoints that mark the boundary with the next entrenched reach. The lower parts of most third-order streams and all fourth- and fifth-order streams are entrenched, cutbanks on meander bends are being undercut, and slumping is active. Thus, in any one stream valley, both erosion and deposition are occurring in different reaches; however, depositional reaches occupy only a small part of the total channel length, and the long-term trend is for erosion and removal of sediment.

Figure 3.6. Aerial view of the Flint Hills of eastern Kansas illustrating the geomorphology of Konza Prairie. (L. Hulbert)

Summary

The geomorphology of Konza Prairie is a complex system with strong control from the bedrock geology (Fig. 3.6). However, the bedrock itself does not determine the character of the landscape, and the modern landscape is the result of surficial processes that have operated for millions of years. Climate, vegetation, grazing, and fire all have varied in the past, and the challenge is to decipher the complex history from a record that has largely been erased. However, the Konza landscape clearly is dynamic, and the rates and magnitudes of surficial processes have changed considerably through time.

On geologic time scales (millions of years), the Konza Prairie landscape is eroding, as shown by the dendritic drainage network that appears to be nearly in equilibrium. On shorter time scales (thousands to hundreds of years), the hillslope and stream systems undergo cyclic periods of erosion and deposition that are recorded as alluvial cut-and-fill cycles in valley bottoms. On even shorter time scales (decades to years), local scour and fill in stream channels and erosion on steep hillslopes and in amphitheater headcuts dominate.

Acknowledgments I am grateful to C. W. Martin, K. L. Ross, G. N. Smith, and W. A. Wehmueller for sharing unpublished data and for fruitful discussions on the geomorphology of Konza Prairie.

4

Soils and Soil Biota

Michel D. Ransom
Charles W. Rice
Timothy C. Todd
William A. Wehmueller

Soils are an integral part of the tallgrass prairie ecosystem. Jenny (1941) suggested that soil formation results from multiple factors: climate, organisms, topography, and parent material, all interacting over time. In the tallgrass prairie-soils ecosystem, the prairie vegetation has exerted great influence on soil formation, but the physical and chemical properties of soil also affect the kinds, amount, and spatial distribution of the vegetation. The relationships between soils and the biota of the prairie are extremely complex.

The soil biota in tallgrass prairie represents a diversity of genetic and physiological traits. Although the aboveground flora and fauna often characterize the prairie, the biomass of the soil biota can be of the same order of magnitude as the visible portion of the prairie (Paul et al. 1979). This invisible prairie is responsible for much of the energy flow and nutrient cycling (Chapters 13 and 14, this volume) in the prairie and contributes to the development of soils. Although some soil organisms use sunlight or inorganic substrates to obtain energy, a vast majority of the soil biota are heterotrophs, using organic compounds to obtain their energy. Photosynthetically fixed C produced by plants decomposes, supporting the microbial community. The ecology and diversity of the soil biota of tallgrass prairie remain relatively unexplored despite its importance in decomposition and nutrient cycling (Clark and Paul 1970; Seastedt et al. 1988a; Seastedt and Hayes 1988).

The goal of this chapter is to describe the soils and the diversity of the soil biota of the tallgrass prairie ecosystem, particularly that associated with Konza Prairie.

Soil Genesis, Morphology, and Classification

Physiography and Previous Work

Konza Prairie is within the Bluestem Hills land resource area of the central Great Plains region (Soil Conservation Service 1981). As noted in Chapter 3 (this volume), the Bluestem Hills consist of uplands locally covered with a thin mantle of loess over Permian limestones and shales. The stream divides are narrow with steep side slopes. The climate is continental with a normal annual precipitation of 835 mm and a mean annual air temperature of about 13°C (Chapter 2, this volume). The soil moisture regime (Soil Survey Staff 1994) for the Bluestem Hills is transitional from udic to ustic. The elevation above sea level at Konza Prairie ranges from 320 to 440 m.

Jantz et al. (1975) conducted a soil survey of Riley County and part of Geary County at a scale of 1:24,000. It included most of Konza Prairie and was an Order 2 soil survey intended for planning purposes, including farm and ranch planning. According to the survey, most soils on upland hillslope positions are formed in residuum weathered from limestones or shales. Smith (1991) mapped the bedrock and Quaternary geology and investigated the geomorphology of Konza Prairie. Much of this material was reviewed in Chapter 3 (this volume).

Glaze et al. (1994) investigated the genesis of three pedons on broad summits in the Bluestem Hills. (A pedon is the smallest volume that can be recognized as an individual soil.) Two of these pedons are within the boundaries of Konza Prairie. The pedons had a mantle of Wisconsin loess ranging from 58 to 72 cm thick over a composite paleosol, or soil formed on a landscape of the past that was welded, or merged, with the modern soil. The paleosol of the two pedons within Konza Prairie formed in hillslope sediment (colluvium) over residuum from Permian cherty limestone. Clusters of pedogenic gypsum occurred in the paleosol formed in hillslope sediment. Although the horizons in which the gypsum occurred had high sodium (Na) contents and electrical conductivity values, the levels of Na were not quite high enough or else occurred too deep in the profile to meet the requirements for classification as natric horizons (Glaze et al. 1994).

Wehmueller et al. (1994) studied the genesis of three pedons occurring on summit and shoulder landscape positions within Konza Prairie using mineralogical and soil thin-section techniques. The pedons had parent material stratigraphy similar to those studied by Glaze et al. (1994). The Bt horizons formed in loess and hillslope sediment exhibited stress features and soil fabrics typical for soils affected by much shrink-swell activity.

The goal of any soil survey is to associate soil map units with landscape positions. Once the soil-landscape associations have been defined, investigators can quickly and accurately predict soil properties based on the landscape position of the area in question. Wehmueller (1996) investigated soil genesis, described soil morphology, and mapped the distribution of soils at a scale of 1:2,000 for a 125-ha watershed, designated N04D, within Konza Prairie. He divided the distribution of soils within Konza Prairie into four groups based on landscape position (Fig. 4.1): (1) soils on the less sloping summits of interfluves

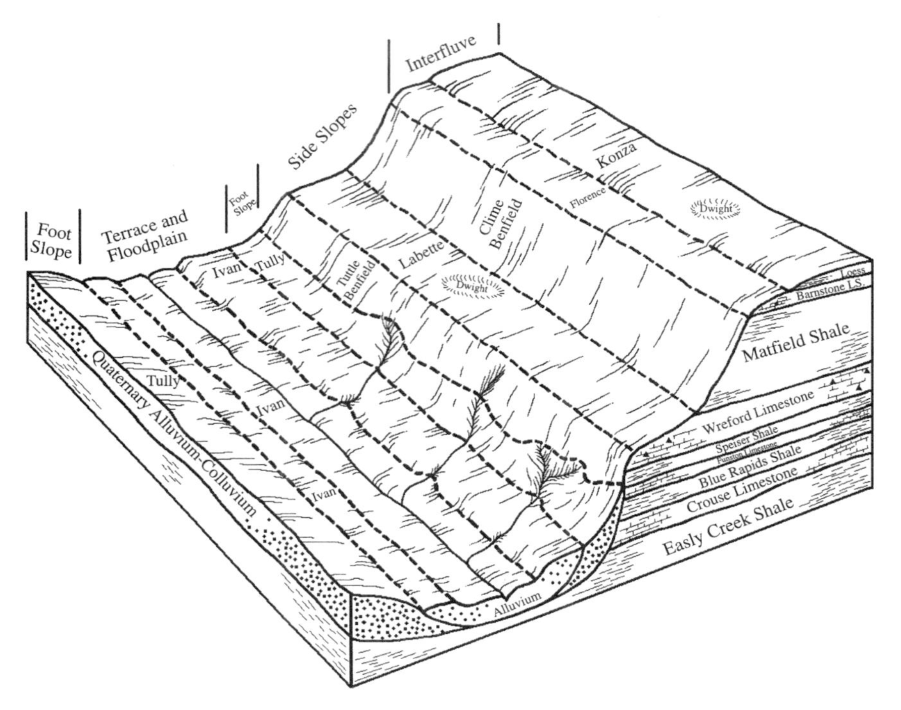

Figure 4.1. Idealized three-dimensional view of soils and landscape relationships on Konza Prairie. The thicknesses of the geologic formations are not to scale.

and benches (Chapter 3, this volume) formed by the Florence Limestone Member of the Barnestone Limestone and the Schroyer Limestone and Threemile Limestone Members of the Wreford Limestone; (2) steeply sloping soils on side slopes; (3) soils on foot slopes; and (4) soils on terraces and floodplains.

In the following discussion, soil and landscape relationships and properties of individual pedons sampled in the N04D watershed are described. Detailed descriptions of the morphology of the pedons, along with complete chemical and physical soil characterization data, are available in Wehmueller (1996). The processes of soil genesis, parent material stratigraphy, and soil and landscape relationships for the N04D watershed are typical for other watersheds within Konza Prairie.

Soils on Interfluves and Benches

Table 4.1 shows selected soil characterization data for Konza, Dwight, and Florence, which are typical soils on interfluves and benches (Fig. 4.1). The Konza soil type is a fine, montmorillonitic, mesic Udertic Paleustoll. This pedon contained high contents of exchangeable Na (Table 4.1), but it was too deep in the profile to meet natric horizon requirements (Soil Survey Staff 1994). The summit on which this pedon occurred was the highest topographic position in the N04D

Table 4.1. Selected physical and chemical characteristics for soils on interfluves and benches, Konza Prairie

Horizon	Depth (cm)	2mm (%)	Sand (%)	Silt (%)	Clay (%)	COLE[a] (cm/cm)	pH 1:1 H_2O	Elec. Cond. (mmhos/cm)	Na Saturation (%)	Organic Carbon (%)	Ext.[b] Fe (%)
Konza (Summit)											
A1	0–10	0	7.2	66.8	26.0	0.04	6.3	0.73	1	4.39	0.8
A2	10–17	<1	6.5	62.9	30.6	0.04	5.6	—	3	2.68	0.9
Bt1	17–33	<1	3.7	47.4	48.9	0.10	6.9	0.53	6	1.50	1.2
Bt2	33-41	<1	2.8	54.1	43.1	0.09	7.6	0.90	8	1.14	1.1
Btk1	41–58	3	4.8	59.3	35.9	0.06	8.2	0.85	10	0.70	1.0
2Btk2	58–71	7	7.1	60.8	32.1	—	8.2	1.29	17	0.54	1.0
2Btky	71–86	3	9.1	56.6	34.3	0.06	7.6	5.73	20	0.31	1.2
2Bt1	86–97	4	9.0	51.9	39.1	0.08	7.7	5.06	20	0.32	1.4
2Bt2	97–107	12	10.5	46.8	42.7	0.09	7.7	3.63	23	0.31	1.5
3Bt3	107–127	58	11.0	34.8	54.2	—	7.7	3.17	26	0.32	1.7
3Bt4	127–148	75	3.8	27.1	69.1	—	7.7	1.34	17	0.35	1.6
3R	148–150	—	—	—	—	—	—	—	—	—	—
Dwight (Summit-depression)											
A	0–9	0	7.7	64.0	28.3	—	5.6	1.37	1	5.77	0.9
Btn1	9–20	0	6.3	51.1	42.6	0.03	6.4	0.69	10	2.43	1.1
Btn2	20–31	0	4.8	46.1	49.1	0.11	7.8	1.17	16	1.62	1.2
Btkn1	31–42	1	4.5	51.5	44.0	0.13	8.2	2.36	27	1.07	1.1
2Btkn2	42–57	1	9.4	48.7	41.9	0.09	8.2	4.23	35	0.70	1.2
2Btkny	57–73	<1	6.0	47.8	46.2	0.14	7.8	8.48	36	0.45	1.3
3Bkn	73–77	39	31.2	37.8	31.0	0.12	8.3	7.74	45	0.44	0.3
3R	77–80	—	—	—	—	—	—	—	—	—	—

(continued)

Table 4.1. continued

Horizon	Depth (cm)	2mm (%)	Sand (%)	Silt (%)	Clay (%)	COLE[a] (cm/cm)	pH 1:1 H_2O	Elec. Cond. (mmhos/cm)	Na Saturation (%)	Organic Carbon (%)	Ext.[b] Fe (%)
Florence (Shoulder)											
A1	0–9	2	11.1	60.0	28.9	—	6.3	—	<1	3.40	1.0
A2	9–18	9	10.8	53.1	36.1	0.04	5.8	—	<1	2.37	1.2
2Bt1	18–28	6	10.0	46.5	43.5	—	5.9	—	<1	1.85	1.5
2Bt2	28–37	19	9.0	45.1	45.9	0.06	6.0	—	<1	1.60	1.6
3Bt3	37–66	84	15.7	33.2	51.1	—	6.0	—	<1	1.41	1.6
3Bt4	66–79	10	4.6	32.6	62.8	0.14	7.1	—	<1	0.76	2.2
3Bt5	79–90	8	3.4	26.4	70.2	—	7.5	—	<1	0.93	2.2
3R	90–93	—	—	—	—	—	—	—	—		—

[a]COLE = Coefficient of Linear Extensibility (Soil Survey Laboratory Staff 1996).
[b]Extractable in Na citrate-dithionite.

watershed. This summit is on a broad interfluve with a thicker loess mantle than the other two pedons. The upper 58 cm was identified as Wisconsin (Peoria) loess based on the high silt and low coarse-fragment contents (Table 4.1). The material from 58 to 107 cm was identified as colluvium or hillslope sediment. The upper horizon of this material from 58 to 71 cm exhibited an increase in coarse-fragment content (Table 4.1) and contained a stone line. Transect work showed that this hillslope sediment could be traced across most of the erosional landscape of the watershed. The high contents of clay and chert fragments suggest that the material from 107 to 148 cm was formed in residuum weathered from the underlying cherty limestone. A strongly developed paleosol occurred in this material.

Konza soils typically have black (10YR 2/1) or very dark brown (10YR 2/2) surface layers with a texture of silt loam or silty clay loam and a range in clay content from 26 to 34%. The subsoil has a clear or abrupt boundary. The upper subsoil is very dark grayish brown (10YR 3/2) or very dark brown (10YR 2/2). It is silty clay, with 36 to 50% clay. The subsoil below the lithologic discontinuity has a hue of 7.5YR, a value of 3 or 4, and a chroma of 3 or 4. It is silty clay loam or silty clay ranging from about 32 to 42% clay. Rock fragments are usually present and make up less than 15% of the whole soil. The paleosol has a hue of 5YR or 7.5YR, a value of 3 or 4, and a chroma of 3 or 4. Texture is silty clay or clay, with a clay content ranging from 50 to 70%. The rock fragment content ranges from 15 to 75%.

The Dwight pedon (Fig. 4.1) was classified as a fine, montmorillonitic, mesic Typic Natrustoll. This pedon had a natric horizon (Soil Survey Staff 1994) as indicated by 15% exchangeable Na (Table 4.1) within 40 cm of the upper boundary of the Btn horizons and columnar structure in the Btn1 horizon. This pedon occurred in a small, irregularly shaped depression on the summit of a bench that was topographically about 25 m lower than the summit where the Konza pedon occurred. The upper 43 cm was identified as Wisconsin loess based on the high silt content and low content of coarse fragments (Table 4.1). The material from 43 to 73 cm was identified as hillslope sediment similar to that of the second parent material of the Konza pedon. The material from 73 to 77 cm was identified as residuum weathered from the underlying limestone.

Dwight soils are moderately deep, moderately well drained, and slowly permeable; they occur in depressions on nearly level summits of benches. They have high amounts of Na that disperse the clay in the upper subsoil. The dispersed clay reduces the permeability of the soil, and the depressions pond water for short periods after rain. The surface layer is generally black (10YR 2/1) silty clay loam, with clay contents ranging from 28 to 35%. It is generally only 4 to 10 cm thick, with an abrupt boundary to the subsoil layer. The upper part of the subsoil has a hue of 10YR, a value of 2 through 4, and a chroma of 2 or 3. It is silty clay, with 40 to 50% clay. The Na saturation percentage is generally lowest near the top of the subsoil and increases with depth. The lower part of the subsoil has a hue of 10YR or 7.5YR, a value of 3 through 6, and a chroma of 2 through 4. The clay content is lower than in the upper subsoil layers and ranges from 35 to 45%. Usually, an accumulation of pedogenic carbonate and gypsum occurs in the lower subsoil.

The Florence pedon (Fig. 4.1) occurred on a shoulder slope below the summit where the Konza pedon occurred and was classified as a clayey-skeletal, montmorillonitic, mesic Udertic Argiustoll. At this location, the interfluve is more sloping and narrower than that of the Konza pedon. The upper 18 cm was identified as Wisconsin loess based on high silt content and low content of coarse fragments (Table 4.1). The second parent material from 18 to 37 cm was identified as hillslope sediment based on subrounded coarse fragments that were mostly chert. An abrupt boundary at 37 cm marked the top of a truncated paleosol formed in residuum similar to the paleosol developed in residuum in the Konza pedon. The limestone and chert fragments in the 3Bt horizon were less angular than those in the 2Bt horizon and were coated with thick clay films.

The surface layer of Florence soils is typically very dark brown (10YR 2/2) or black (10YR 2/1). It is silt loam or silty clay loam, ranges from 26 to 32% clay, and generally does not have many rock fragments. The subsurface layer has a hue of 7.5YR or 10YR, a value of 2 or 3, and a chroma of 2. It is silty clay loam, ranges in clay from 34 to 40%, and generally has less than 10% rock fragments. The subsoil, below the lithologic discontinuity, has a hue of 7.5YR or 5YR, a value of 3 or 4, and a chroma of 3 or 4. The texture is silty clay or clay and ranges from 40 to 70% clay. Chert and limestone fragments range from 10% to more than 50%.

Labette soils (Fig. 4.1) are moderately deep and well-drained, and they occur on gently sloping benches. These soils formed in thin loess over residuum from the underlying limestone. The depth to limestone ranges from 60 to 100 cm. The surface layer is very dark brown (10YR 2/2) or black (10YR 2/1) silt loam or silty clay loam. This layer ranges from 12 to 30 cm thick and has an abrupt boundary to the subsoil. The upper subsoil has a hue of 10YR, a value of 2 or 3, and a chroma of 1 or 2. The texture is silty clay. The lower subsoil, formed in residuum, has a hue of 7.5YR or 5YR, a value of 3 through 5, and a chroma of 3 or 4. The texture is silty clay or clay and ranges from 50 to 70% clay.

Soils on Side Slopes

Soils on side slopes (Fig. 4.1) have a wide range in depth, amount of carbonate, and content of rock fragments. Soils that are calcareous to the surface are on the steepest slopes and are usually below outcrops of limestone. The Tuttle pedon (Figs. 4.1, 4.2) is a loamy-skeletal, mixed, mesic Pachic Calciustoll. The pedon has a calcic horizon as indicated by $CaCO_3$ contents over 15% (Table 4.2). This pedon formed in hillslope sediment over residuum weathered from shale. The depth distribution of the clay content (Table 4.2) does not show evidence of clay translocation. This lack of clay translocation and abundance of carbonates indicate that these soils on steep side slopes are not highly weathered. The dominant pedogenic process has been translocation and redistribution of carbonate, which suggests that the hillslope sediment has not been in place very long or that, because of the steep slopes, water movement through the soil is limited. Soils with similar parent materials on adjoining areas with less slope contain translocated clay and have argillic horizons. Hence, a combination of slope and external ad-

Figure 4.2. Although some soils on Konza Prairie can be very deep (over 2 m), most are shallower and contain numerous rocks and small fragments. These characteristics are illustrated by this profile of a Tuttle silty clay loam. (M. Ransom)

ditions appears to have influenced the leaching environment and limited pedogenesis to redistribution of carbonate.

Tuttle soils typically have a surface layer that is black (10YR 2/1) or very dark gray (10YR 3/1). Texture is very gravelly or channery silty clay or silty clay loam. The subsurface layers have similar colors but usually have greater than 25% limestone fragments. The subsoil has a hue of 10YR or 2.5Y, a value of 3 through 6, and a chroma of 2 through 4. It is flaggy or very flaggy silty clay loam, with the flagstones making up as much as 60% by volume. Below the lithologic discontinuity, there are few rock fragments, and the texture is silty clay or silty clay loam.

Soils on less sloping areas of side slopes are leached more deeply than strongly sloping soils on side slopes. The Benfield pedon (Fig. 4.1) is typical on this landscape. It is classified as a very fine, montmorillonitic, mesic Uder-

Table 4.2. Selected physical and chemical characterization data for soils on side slopes, footslopes, terraces, and floodplains, Konza Prairie

Horizon	Depth (cm)	2mm (%)	Sand (%)	Silt (%)	Clay (%)	COLE[a] (cm/cm)	pH 1:1 H_2O	Organic Carbon (%)	$CaCO_3$ <2 mm (%)	Ext.[b] Fe (%)
Tuttle (Side Slopes)										
A1	0–8	20	6.4	52.3	41.3	0.049	7.5	5.10	6.0	0.8
A2	8–17	22	6.5	49.9	43.6	—	7.6	4.40	7.5	0.7
A3	17–32	30	8.3	48.1	43.6	—	7.3	3.78	10.3	0.8
2A4	32–57	72	9.5	48.4	42.1	0.025	7.8	2.62	22.7	0.8
2Bk1	57–70	78	18.7	50.1	31.2	—	7.9	0.86	50.6	0.8
2Bk2	70–109	72	16.6	51.6	31.8	0.031	8.1	0.40	52.0	0.7
3Bk3	109–154	10	4.7	47.7	47.6	0.076	8.1	0.17	21.2	1.6
Benfield (Side Slopes)										
A1	0–9	—	2.7	57.0	40.3	0.053	6.6	6.06	—	1.0
A2	9–22	1	3.7	51.9	44.4	—	5.9	4.12	—	1.1
A3	22–42	25	5.2	40.4	54.4	—	6.1	2.70	—	1.3
2Bw	42–58	1	1.5	30.5	68.0	0.143	6.8	1.06	—	1.5
2Bss	58–80	2	1.1	31.0	67.9	0.129	7.0	0.71	—	1.6
3Btss	80–98	11	3.3	37.6	59.1	0.130	7.9	0.39	3.1	1.6
3Btk	98–111	73	10.4	60.6	29.0	0.067	8.1	0.28	53.1	0.8
3Crk1	111–135	7	6.1	64.0	29.9	0.047	8.2	0.15	47.4	0.9
3Crk2	135–181	16	2.8	60.5	36.7	0.054	8.1	0.10	32.6	1.1
3Crk3	181–190	2	0.9	59.7	39.4	0.058	8.1	0.06	26.7	1.0
Ivan (Low Terrace–Floodplain)										
A1	0–12	<1	7.0	56.9	36.1	—	6.7	5.94	—	0.8
A2	12–33	—	6.9	53.3	39.8	0.088	7.1	2.99	—	0.8
A3	33–43	4	11.4	50.6	38.0	—	7.6	2.00	3.2	1.0

Bw1	43–53	8	20.8	47.7	31.5	—	7.8	1.09	11.8	1.0
Bw2	53–68	69	32.6	41.7	25.7	—	7.9	0.69	19.0	0.9
2Bwb	68–86	15	26.3	46.0	27.7	0.049	8.0	0.63	18.9	0.9
2C	86–121	81	39.6	37.1	23.3	0.042	8.0	0.45	26.8	0.9
3Bwb1	121–140	5	12.6	48.9	38.5	0.087	8.0	0.71	5.1	0.8
3Bwb2	140–155	2	11.6	50.0	38.4	—	8.0	0.63	4.0	0.7
4R	155–156	—	—	—	—	—	—	—	—	0.1
No Series Designated (High Terrace)										
A1	0–8	—	5.7	57.3	37.0	—	7.4	6.14	14.1	0.3
A2	8–25	—	9.0	56.6	34.4	—	7.7	3.35	18.2	0.3
Bw1	25–46	<1	8.9	54.7	36.4	0.069	7.9	1.70	15.3	0.3
Bw2	46–77	1	11.3	52.8	35.9	0.059	8.0	0.84	7.5	0.5
2Bw3	77–85	48	18.2	47.8	34.0	—	7.9	0.54	4.4	0.8
2Bw4	85–102	<1	12.1	57.5	30.4	—	8.0	0.46	1.1	0.6
3Ab	102–125	1	3.5	54.7	41.8	—	7.7	0.59	0.1	0.6
3Bwb1	125–137	1	4.0	50.8	45.2	—	7.7	0.43	—	0.9
3Bwb2	137–152	<1	6.8	49.1	44.1	0.108	7.8	0.25	0.2	1.0
3Bwb3	152–182	1	6.2	56.6	37.2	0.092	7.9	0.13	0.3	0.9
3Bkb1	182–194	10	7.2	61.4	31.4	0.069	8.1	0.12	8.6	0.5
3Bkb2	194–206	9	5.9	63.6	30.5	0.049	8.1	0.08	7.8	0.6
3BC	206–222	<1	4.7	64.4	30.9	0.052	8.1	0.05	2.9	0.8

[a]COLE = Coefficient of Linear Extensibility (Soil Survey Laboratory Staff 1996).
[b]Extractable in Na citrate-dithionite.

tic Argiustoll. It was leached of carbonates to a depth of 80 cm (Table 4.2). This pedon typically has high shrink-swell characteristics as indicated by COLE values over 0.06. High shrink-swell activity disrupts the formation of clay films in this soil. This pedon is on a backslope of a noseslope and is formed in two layers of hillslope sediment over residuum from calcareous shale. In other less sloping areas of sideslopes, the underlying bedrock is limestone instead of shale.

Benfield and Clime are the most extensive soils on side slopes that are less sloping (Fig. 4.1). Benfield soils typically have a surface layer that is black (10YR 2/1) or very dark brown (10YR 2/2) silty clay loam. The surface soil layers generally have between 40 and 55% clay, and the rock fragment content ranges from about 15 to 30%, with a lithologic discontinuity immediately below. This discontinuity is marked by an abrupt decrease in the quantity of rock fragments and organic C content. The upper part of the subsoil also formed in hillslope sediment. This layer usually has more than 60% clay. The colors have a hue of 10YR, 7.5YR, or 2.5Y, with values of 3 through 5 and a chroma of 2 or 3. The lower subsoil, occurring below the discontinuity, formed in residuum weathered from shale. It generally is calcareous and has 25 to 40% clay. The hue is 2.5Y or 10YR, value ranges from 4 through 6, and chroma ranges from 2 through 4. The depth to shale or limestone bedrock ranges from 95 to 115 cm.

Clime soils formed in hillslope sediment, 20 to 40 cm thick, over residuum weathered from shale. Depth to shale ranges from 50 to about 110 cm. Clime soils typically have a surface layer that is gravelly silty clay loam or gravelly silty clay. The clay content ranges from 35 to 50%, and the rock fragment content ranges from 15 to 35%. It is black (10YR 2/1) or very dark brown (10YR 2/2). The subsoil below the lithologic discontinuity has a hue of 2.5Y, a value of 4 to 6, and a chroma of 2 through 4. This layer is generally less clayey than the surface soil and ranges from 20 to 35% clay.

Soils on Footslopes

Soils on the footslopes (Fig. 4.1) are generally thicker than the soils on side slopes and are deeper to bedrock. The soils in the upper reaches of the watershed are usually more than 125 cm but less than 200 cm thick over shale or limestone bedrock. Typically, the soils on footslopes in the upper reaches of the watershed formed in material that Smith (1991) identified as undifferentiated Quaternary alluvial-colluvial deposits. Most of these soils are formed in two units of hillslope sediment over residuum. The residuum exhibits an accumulation of pedogenic carbonates; development of soil structure; and, in some pedons, illuvial clay accumulations.

In the lower part of the watershed, the soils on footslopes are formed in hillslope sediment over alluvium, and the thickness of the valley fill ranges from 200 to more than 400 cm. In these soils, the hillslope sediment has been in place long enough for the material to be completely leached of carbonates. The argillic horizon in these soils is not as clayey or as thick as the argillic horizon formed in soils on footslopes in the upper reaches of the watershed. Usually, soil formation is evident in the alluvium below the discontinuity, which suggests that the alluvium was weathered before burial by hillslope sediment.

Tully soils (Fig. 4.1) are the most extensive soils on footslopes. They are fine, mixed, mesic Pachic Argiustolls. Typically, the surface layer is black (10YR 2/1) or very dark brown (10YR 2/2) silty clay loam with 35 to 40% clay. Although rock fragments are common throughout the surface layers, they rarely make up more than 15% by volume. Areas of this soil that are next to the drainage ways generally have more rock fragments. The subsoil above the lithologic discontinuity has a hue of 10YR or 7.5YR, a value of 3 through 4, and a chroma of 2 or 3. The texture is silty clay or clay. The lower part of the subsoil is the most clayey and generally will have the most rock fragments. The discontinuity is marked by an abrupt decrease in rock fragments. Clay content in the lower subsoil ranges from 45 to 60%, and rock fragments range from 5 to 20%. The subsoil below the discontinuity is formed in residuum from shale or limestone. It is less clayey than the subsoil above the discontinuity and generally contains pedogenic carbonate in the form of nodules or soft masses.

Soils on Terraces and Floodplains

Multiple levels of terraces and floodplains occur on Konza Prairie (Fig. 4.1). The floodplains are narrow and contain recent alluvial deposits, point bar deposits, and a stream channel. The alluvial deposits have experienced little if any pedogenesis and have a wide range in properties.

The Ivan pedon occurred on the lowest terrace about a meter above the floodplain. It is classified as a fine-loamy, mixed, mesic Cumulic Hapludoll. The organic C content (Table 4.2) decreases irregularly with depth. The parent material is alluvium deposited in three episodes. The lower two alluvial units contain buried soils that represent periods of stability in the landscape. These periods were followed by truncation of the soil, then another period of deposition, followed by stability and soil formation.

Ivan soils are dominant on low terraces and floodplains (Fig. 4.1). They are deep, moderately well-drained, and moderately permeable soils. The content of rock fragments varies from few to over 50%. In general, the content of rock fragments is higher for soils next to the stream channel. These soils also commonly have buried soils within a depth of 1 m. Pedons that are calcareous to the surface are mixed intricately with those in which the carbonates are deeper. The surface layers have a hue of 10YR, a value of 2 or 3, and a chroma of 1. They are typically silty clay loam, ranging from 35 to 40% clay. Rock fragments are generally less than 15%. The subsoil horizons have similar colors and textures. They generally are higher in carbonates than the surface horizons. Layers with greater amounts of chert and limestone mark the boundaries of different alluvial periods.

A second terrace occurs about 2.5 to 3.5 m above the floodplain (Fig. 4.1). The pedon sampled in the N04D watershed is an example of soils occurring on this higher terrace (Table 4.2). This soil does not have an assigned series name; it classified as a fine, mesic Typic Calciaquoll. This pedon was developed in three layers of alluvium and contained a buried paleosol. The carbonate accumulation in the modern soil (Table 4.2) resulted from springs that saturate this site periodically. The areas of this terrace that are not below springs generally have soils that classify as fine, mixed, mesic Cumulic Hapludolls. These soils are formed in similar alluvial deposits but have different drainage characteristics because of differences in the hydrology of the sites.

Biology of Tallgrass Prairie Soils

The previous section showed that soils on Konza Prairie vary according to landscape position in many properties, such as soil depth, parent material stratigraphy, and clay content. However, landscape position does not have a major effect on the organic C content of soils on Konza Prairie. The amount of organic C in the surface horizon and the depth distribution of organic C (Tables 4.1, 4.2) are quite similar for all soils regardless of the landscape position. The only exception is that soils on terraces and floodplains tend to exhibit a higher organic C content with depth than soils on other landscape positions. The prairie vegetation exerts such a major influence on soil formation and organic matter accumulation that it overcomes any effects on organic C content caused by topographic variation. The organic C in the soils of the tallgrass prairie is a result of the input of C and N by the vegetation. This input of C supplies the energy needed by the organisms living in the soil. A portion of the plant material decomposed by the microorganisms forms soil organic matter. The allocation of C and nutrients belowground supports a diversity of organisms in the soil. Several groups of these organisims are discussed in the following.

Bacteria and Fungi

The abundance of most soil organisms varies significantly with depth. The aerobic heterotrophic bacteria are concentrated near the soil surface, where most of the organic matter is present (Fig. 4.3; Dodds et al. 1996a). Populations of bacteria steadily decline to the water table, then increase and remain constant in the water-saturated zone. Numerically, bacteria and actinomycetes make up a larger portion of the microbial community than fungi. Garcia (1992) reported that population levels of bacteria and actinomycetes were 10^5 to 10^6 cells g^{-1} soil in the surface 5 cm, whereas populations of fungi were an order of magnitude lower. Burning did not significantly affect the composition of the microbial community. Rice and Parenti (1978) reported that burning increased fungal populations. No information is currently available on the diversity of the microbial community in tallgrass prairie.

We have made some estimates of different physiological groups of the microbial community. Denitrifiers represent a group of bacteria capable of growth under aerobic and anaerobic conditions. Under anaerobic conditions, this group of bacteria reduces NO_3^- to N gases, a process that results in a loss of N from the soil to the atmosphere. Most of these bacteria are heterotrophs and are involved in the decomposition of organic matter. Within one soil pro-

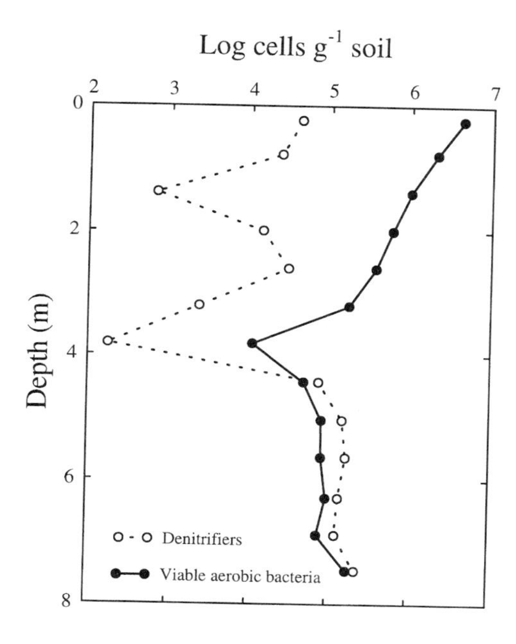

Figure 4.3. Profile distribution of aerobic heterotrophic bacteria and denitrifiers in a Reading silt loam soil on Konza Prairie. From Dodds et al. (1996a); Sotomayor and Rice (1996).

file on Konza Prairie, denitrifier populations decreased with depth similar to the aerobic bacteria (Fig. 4.3) (Sotomayor and Rice 1996). *Pseudomonas fluorescens* was the dominant species of the denitrifier community, although *Pseudomonas* sp., *P. putida*, and *P. mephitica* also have been identified from the denitrifier isolates (Halda-Alija 1996).

Nitrifiers are a specific group of bacteria that oxidize NH_4^+ to NO_3^-. This oxidation process is carried out in two steps by two distinct groups. One group oxidizes NH_4^+ to NO_2^-, and the other oxidizes NO_2^- to NO_3^-. Early investigations reported that nitrifiers or their activity were inhibited in grassland soils; however, more recent studies suggest that competition for NH_4^+ limits nitrification (Strauss 1995; Dodds et al. 1996a). In one study on Konza Prairie, populations of NH_4^+ oxidizers were 10^3 cells g^{-1} soil, 10 times lower than populations in a similar soil under cultivation. Populations of NO_2^- oxidizers were 10^3 cells g^{-1} and were not affected by cultivation. When excess NH_4^+ was added, nitrification was observed in the prairie soil (Strauss 1995).

Mycorrhizal Fungi

Extensive research has been conducted on Konza Prairie involving the symbiotic association of tallgrass prairie plant roots with arbuscular mycorrhizal (AM) fungi. Much of the research involving the symbiosis with specific plant species and plant responses to AM fungi is reported in Chapter 6 (this volume), and fire, mowing, and nutrient effects are reported in Chapter 13 (this volume). On various sample sites, spores of up to 20 species are found, with most belonging to the

Table 4.3. Density, relative abundance, and frequency of arbuscular mycorrhizal fungi isolated from tallgrass prairie plots under various management regimes

AM Fungal Species	Mean Density (Spores/10 g soil)[a]	RA (%)[b]	Frequency (%)[c]
Acaulospora longula Spain & Schenck	9.94	2.91	57.82
Entrophospora infrequens (Hall) Ames & Schneider	5.78	1.69	81.25
Gigaspora gigantea (Nicol. & Gerd.) Gerdemann & Trappe	1.25	0.37	35.93
Glomus aggregatum Schenck & Smith emend. Koske	31.06	9.10	48.44
Glomus claroideum Schenck & Smith	0.53	0.156	4.69
Glomus clarum Nicolson & Schenck	0.91	0.27	3.12
Glomus constrictum Trappe	10.83	3.17	73.43
Glomus etunicatum Becker & Gerdemann	127.94	37.49	73.43
Glomus fecundisporum Schenck & Smith	9.30	2.72	10.94
Glomus heterosporum Smith & Schenck	86.56	25.37	56.25
Glomus macrocarpum Tulasne & Tulasne	4.36	1.28	3.12
Glomus mortonii Bentivenga & Hetrick	11.30	3.31	25.00
Glomus mosseae (Nicol. & Gerd.) Gerdemann & Trappe	2.10	0.61	32.81
Glomus rubiforme (Gerd. & Trappe) Almeida & Schenck	35.61	10.44	9.38
Glomus sinuosum (Gerd. & Bakshi) Almeida & Schenck	0.75	0.22	4.69
Glomus sp. (white)	2.78	0.82	21.87
Scutellospora calospora (Nicol. & Gerd.) Walker & Sanders	0.23	0.07	5.00

[a]Mean of the total spore counts from the subplots in which the species was observed.
[b]Relative abundance = (no. of spores of a species/total spores) × 100.
[c]Percentage of the 64 subplots in which the species was observed.

genus *Glomus*. Spore numbers and species composition vary significantly between annually burned and unburned prairie and with changes in topography (Gibson and Hetrick 1988). In 1991, 11 species were identified in the Belowground Plots (Chapter 14, this volume) on Konza Prairie (Table 4.3). *G. ambisporum* was clearly the dominant species, with 70% relative abundance. *Glomus mortonii* was next, with 14% relative abundance (Bentivenga and Hetrick 1992a). A 1994 sampling of the same sites found *G. etunicatum* and *G. heterosporum* to be the most common species, with relative abundances of 37% and 25%, respectively (Eom and Hartnett unpublished data).

Invertebrates

The soil fauna in grasslands is diverse and abundant, encompassing microfauna (protozoa), mesofauna (nematodes and microarthropods), and macrofauna (macroarthropods and earthworms). Studies of tallgrass prairie have emphasized members of the latter two groups (Risser et al. 1981; Seastedt et al. 1987, 1988a;

Table 4.4. Autumn density and biomass estimates for the major invertebrate groups in upland tallgrass prairie

Invertebrate Group	Sample Depth	Density (no. m^{-2})	Biomass (g/m^2)
Earthworms	30 cm	300	6.0
Macroarthropods	30 cm	100	1.5
Microarthropods	5 cm	50×10^3	ND
Nematodes	20 cm	5.5×10^6	0.5[a]

Sources: From Seastedt et al. (1987, 1988a) and Todd et al. (1992).

[a]Estimated from biomass values published in Smolik and Lewis (1982).

Todd et al. 1992), for which total biomass has been estimated to be 10 times that of aboveground invertebrates (Risser et al. 1981). Despite the significance of this observation, only limited information on invertebrate populations in tallgrass prairie soil or their impact on ecosystem processes was available before the Konza Prairie LTER Program. Previous studies cataloged the taxonomic or trophic composition of various soil invertebrate communities in tallgrass prairie (Orr 1965; Schmitt and Norton 1972; Lussenhop 1976, 1981; Norton and Schmitt 1978; Risser et al. 1981). Few of these studies included ecosystem processes as a component, and only research from the Osage Site in Oklahoma (Risser et al. 1981) included multidisciplinary studies of tallgrass prairie invertebrates. Since 1982, soil invertebrate communities on Konza Prairie have been characterized in numerous experiments, the results of which will be summarized in this chapter and Chapter 14 (this volume). A major goal of this research is to provide long-term records of invertebrate population patterns and interactions in tallgrass prairie.

Among the soil invertebrate communities on Konza Prairie, nematodes dominate in terms of abundance, although earthworms constitute the majority of the biomass (Table 4.4). Orr (1965) cataloged 82 genera and 238 species of nematodes from a tallgrass prairie in Kansas. The most abundant taxa on Konza Prairie represent fewer than 30 genera in 20 families and 7 orders (Table 4.5). The Tylenchida constitute more than 65% of the total nematode community, followed by the Rhabditida and Dorylaimida, which each contribute 10 to 15% of the total density (Seastedt et al. 1987). Fungivores (hyphal feeders) are the dominant trophic group (many of these species may also feed on lower plants; Yeates et al. 1993), constituting approximately 40% of the nematode community. Herbivores (root feeders) and microbivores (bacterial and unicellular eukaryote feeders) account for 30% and 20% of the total population, respectively, with the remaining 10% consisting of omnivores and predators. These proportions are similar to those reported by Risser et al. (1981) for a tallgrass prairie site in Oklahoma. Overall densities of predaceous nematodes are low, but they typically represent a much higher proportion of the total nematode biomass (Risser et al. 1981).

The arthropod community displays the most diversity among the invertebrate groups in tallgrass prairie soils, with 15 orders and 46 families identified by Risser et al. (1981). The most abundant taxa found on Konza Prairie and their

Table 4.5. Trophic classification of the most abundant invertebrate fauna in Konza Prairie soil

Trophic Level	Group	Taxa
Herbivores	Macroarthropods	Coleoptera (Scarabaeidae: *Phyllophaga*; Elateridae)
		Homoptera (Cicadidae)
	Nematodes	Tylenchida (*Helicotylenchus*, Paratylenchinae, *Tylenchorhynchus*)
		Dorylaimida (*Xiphinema*)
Detritivores	Earthworms	Lumbricidae (*Aporrectodea*, *Bimastos*)
		Megascolecidae (*Diplocardia*)
	Macroarthropods	Diplopoda
	Microarthropods	Oribatei
Fungivores	Microarthropods	Oribatei
		Prostigmata
		Collembola
	Nematodes	Tylenchida (*Aglenchus*, *Aphelenchoides*, *Aphelenchus*, Boleodorinae, *Ditylenchus*, *Filenchus*)
		Dorylaimida (*Diptherophora*, *Dorylaimellus*, *Tylencholaimellus*, *Tylencholaimus*)
Microbivores (bacterivores)	Nematodes	Rhabditida (*Acrobeles*, *Chiloplacus*, *Eucephalobus*)
		Araeolaimida (*Plectus*, *Wilsonema*)
		Monhysterida (*Monhystera*)
		Enoplida (*Prismatolaimus*)
Predators	Macroarthropods	Chilopoda
		Coleoptera (Carabidae)
		Diptera (Asilidae, Tabanidae)
	Microarthropods	Mesostigmata
		Prostigmata
	Nematodes	Dorylaimida (*Aporcelaimellus*, *Eudorylaimus*, *Mesodorylaimus*)
		Mononchida (*Monochus*, *Mylonchulus*)

Sources: Summarized from Orr (1965); Seastedt (1984a, 1984b); Seastedt et al. (1987).

trophic classification are summarized in Table 4.5. Nearly 60% of macroarthropod densities and biomass occurs as herbivorous species (mainly June beetle larvae, wireworms, and cicada nymphs), with the remaining 40% nearly equally divided between detritivores (millipedes) and predators (centipedes, carabid beetle, and Diptera larvae) of other soil arthropods (Seastedt et al. 1987; Todd et al. 1992). Microarthropod trophic groups are less discernible, but trophic categories consist primarily of fungivore-detritivores and predators (Seastedt 1984b). The predaceous microarthropods in grassland soils are dominated by prostigmatic mites (e.g., Rhagidiidae), which are mostly specialized predators of arthropods, and mesostigmatic mites, which are primarily nematophagous (Walter and Ikonen 1989).

The earthworm community of tallgrass prairie in the Flint Hills region con-

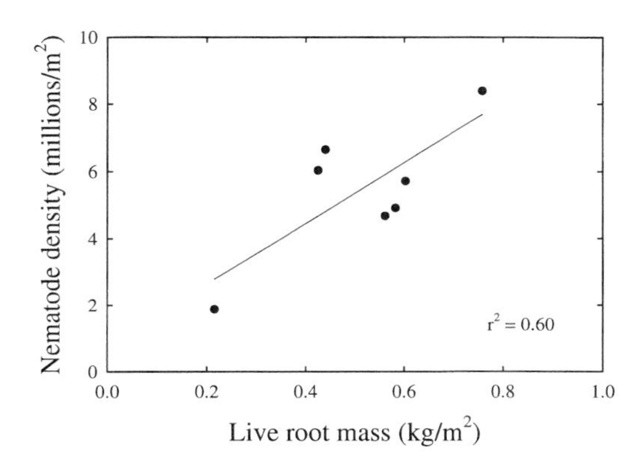

Figure 4.4. Relationship between autumn nematode density and live root biomass in annually burned tallgrass prairie. From Seastedt et al. (1987); Todd et al. (1992); Todd (1996).

sists of a mixture of North American and introduced European species (James 1984). On Konza Prairie, the European Lumbricid *Aporrectodea turgida* Eisen can account for 50% of the total biomass, although native Megascolecidae (*Diplocardia* spp., primarily *D. smithii* McNab and McKey-Fender) constitute 60 to 70% of total numbers (Seastedt et al. 1987; Todd et al. 1992). All of the nine species identified from Konza soils are detritivores, but native and introduced species exhibit some differences in feeding and burrowing strategies (James and Seastedt 1986; James and Cunningham 1989).

Spatial and temporal patterns in soil invertebrate communities on Konza Prairie have been summarized previously (James 1982; Seastedt 1984a, 1984b; Seastedt et al. 1987; Todd et al. 1992). Nearly 70% of the nematode population occurs in the surface 20 cm of soil (Todd unpublished data), and this appears to be representative of the vertical distribution of nematodes in tallgrass prairie (Risser et al. 1981). Topographic variation is minimal, with average nematode densities in the surface 20 cm of both upland and lowland soils increasing from approximately 3 million m^{-2} in the spring to nearly 6 million m^{-2} in autumn, largely because of increases in the herbivore component of the nematode community (Seastedt et al. 1987; Todd et al. 1992; Todd unpublished data). Maximum autumn densities in annually burned prairie range from 2 to 9 million m^{-2} depending on the quantity of root biomass and growing-season precipitation (Fig. 4.4).

Soil arthropods and earthworms in tallgrass prairie are stratified in the soil profile and across topographic gradients (Risser et al. 1981; James 1982; Seastedt 1984b; O'Lear 1996). Maximum microarthropod densities on Konza Prairie are found at depths ranging from the surface 5 cm of soil to below 20 cm, depending on soil moisture conditions, and tend to be higher in upland versus lowland sites (Seastedt 1984b; O'Lear 1996). The occurrence of large numbers of microarthropods below 20 cm in tallgrass prairie soil also was reported by Risser et al. (1981). Earthworm species exhibit differences in depth distribution related to feeding habits. Native *Diplocardia* spp. often are concentrated among the dense roots in the upper 10 cm of soil, whereas the introduced species *A. turgida* feeds and casts on the soil surface (James 1982; James and Seastedt 1986; James and Cunningham 1989). In contrast to microarthropods, earthworms are more abundant in the deeper soils found lower on slopes (James 1982).

Microarthropod numbers generally increase during the growing season, with maximum densities in the upper 5 cm of soil typically between 50 and 100 thousand m^{-2} (Seastedt et al. 1987; O'Lear 1996). Densities of the native earthworm *D. smithii* also increase during the growing season when conditions are favorable, although biomass of adults may actually decrease (James 1992b). Reproduction by most native and introduced earthworm species occurs primarily in the spring, with some species displaying a secondary reproductive period in autumn.

Summary

Soils are important components of the tallgrass prairie ecosystem. On Konza Prairie, soil genesis is affected by the native tallgrass vegetation, but soil properties also affect the kinds, amounts, and spatial distribution of vegetation. The soils on Konza Prairie vary according to landscape position. Although landscape position affects many soil properties on Konza Prairie such as depth, parent material stratigraphy, and texture, it does not significantly affect organic C content. The prairie vegetation exerts more of an influence on soil formation and organic matter accumulation than does topographic variation on Konza Prairie.

The soils on Konza Prairie contain a diverse assemblage of organisms. The microorganisms are essential in the recycling of nutrients and formation of soil organic matter and are supported by the large amount of C allocated belowground. It is difficult to assess all bacterial species, but two specific groups of bacteria have been studied in tallgrass prairie. The denitrifiers are numerous and probably function in multiple physiological capacities. The nitrifiers are a less diverse group and are limited by the availability of ammonium. Research on arbuscular mycorrhizal fungi demonstrates the dependency of many warm-season grasses on these fungi, which enhance nutrient uptake. Soil invertebrates indirectly stimulate decomposition of plant material by microbes and help mix the soil. All these biological components facilitate the formation of the soils on Konza Prairie by formation of soil organic matter, biological mixing of the soil, and formation of aggregates. These processes sustain the tallgrass prairie vegetation, while aiding in formation of the underlying soils; however, much remains to be studied in this unseen world.

Acknowledgments This is contribution no. 97-192-B from the Kansas Agricultural Experiment Station, Kansas State University, Manhattan. We gratefully acknowledge the USDA Natural Resources Conservation Service for their help with field sampling and for making the soil survey. Partial financial support was provided by the Department of Agronomy. We also thank Judy Adams for her careful editing and typing of the manuscript.

Part II

Terrestrial Populations and Communities

5

The Flora of Konza Prairie
A Historical Review and Contemporary Patterns

Craig C. Freeman

The contemporary flora of Konza Prairie reflects the spatial and temporal dynamics of its vegetation as affected by a host of complex and interacting factors, including geology, climate, fire, grazing, and human activities (Chapter 6, this volume). Located near the western edge of the tallgrass prairie ecosystem, Konza Prairie is dominated by grassland vegetation characterized by a few native, dominant, perennial, warm-season grasses and an assortment of other species with decreasing abundance and influence in the prairie. The flora is decidedly calciphilous and has a preponderance of widespread species. Forest and wetland natural communities typical of eastern Kansas are limited in area, but they contribute significantly to the site's total species richness.

Historical Vegetation Patterns

The principal geohistorical events that gave rise to the temperate grasslands of central North America are summarized by Bare and McGregor (1970), Wright (1970), Risser et al. (1981), Axelrod (1985), Graham (1993), and Delcourt and Delcourt (1993). These overviews describe some of the major vegetational changes that likely have affected Konza Prairie during the past 10 million years.

From the mid-Tertiary to late Tertiary (35–10 million yr B.P.), broad-leaved deciduous forests of the Arcto-Tertiary geoflora held sway through the temperate latitudes of North America (Graham 1993). Widespread cooling throughout North America and reduced summer rains in the midcontinent region during the late Tertiary (10–2 million yr B.P.) promoted the development of extensive grasslands at the expense of deciduous hardwood forests, but paleofloras from the central and southern Great Plains are inadequate to permit detailed reconstructions of the vegetation during this interval (Graham 1993). Macrofossils from the middle Miocene to early Pliocene suggest that forests and woodlands were present in valleys and mesic localities in the Great Plains, whereas uplands supported scattered grassland and parkland (Axelrod 1985).

Extensive grasslands probably arose initially in central North America 7 to 5 million yr B.P. following the Miocene uplift of the Rocky Mountains (Axelrod 1985). The ensuing rain shadow and lowered minimum winter temperatures in the midcontinent region during the Pliocene promoted the spread of prairie vegetation and further restricted woody elements to more mesic and sheltered sites (Graham 1993). Many extant plant genera and species evolved before the close of the Tertiary (Delcourt and Delcourt 1993).

Glacial and interglacial episodes during the Quaternary caused dramatic shifts in the distribution of the Great Plains vegetation (Bare and McGregor 1970; Wright 1970; Axelrod 1985; Delcourt and Delcourt 1993). Moister climates returned during the early and mid-Quaternary, favoring the expansion of forests and woodlands at the expense of prairie. Fossil pollen and macrofossils indicate that boreal forest covered most of the eastern and central United States during maximum Wisconsin glaciation 20,000 to 14,000 yr B.P. (Wright 1970; Delcourt and Delcourt 1993), extending as far west as northeastern Kansas (McGregor 1968). Boreal forest was replaced by deciduous forests in the southern Great Plains about 12,500 yr B.P. (Wright 1970; Axelrod 1985). By 10,000 yr B.P., prairie vegetation dominated the central and southern Great Plains (Delcourt and Delcourt 1993).

During the middle Holocene (9,000–6,000 yr B.P.), a wedge of prairie vegetation reached its easternmost limit in the eastern Midwest (Delcourt and Delcourt 1993). Global climate cooling during the last 5,000 years resulted in the westward retreat of prairie and the readvance of deciduous forest along valleys well into the grasslands. This migration is recorded in pollen profiles from Muscotah Marsh (ca. 95 km ENE of Konza Prairie in Atchison County, Kansas) around 5,100 yr B.P. (Wright 1970). These global climatological changes and the influence of Native Americans on the biota, particularly during the past 5,000 years, ultimately gave rise to the modern prairie flora (Axelrod 1985; Delcourt and Delcourt 1993).

Initial settlement of the Flint Hills by Europeans commenced in the 1840s (Malin 1942). General Land Office surveys of the Kansas Flint Hills were conducted from the 1850s until the early 1870s (Barker 1969), with the Konza Prairie area being surveyed from 1855 to 1856 (Abrams 1986). Surveyors' field notes offer sketchy views of the region's vegetation because they are biased toward areas of timber and potential farmland. Details about the prairie are uniformly lacking. Valleys along rivers, streams, and creeks often are described as "well timbered," and the woody species frequently reported correspond well with species seen today in remnant forests and woodlands (Barker 1969).

Settlements were established in the Kansas River valley as far west as the

Konza Prairie vicinity soon after the Kansas Territory was established in 1854 (Malin 1942). By the 1870s, diversified farming was widespread on bottomlands (Barker 1969; Isern 1983), but agricultural growth statewide was curtailed because of two droughts during the decade (Madden 1973). Limited attempts were made to farm the uplands, but these areas continued to be used primarily for livestock grazing. Barbed wire and herd laws ended the era of free rangeland by the mid-1880s (Malin 1942; Isern 1983). By the end of the decade, a major livestock industry had developed in the Flint Hills; cattle were shipped to Kansas by rail in the spring to be fattened and then shipped to eastern markets by late summer (Barker 1969; Isern 1983).

Declines in the area of original tallgrass prairie exceed those of all major ecosystems on the continent (Samson and Knopf 1994), but the shallow, rocky soils of the Flint Hills spared most of Konza Prairie from the plow. The native vegetation of the original Konza Prairie generally was in good to excellent condition when it was acquired in 1971; conditions on the Dewey Ranch at the time of its purchase varied from fair to excellent (Freeman and Hulbert 1985).

Contemporary Vegetation Patterns

Tallgrass prairie dominates the Konza Prairie landscape, but other natural communities with limited representation contribute significantly to the overall species richness. Major natural community types are tallgrass prairie, deciduous forest, wetlands, and aquatic communities. Anthropogenic disturbances have altered parts of Konza Prairie, providing repeated opportunities for the introduction and establishment of nonnative species.

Tallgrass Prairie

More than 90% of the vegetation of Konza Prairie is unplowed native, tallgrass prairie (Freeman and Hulbert 1985). Küchler (1974a) mapped the potential natural vegetation of the area as bluestem prairie—a tallgrass prairie association. Lauver (1989) classified the natural vegetation as Flint Hills tallgrass prairie. Most tallgrass prairie on Konza Prairie is mesic to dry-mesic. Wet-mesic associations are highly localized along upland drainages and around seeps on slopes.

The grassland vegetation of Konza Prairie is characterized by a few dominant C_4 grasses that have wide ranges and broad ecological amplitudes. The dominant species are *Andropogon gerardii* Vitman, *A. scoparius* Michx., *Sorghastrum nutans* (L.) Nash, and *Panicum virgatum* L. Grasses that are common on mixed-grass and shortgrass prairies farther west (*Bouteloua gracilis* [H.B.K.] Lag. ex Griffiths, *B. curtipendula* [Michx.] Torr., and *Buchloe dactyloides* [Nutt.] Engelm.) occur in xeric sites.

Nearly 250 native species grow largely in prairie habitats on Konza Prairie, 70% of which are perennials and 26% of which are annuals. Forbs (nongraminoid herbs) are conspicuous elements of the prairie flora. Common species on mesic sites include *Aster ericoides* L. subsp. *ericoides*, *Erigeron strigosus* Muhl. ex Willd., *Psoralea tenuiflora* Pursh, *Salvia azurea* Lam., *Solidago missouriensis* Nutt. var. *fasciculata* Holz., and *S. rigida* L. var. *humilis* Porter. Drier sites support populations of *Ambrosia psilostachya* DC., *Artemisia ludoviciana* Nutt. var. *ludoviciana*, *Aster oblongifolius* Nutt. var. *oblongifolius*, *Echinacea angustifolia* DC., *Hedyotis nigricans* (Lam.) Fosb., *Liatris aspera* Michx., *L. mucronata* DC., *L. punctata* Hook., *Tragia betonicifolia* Nutt., and *Vernonia baldwinii* Torr. subsp. *interior* (Small) Faust. Woody species, primarily shrubs, are more common in unburned tallgrass prairie. Common species are *Amorpha canescens* Pursh, *Cornus drummondii* C.A. Mey., *Rhus aromatica* Ait. subsp. *serotina* (Greene) Rehd., *Rhus glabra* L., and *Rosa arkansana* Porter.

About 30% of the bryophytes reported on Konza Prairie occur exclusively or occasionally in prairie habitats (Merrill 1991a). *Ephemerum spinulosum* Bruch & Schimp. may be the most abundant species, and *Astomum muehlenbergianum* (Sw.) Grout, *Phascum cuspidatum* Hedw., and *Weissia controversa* Hedw. are encountered commonly.

Deciduous Forest

Nearly continuous, 10- to 300-m-wide bands of deciduous forest extend onto Konza Prairie along tributaries to the Kansas River (Abrams 1986), covering about 7% of the site in 1985 (Knight et al. 1994). These forests, which are best developed along Kings Creek and Shane Creek, have expanded in area since the mid-1800s (Abrams 1986; Knight et al. 1994). The forests on Konza Prairie are western extensions of the deciduous forests that historically dominated the eastern half of the United States.

Roughly 100 species, 80% of which are perennials, are restricted to forest habitats on Konza Prairie. Forests on Konza Prairie exhibit lower species richness than do similar natural communities farther east. Still, they support many woody and herbaceous species at or near the western limit of their ranges. Dominant trees are *Quercus macrocarpa* Michx., *Q. muehlenbergii* Engelm., *Celtis occidentalis* L., and *Ulmus americana* L., with *Q. muehlenbergii* Engelm. being prevalent in the most xeric sites (Abrams 1986). Overstory associates are *Acer negundo* L. var. *violaceum* (Kirchn.) Jaeg., *Carya cordiformis* (Wang.) K. Koch, *Cercis canadensis* L., *Fraxinus pennsylvanica* Marsh. var. *subintegerrima* (M. Vahl) Fern., *Gleditsia triacanthos* L., *Gymnocladus dioica* (L.) K. Koch, *Juglans nigra* L., *Juniperus virginiana* L., *Morus alba* L., *Platanus occidentalis* L., *Populus deltoides* Marsh. subsp. *monilifera* (Ait.) Eckenw., and *U. rubra* Muhl. Understory shrubs include *Cornus drummondii*, *Prunus americana* Marsh., *Ribes missouriense* Nutt. ex Torr. & A. Gray, *Rhus aromatica* subsp. *serotina*, *R. glabra*, *Symphoricarpos orbiculatus* Moench, and *Zanthoxylum americanum* P. Mill. *Symphoricarpos* is particularly common in areas that have a history of disturbance. Common vines are *Menispermum canadense* L., *Parthenocissus quinquefolia* (L.) Planch., *Toxicodendron radicans* (L.) O. Ktze. subsp. *negundo* (Greene) Gillis, and *Vitis riparia* Michx. A depauperate assemblage of spring ephemerals along gallery forests includes *Arisaema dracontium* (L.) Schott, *Chaerophyllum procumbens* (L.) Crantz, *Cardamine concatenata* (Michx.) O. Schwarz, *Dicentra cucullaria* (L.) Bernh., *Erythronium albidum* Nutt., *Phlox divaricata* L. subsp. *laphamii* (Wood) Wherry, and *Ranunculus abortivus* L.

Based on data from Merrill (1991a), over 60% of the bryophytes on Konza Prairie are reported in forests and shrub thickets, but fewer than 30% of these are restricted to such habitats. The forest bryoflora comprises widespread species typical of eastern deciduous forests. *Leskea gracilescens* Hedw., *Brachythecium acuminatum* (Hedw.) Aust., *Entodon seductrix* (Hedw.) C. Müll., *Anomodon minor* (Hedw.) Furnr., *Lindbergia bracyptera* (Mitt.) Kindb., *Orthotrichum pusillum* Mitt., *O. strangulatum* P. Beauv., and *Frullania inflata* Gott. are the most common species. Shaded soil banks support *Brachythecium oxycladon* (Brid.) Jaeg., *Campylium hispidulum* (Brid.) Mitt., and *Plagiomnium cuspidatum* (Hedw.) T. Kop.

Shrub thickets are common at the upper reaches of draws and on slopes around limestone outcrops. These sites typically are dominated by *Cornus drummondii*, *Rhus aromatica* subsp. *serotina*, *R. glabra*, and *Symphoricarpos orbiculatus*. *Schistidium agassizii* Sull. & Lesq. is a common moss on shaded and exposed limestone outcrops in gallery forests and shrub thickets (Merrill 1991a).

Wetlands and Aquatic Sites

Wetland and aquatic natural communities are poorly represented, but they contribute significantly to total species richness of Konza Prairie. More than 28% of the 311 vascular plant species on Konza Prairie assigned a wetland indicator status in the Central Plains region (Reed 1988) are facultative wetland or obligate wetland species. Hydrophytes in and along perennial and intermittent streams are *Amorpha fruticosa* L., *Bidens polylepis* Blake, *Carex annectens* (Bickn.) Bickn. var. *xanthocarpa* (Bickn.) Wieg., *Cyperus odoratus* L., *Eleocharis erythropoda* Steud., *Glyceria striata* (Lam.) Hitchc., *Juncus* spp., *Leucospora multifida* (Michx.) Nutt., *Lippia lanceolata* Michx., *Lobelia cardinalis* L., *Lysimachia ciliata* L., *Polygonum* spp., *Rorippa palustris* (L.) Bess. subsp. *glabra* (Schulz) Stuckey, *Rumex altissimus* Wood, *Spartina pectinata* Link, *Verbena hastata* L., and *Veronica catenata* Penn. var. *catenata*. Typical bryophytes are *Didymodon tophaceus* (Brid.) Lisa and *Hygroamblystegium tenax* (Hedw.) Jenn., with *Bryum pseudotriquetrum* (Hedw.) Gaertn. et al. forming cushions along stream margins (Merrill 1991d).

Seeps are localized wetlands that have soils or substrates saturated by seeping groundwater (Lauver 1989). Throughout the Flint Hills, these natural communities usually occur on slopes at contacts between limestone and shale bedrock layers. Frequent hydrophytes around seeps are *Eleocharis* spp., *Equisetum laevigatum* A. Br., *Erigeron philadelphicus* L., *Lythrum alatum* Pursh, *Juncus* spp., *Tripsacum dactyloides* (L.) L., and *Typha latifolia* L. The constant or near-constant flow of water from springs provides habitat for populations of *Nasturtium officinale* R. Br., *Mimulus glabratus* H.B.K. var. *fremontii* (Benth.) A. L. Grant, and *Veronica catenata* var. *catenata*. *Didymodon tophaceus*, *Drepanocladus aduncus* (Hedw.) Warnst., *Philonotis marchica* (Hedw.) Brid., and

Aneura pinquis (L.) Dum. are bryophytes often found around seeps and springs on Konza Prairie (Merrill 1991a).

A few submerged and floating-leaved hydrophytes grow in the dozen scattered, man-made ponds on Konza Prairie. Included are *Najas guadalupensis* (Spreng.) Morong, *Potamogeton foliosus* Raf., *P. nodosus* Poir., and *P. pusillus.* L. *Ammannia coccinea* Rottb., *Sagittaria latifolia* Willd., and *Typha latifolia* are common emergent species.

Disturbed Areas

Humans have altered greatly some parts of Konza Prairie. We can only speculate about the exact influence of Native Americans on the Konza Prairie flora, but many anthropogenic changes since the 1850s are obvious. Farmsteads, roadsides, ditches, cultivated ground, cool-season grasslands, and assorted other disturbed sites are found in many parts of Konza Prairie. Lowlands near the headquarters include roughly 50 ha that were cultivated or converted to brome pasture (Zimmerman 1985). Other significant anthropogenic disturbances have resulted from selective logging, cattle grazing, and herbicide spraying (Abrams 1986).

Common, nonnative weeds are *Bromus japonicus* Thunb. ex Murr., *Capsella bursa-pastoris* (L.) Medic., *Digitaria sanguinalis* (L.) Scop., *Echinochloa crus-galii* (L.) Beauv. var. *crusgalii*, *Erysimum repandum* L., *Festuca arundinacea* Schreb., *F. pratensis* Huds., *Hibiscus trionum* L., *Ipomoea hederacea* Jacq., *Kochia scoparia* (L.) Schrad., *Melilotus alba* Medic., *M. officinalis* (L.) Pall., *Taraxacum officinale* Weber, *Thlaspi arvense* L., *Torilis arvensis* (Huds.) Link, and *Tribulus terrestris* L. The mosses *Bryum argenteum* Hedw. and *Ceratodon purpureus* (Hedw.) Brid. grow on compacted soil near the Konza Prairie headquarters, and *Physcomitrium pyriforme* (Hedw.) Hampe occurs in lowland brome fields (Merrill 1991a).

Localized disturbances caused by various activities of animals provide sites for the establishment of many native and naturalized species (Chapter 6, this volume). Gibson (1989) found that annuals are important components of the flora of badger mounds, whereas common perennials are more abundant on pocket gopher mounds and anthills. Relict bison wallows on uplands tend to support plants commonly encountered on mesic lowlands and therefore are floristically distinct from the surrounding upland prairie. Most species reported by Gibson (1989) at sites disturbed by native animals are indigenous.

Floristic Studies

Organized efforts to determine which vascular plants occur on Konza Prairie began almost immediately after the original Konza Prairie was acquired. Marple (1975) compiled a checklist of 270 species of vascular plants based on collections and observations restricted to the original Konza Prairie. Hulbert (1976) revised and expanded this list. The acquisition of the Dewey Ranch in 1976 increased the area by more than 3,000 ha, capturing a variety of natural community types unrepresented on the original Konza Prairie. Freeman (1977) conducted a floristic survey of the ranch, expanding the list for the entire site to 363 species. Periodic updates from 1980 to 1987 raised the number to 467 (Freeman and Hulbert 1985; Freeman and Gibson 1987). Continued research since 1987 has added many previously unknown species to the Konza Prairie vascular flora (Kazmaier 1993). Presently, 529 vascular plant species (531 taxa) are reported from Konza Prairie.

Systematic bryological surveys did not begin until late 1987 (Merrill 1991a). By 1991, 62 species of bryophytes had been reported (Merrill 1991a). Nomenclatural adjustments (Anderson et al. 1990) brought the number to 60 species (61 taxa). Voucher specimens for most vascular plants and bryophytes reported from Konza Prairie have been deposited in the Kansas State University Herbarium.

Characteristics of the Konza Prairie Flora

Vascular Plants

Five hundred twenty-nine species of vascular plants in 97 families and 324 genera have been reported on Konza Prairie since 1975. A list of plants is available at http://climate.konza.ksu.edu. These numbers represent 88%, 71%, and 54% of the families, genera, and species, respectively, in the Flint Hills based on data provided by Barker (1969). Compared with the entire Kansas flora, the figures are 64%, 42%, and 24%, respectively.

The 10 most species-rich families in descending order are Asteraceae (AST); Poaceae (POA); Fabaceae (FAB); Brassicaceae (BRA); Euphorbiaceae (EUP); Cyperaceae (CYP); Lamiaceae (LAM); Scrophulariaceae (SCR); Polygonaceae (POL); and Rosaceae (ROS) (Figs. 5.1, 5.2). These families collectively account for 57% of all species known from Konza Prairie. The same 10 families are identified by Barker (1969) as the most species-rich in the Flint Hills. The rank order of families of the two floras is similar (but not identical), suggesting that the Konza Prairie flora is a taxonomically and ecologically representative subset of the entire Flint Hills flora. Among the 10 largest families on Konza Prairie, introduced species are proportionally highest in the Brassicaceae (63%), Lamiaceae (31%), Poaceae (28%), and Scrophulariaceae (25%).

Provenance, longevity, life-form, and habitat data for Konza Prairie plants are summarized in Figures 5.3 through 5.5. Provenance and longevity data are derived primarily from *Flora of the Great Plains* (Great Plains Flora Association 1986). Raunkiaer life-form codes are assigned based on descriptions in that publication. Habitat codes are based on Freeman and Hulbert (1985), Freeman and Gibson (1987), and the Great Plains Flora Association (1986).

Nearly 100 introduced species have been reported on Konza Prairie—18% of the total number. Nearly identical percentages are reported for the Flint Hills (Barker 1969) and for Kansas (Stuckey and Barkley 1993). Human activities have resulted in the intentional and accidental introduction of many species into the grassland. Problems associated with exotic species frequently are permanent

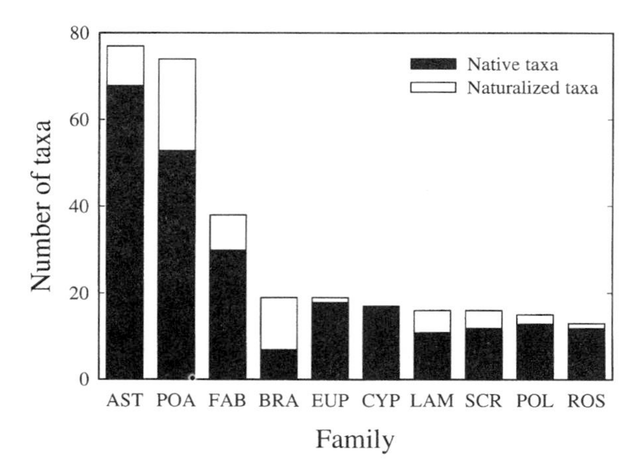

Figure 5.1. Native and naturalized taxa in the 10 most species-rich families on Konza Prairie.

and may be among the most pervasive human-induced influences affecting biodiversity (Coblentz 1990).

Perennial plants constitute 65% of all the species on Konza Prairie (Fig. 5.3). Most of the remaining species are annuals (29%). Biennials and species with variable longevity states are relatively rare. If the longevity of only indigenous species is considered, perennials, annuals, biennials, and species with variable longevity represent 70%, 26%, 2%, and 2% of species, respectively. Annuals are more common among introduced species, accounting for 43% of this group.

Life-form classes provide useful functional criteria for describing vegetation (Goldsmith et al. 1986), allowing elaboration of simple longevity data. Raunkiaer's (1934) system, which groups seed plants based on the position of perennating buds relative to the soil surface, distinguishes five classes: phanaerophytes

Figure 5.2. The most species-rich family on Konza Prairie is the Asteraceae. *Rudbeckia hirta* is a common and conspicuous member of this family. (C. Freeman)

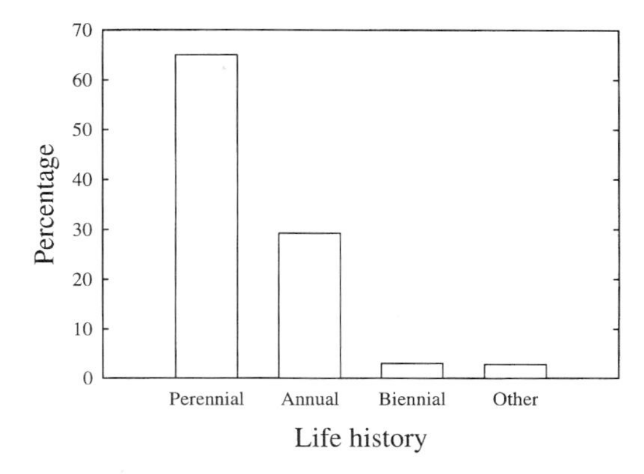

Figure 5.3. Life histories of Konza Prairie vascular plants. Other = various combinations of life history states.

(trees and tall shrubs with buds at least 25 cm above the ground); chamaephytes (low shrubs and perennial herbs with buds less than 25 cm above the ground); hemicryptophytes (perennials and biennials with buds at ground level or within the surface layer of soil); geophytes (= cryptophytes) (perennials with rhizomes, tubers, or bulbs located well below the ground); and therophytes (annuals, which survive unfavorable periods as seeds).

Frequencies of life-form classes for Konza Prairie plants are 11% phanaerophytes; 1% chamaephytes; 38% hemicryptophytes; 19% geophytes; 29% therophytes; and 2% with variable life-forms (Fig. 5.4). This spectrum is very similar to that reported by Stalter et al. (1991), which was based on the 441 species listed in Freeman and Hulbert (1985). Both spectra reveal that hemicryptophytes constitute the most abundant life-form on Konza Prairie, followed by therophytes, geophytes, phanaerophytes, and chamaephytes. This pattern is congruous with

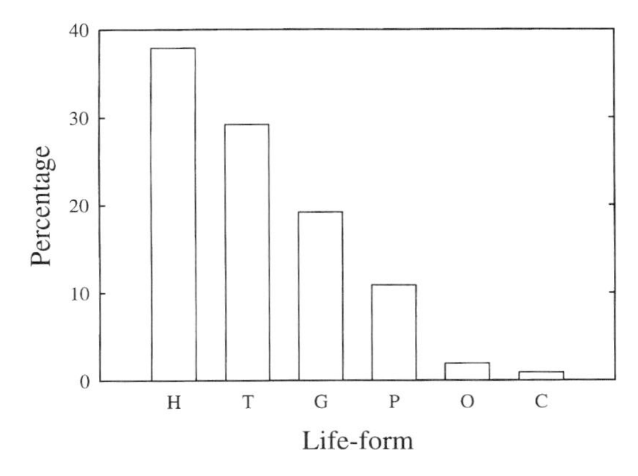

Figure 5.4. Life-forms of Konza Prairie vascular plants. H = hemicryptophytes; T = therophytes; G = geophytes; P = phanaerophytes; O = various combinations of these states; C = chamaephytes.

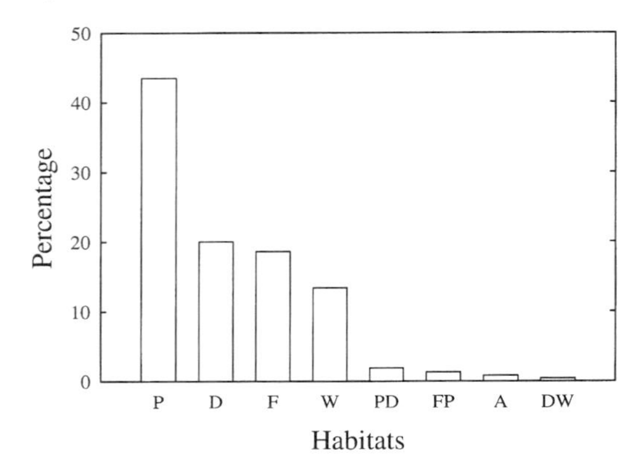

Figure 5.5. Habitats of Konza Prairie vascular plants. P = prairie; D = disturbed sites; F = forest; W = wetland; A = ponds and pools in streams (aquatic); combinations of letters are for taxa regularly found in more than one habitat type.

temperate vegetation types as a whole (Polunin 1960) and matches closely the prairie studied by Stalter et al. (1991).

The assignment of species to broad habitat types on Konza Prairie allows some generalization about the flora, even though these types may not be mutually exclusive (e.g., wetlands occur in both prairies and forests). Also, many species regularly occur in two or more habitat types.

A majority of the species occur primarily in prairie habitats (Fig. 5.5). This observation is expected because prairie occupies more than 90% of the area. Forests cover roughly 7% of Konza Prairie, but they contribute 19% of the total species. Wetlands and aquatic habitats are even more limited, but more than 14% of the species occur in these habitats. Species associated with disturbed sites include disturbance-dependent native species and introduced species. Slightly more than 70% of the species that regularly occupy disturbed areas of Konza Prairie are introduced.

Wells (1970a) suggested that the flora of the central Great Plains is derived recently from surrounding biomes, citing as evidence the extremely low level of endemism, dominant grasses that have wide ranges and occur throughout much of their ranges as synusial components of woodland communities, and the inherent instability of Great Plains grassland ecosystems. A variety of complementary evidence seems to support this claim (Axelrod 1985). Also, species richness and endemism generally are correlated positively (Noss and Cooperrider 1994). Only about 20% of all native North American vascular plants north of Mexico occur in the prairie province, and less than 2% of the grassland flora is endemic (Thorne 1993). Figures for biomes peripheral to the grasslands consistently show greater species richness and higher levels of endemism (Gentry 1986; Thorne 1993).

Based on distribution data in the *Atlas of the Flora of the Great Plains* (Great Plains Flora Association 1977), 56% of the native plants on Konza Prairie are widespread, occurring throughout more than half of the Great Plains region. Most of the remaining native species also are widespread, but the majority of their ranges lie outside the Great Plains. Within the Great Plains, 20% are limited to the southeastern quarter, 16% are confined to the eastern half, and 9% have distributions restricted to the southern half. Approximately 22% of the native species on Konza Prairie (95 of 438) reach the limit of their range in the Flint Hills region. Only a few species on Konza Prairie truly can be considered grassland endemics (e.g., *Astragalus plattensis* Nutt. ex Torr. & A. Gray), and no vascular plants are endemic to the site. The widespread occurrence of many tallgrass prairie plants undoubtedly has buffered most species from staggering habitat losses. No federally threatened or endangered plant species occur on Konza Prairie, and only one state-rare species (*Chenopodium pallescens* Standl.) occurs on the site.

Bryophytes

Sixty species of bryophytes in 21 families and 41 genera are reported from Konza Prairie (see http://climate.konza.ksu.edu). Mosses (56 species, 18 families, 38 genera) constitute the largest class, with the remainder being liverworts (4 species, 3 families, 3 genera). No hornworts are known from Konza Prairie. These numbers represent 66%, 64%, and 57% of the families, genera, and species, respectively, in the Flint Hills and 37%, 35%, and 25% of the same groups in the Kansas bryoflora based on data in Churchill (1985); McGregor (1955); Merrill (1991a, 1991b, 1991c, 1991d); and Merrill and Timme (1991).

The five most species-rich families on Konza Prairie are the Amblystegiaceae (8 species), Pottiaceae (8 species), Brachytheciaceae (6 species), Bryaceae (5 species), and Orthotrichaceae (5 species). Collectively, these families include 53% of all species. The Pottiaceae, Amblystegiaceae, Bryaceae, and Brachytheciaceae also rank among the five largest families in the Flint Hills.

The Konza Prairie bryoflora is characterized by species that are more or less widespread in eastern North America (Merrill 1991a), an observation that applies also to the entire moss flora of Kansas (Churchill 1985). Nearly two-thirds of the species have been found in gallery forests or shrub thickets, but only about 30% are restricted to these habitats. Less than one-third of Konza Prairie bryophytes have been found on the prairie, and just five species are known only from grassland sites. Absent from the Konza Prairie bryoflora are any elements from the western and southwestern Great Plains (Merrill 1991a).

No endemic or globally rare species are known from Konza Prairie, but species on the site currently reported from five or fewer counties in Kansas are *Astomum phascoides* (Hook. ex Drumm.) Grout (4), *Brachythecium salebrosum* (Web. & Mohr) Schimp. (1), *Drepanocladus aduncus* (Hedw.) Warnst. (4), *Fissidens obtusifolius* Wils. var. *kansanus* Ren. & Card. (2), *Homomallium mexicanum* Card. (1), *Orthotrichium ohioense* Sull. & Lesq. (4), *Pleuridium subulatum* (Hedw.) Rabenh. (3), and *Schistidium apocarpum* (Hedw.) Bruch & Schimp. (2). Given our relatively fragmentary knowledge of the state's bryoflora, the apparent rarity of most of these species may be a collection artifact.

Summary

The contemporary flora of Konza Prairie has evolved over the past few million years in response to geology, climate, fire, grazing, human activities, and other influences. Located near the western edge of the tallgrass prairie ecosystem, Konza Prairie is dominated by grassland vegetation characterized by a few dominant perennial, warm-season grasses and a wide variety of other graminoids and forbs. A majority of the native species have broad ranges. No vascular plants or bryophytes are endemic to Konza Prairie, and few of its species are grassland endemics. No federally protected plants and less than a dozen species that are rare in Kansas occur on Konza Prairie.

The vascular flora comprises 529 species and is well known. Ten percent of the families account for more than half of all species on the site. Introduced species make up 18% of the total. Nearly two-thirds of the species are perennial, and most of the remaining species are annuals. The life-form spectrum of Konza Prairie is typical of most temperate ecosystems; hemicryptophytes and geophytes constitute a large percentage of the species, and chamaephytes are quite rare. Nearly 250 species occur primarily in prairie habitats; however, forests, wetlands, and aquatic natural communities support many nongrassland species, greatly increasing the overall species richness. More than half of the native vascular plants on Konza Prairie are widespread in the Great Plains.

Although less thoroughly studied, the bryoflora of Konza Prairie includes 60 species and is representative of the region. The five largest families (out of a total of 21) include slightly more than half of all species reported from the site. Most mosses and liverworts known from Konza Prairie have large ranges and floristic affinities with forests and woodlands in the eastern United States. Nearly 60% of the species are associated with habitats in gallery forests and shrub thickets.

6

Plant Populations
Patterns and Processes

David C. Hartnett
Philip A. Fay

Long-term research on Konza Prairie has greatly increased our understanding of plant population structure, dynamics, and interactions in tallgrass prairie. It is well recognized that periodic fire, grazing, and a variable climate historically have been integral factors in shaping the structure of plant communities and directing the evolution and coevolution of their component species (Chapter 1, this volume). On Konza Prairie, landscape heterogeneity associated with the Flint Hills topography also interacts with these three primary factors to influence the distribution and abundance of plant species. These key factors have important direct, indirect, and interacting effects on tallgrass prairie plant life histories, population dynamics, species interactions, and community structure (Fig. 6.1). The structure and composition of the vegetation, in turn, influence populations of birds, mammals, and invertebrate consumers (Chapters 7, 8, this volume). Ultimately, these interactions drive local patterns of species abundance, migration, and extinction, imparting patterns to the distribution and abundance of species at the watershed and larger scales.

In this chapter, we focus on plant populations of the tallgrass prairie. We provide a brief overview of our understanding of the population ecology of prairie plants before the LTER Program; a description of spatial distributions and temporal patterns of abundances of these populations derived from long-term studies on Konza Prairie; and a summary of their responses to fire, grazing, climate, and their interactions. We explore the numerous interactions and feedbacks involving plants depicted in Figure 6.1 and attempt to weave together our understanding of population processes and the abundances and dynamics of various taxa into a broader picture, highlighting the generalities and further questions that have emerged from studies of these various populations over the past 15 years.

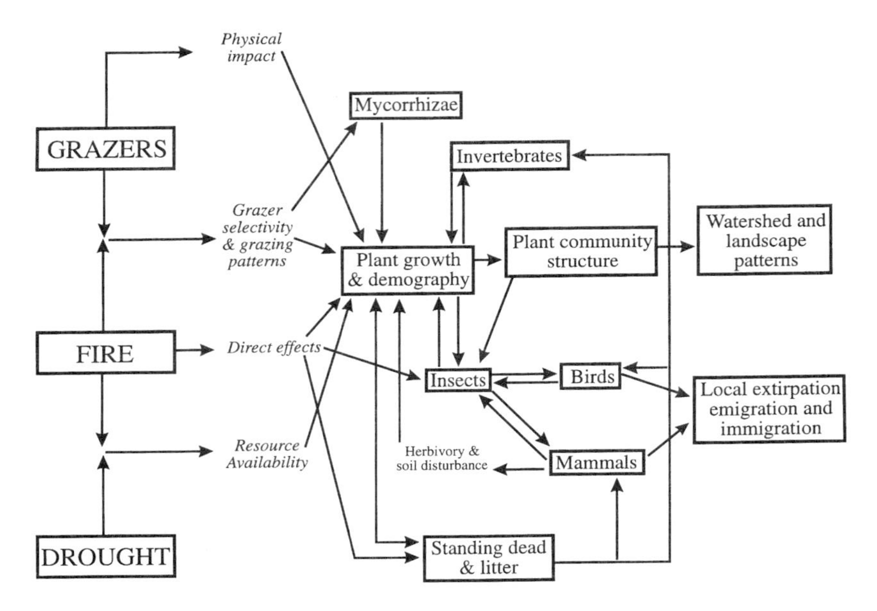

Figure 6.1. Relationships between the core abiotic and biotic components affecting plant populations in tallgrass prairie. Arrows indicate known and hypothesized interactions among components.

Plant ecology in the United States has strong historical roots in the midwestern prairies where F. E. Clements, J. E. Weaver, H. A. Gleason, and others conducted numerous studies during the first half of this century (Chapter 1, this volume). A wealth of information exists on species composition and successional dynamics of tallgrass prairie vegetation and its response to drought and grazing (Weaver and Albertson 1944; Weaver 1954, 1968). Tallgrass prairie studies associated with the International Biological Program (IBP) were begun in 1969 at the Osage Site in northeastern Oklahoma. Although the ecosystems approach was the primary focus of these studies, various floristic analyses provided general patterns of life history and morphological adaptations of prairie plants and patterns of diversity and abundances (Risser et al. 1981). The IBP studies also addressed effects of grazing, fire, nutrients, irrigation, and pesticides, thus contributing to our understanding of effects of these factors on plant community structure.

Most early studies focused primarily on describing plant species composition and abundances in tallgrass prairie. Later studies examined underlying processes organizing prairie plant communities, emphasizing niche relations and competition (e.g., Platt and Weis 1977; Parrish and Bazzaz 1976, 1979; Bazzaz and Parrish 1982) and the potential role of disturbance and nonequilibrium processes (Platt 1975; Pickett 1980). Only a small number of studies have examined in detail the life histories and demography of prairie grasses and forbs (e.g., Platt et al. 1974; Rabinowitz 1978; Rabinowitz and Rapp 1980; Rapp and Rabinowitz 1985; Briske and Butler 1989; Louda et al. 1990; Keeler 1991; Hartnett and Keeler 1995). Many characteristic features of the perennial herbs that dominate the tall-

Figure 6.2. View from an upland portion of Konza Prairie across a watershed dissected by gallery forest. Population and community data are collected from upland, hillside, and lowland topographic positions in a variety of watersheds on Konza Prairie. (A. Knapp)

grass prairie make study of their demography extremely difficult. Their extensive vegetative reproduction via rhizomes and intermingling of clones makes nondestructive identification, aging, and censusing of individuals (genets) unfeasible. In addition, the high and variable local population densities and the large number of species that co-occur make accurate estimation of populations a highly labor-intensive task. Yet this spatial and temporal variability is not simply a hindrance to sampling and estimation of plant population and community patterns but rather an important characteristic of these grasslands of direct ecological interest.

Patterns of Species Abundance and Distribution

The cover and frequencies of plant species on Konza Prairie have been measured annually since 1981 via sampling along permanent transects stratified across a topographic gradient (Fig. 6.2) from shallow soil upland (Florence cherty silt loam) to lowland sites (Tully silt clay loam) in watersheds burned in the spring at various intervals and since 1988 in watersheds grazed by *Bos bison*. In addition, the tiller densities and flowering of selected dominant grasses and the growth and reproduction of selected forbs are measured annually.

Overall, both spatial and temporal distributions of plant species show bimodal patterns (Fig. 6.3), conforming closely to predictions of Hanski's "core-satellite" hypothesis (Hanski 1982; Collins and Glenn 1991). The vegetation consists of a matrix of widely distributed, temporally stable, and abundant "core" species (mostly dominant warm-season grasses) and a second group of localized, less abundant, and temporally dynamic and unpredictable "satellite"

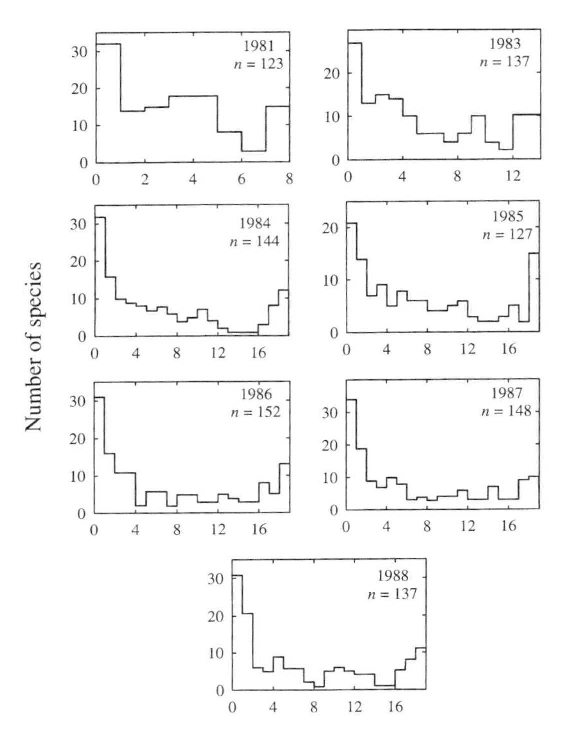

Figure 6.3. Plant species distributions on LTER species composition transects, 1981–1988. Data represent a composite sample from upland and lowland sites in watersheds with various burn frequencies. Eight sites were sampled in 1981, 14 sites in 1983, and 19 sites for 1984–1988; n = total number of species among all sites. Data for 1986 modified from Collins and Glenn 1991, with permission of the Ecological Society of America.

species (subdominant grasses and forbs) within the grass matrix. These two distinct modes of species abundances and dynamic behavior suggest two distinct sets of processes influencing the population dynamics and community structure of prairie plants.

Glenn and Collins (1990) showed that patch structure of the plant communities on Konza Prairie, as defined by satellite species, was not different from a simulated random patch structure. Furthermore, plant community heterogeneity (dissimilarity in species composition from one point to another in the plant community) and local rates of extinction of satellite species were correlated positively with total plant species richness and correlated negatively with cover of the dominant matrix grass *Andropogon gerardii* Vitman (Collins 1992; Glenn and Collins 1992). These patterns suggest that patch structure and species richness in tallgrass prairie vegetation are determined principally by the dynamic behavior of the satellite species growing within the grass matrix, and that stochastic dispersal and extinction may be the primary processes influencing plant community

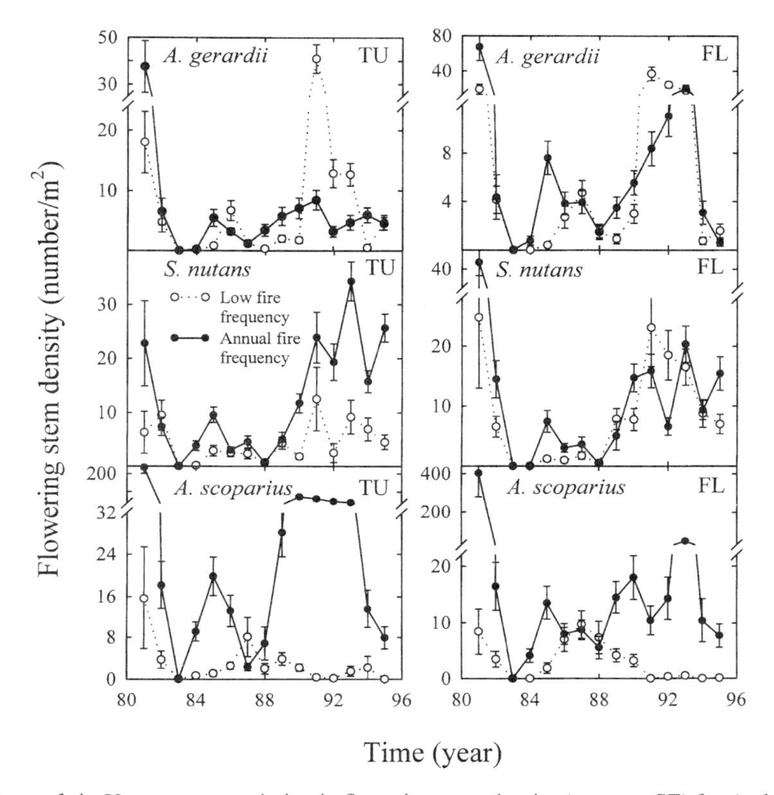

Figure 6.4. Year-to-year variation in flowering stem density (mean ± SE) for *Andropogon gerardii, A. scoparius,* and *Sorghastrum nutans* on lowland (Tully, TU) and upland (Florence, FL) soils in annually burned versus infrequently burned LTER watersheds.

structure. These nonequilibrium patch dynamics of satellite species were suggested by earlier studies in other tallgrass prairie sites. For example, Rabinowitz (1978) found a strong positive correlation between mean abundance and mean propagule weight of eight prairie grasses (i.e., rarer species were characterized by smaller seeds), suggesting that sparser species persist by having greater seed dispersal capabilities and higher immigration rates into small empty patches.

By contrast, the spatial and temporal stability of matrix grasses suggests equilibrium, and that competition and other biotic interactions among these "core" species may be important influences. Indeed, evidence from Konza Prairie populations indicates that competition is an important factor regulating the performance and abundance of dominant grasses and some forbs (Hartnett 1991, 1993; Hartnett et al. 1993; Fahnestock and Knapp 1993). The relative importance of different resources in tallgrass prairie plant competition and bottom-up controls on community dynamics are explored in Chapter 9.

Although the dominant matrix grasses show significantly less spatial and temporal variation than satellite species, their populations are far from static. A 10-year record of flowering stem densities of *A. gerardii*, *A. scoparius* Michx.,

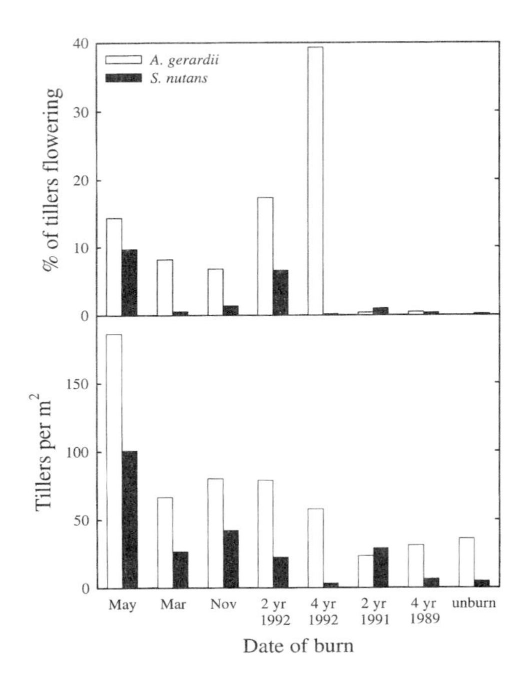

Figure 6.5. Temporal variation in flowering rates and tiller densities for *Andropogon gerardii* and *Sorghastrum nutans*. From Towne (1995).

and *Sorghastrum nutans* (L.) Nash shows high temporal variation associated with annual weather variation (Fig. 6.4). However, in herbaceous perennials, reproduction generally is more sensitive to variation in resource availability and abiotic conditions than is plant growth or population density, and only 2 to 15% of these grass tillers flower in most years (Fig. 6.5). Thus, temporal variation in flowering may be much greater than variation in tiller or plant (genet) density. Variation in reproduction and densities of some forbs on Konza Prairie is related strongly to fire regimes (Fig. 6.6). Their populations decrease under conditions that favor the core grass species (e.g., frequent burning or lack of grazing), further supporting the notion that grass-forb competition is an important factor influencing relative abundances of plant species. Overall, results of Konza Prairie LTER research and previous studies (Bazzaz and Parrish 1982) clearly indicate that the interplay between plant competition and nonequilibrium processes (disturbance and patch dynamics) regulates plant species coexistence in tallgrass prairie, and that the relative importance of these processes varies both among taxa and with time. These patterns clearly reflect our central thesis that multiple limiting factors vary in importance in time and space in tallgrass prairie (Chapter 1, this volume).

Population Responses to Fire

A substantial amount of research has focused on plant responses to fire in tallgrass prairie (Collins and Wallace 1990). However, much of this previous work (including Konza Prairie studies conducted from 1980 through 1987) was done

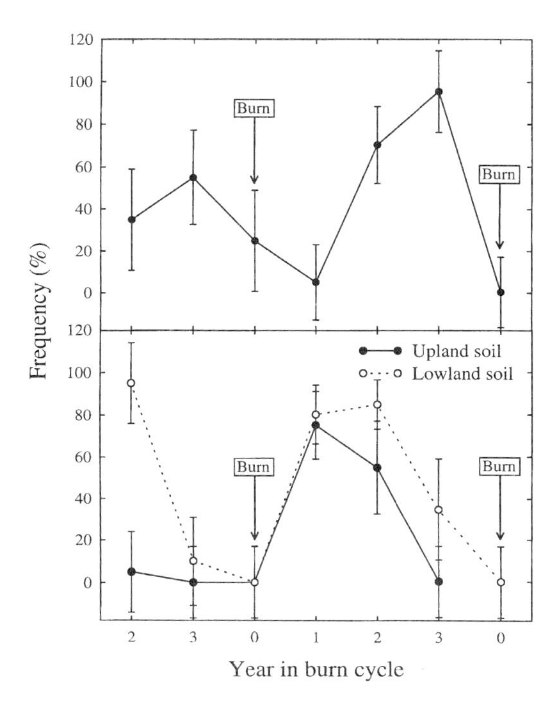

Figure 6.6. Year-to-year variation in frequencies (mean ± SE) of *Erigeron strigosus* var. *strigosus* (upper panel) and *Viola rafinesquii* (lower panel). *Erigeron* was absent in lowlands. From Gibson (1988), with permission of the Bulletin of the Torrey Botanical Club.

on ungrazed prairie, and recent work indicates significant interactions among fire, large grazers, and plant communities (Vinton et al. 1993). Here we summarize results of these recent studies and attempt to provide a general synthesis of the interacting roles of fire and grazing on population processes.

Perhaps the most conspicuous and significant effect of frequent fire on tallgrass prairie is the reduction of populations of woody species. Woody plant densities on Konza Prairie decrease slowly under annual spring burning and increase rapidly with longer fire-return intervals (Fig. 6.7; Briggs and Gibson 1992; Chapter 9, this volume). Fire-induced mortality of woody plants is related to their morphology and life history traits (e.g., unprotected aboveground meristems, timing of active aboveground growth, and translocation of carbohydrate reserves). Variation in population responses to fire among the common woody species, *Rhus glabra* L., *Cornus drummondii* C.A. Mey, *Juniperus virginiana* L., *Ceanothus herbaceous* L., and *Symphoricarpos orbiculatus* Moench, is associated with interspecific variation in these traits. In the absence of fire, accumulation of detritus, successful establishment of woody plants, and altered competitive interactions lead to a fairly rapid increase in shrub canopy cover (over 1% per year) and succession toward a woody-dominated community (Bragg and Hulbert 1976). Our observations on Konza Prairie indicate clearly that the activities of *B. bison* also contribute significantly to increased adult mortality and decreased establishment of woody plants, particularly *J. virginiana*.

Responses of grasses and forbs to fire on tallgrass prairie also are influenced by their morphologies and phenologies. The convergent selection pressures as-

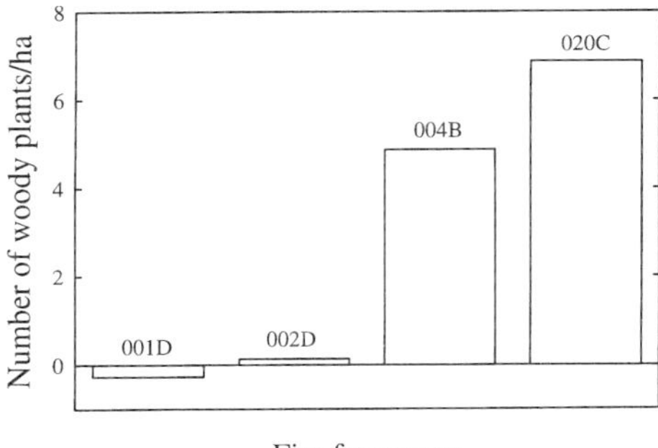

Figure 6.7. Change in the number of individual woody plants per ha from 1981 to 1986 on four Konza Prairie watersheds under different burn frequencies. 001D = annually burned, 002D = biannually burned, 004B = burned every 4 years, and 020C = long-term unburned. From Briggs and Gibson (1992), with permission of the Bulletin of the Torrey Botanical Club.

sociated with drought, fire, and grazing have led to traits such as a large mass of perennating organs and allocation of resources belowground (e.g., large root: shoot ratios) and the maintenance of meristems at or below the soil surface. In the dominant C_4 grasses *A. gerardii* and *S. nutans*, fire stimulates rhizome development and tillering early in the season, resulting in high tiller densities and percentage of tillers flowering (Knapp and Hulbert 1986; Hulbert 1986; Hulbert and Wilson 1983; Peterson 1983). The magnitude of these responses varies considerably among years, with the season of fire, and among species (Fig. 6.5; Gibson 1988; Towne 1995; Svejcar 1990). Most cool-season C_3 grasses show decreased flowering and tillering in response to frequent spring fires. These demographic responses to fire occur in the context of high variability associated with annual and seasonal weather variation (Fig. 6.4).

An understanding of tiller dynamics and densities of the dominant grasses is essential not only for understanding their population biology and the role of fire but also because tillers represent a crucial link between population and ecosystem processes in tallgrass prairie. Grasses constitute most of the plant biomass on tallgrass prairie (Chapter 12, this volume), and more of the variation in aboveground production can be accounted for by variation in tiller (shoot) numbers than in individual tiller size. Thus, spatial and temporal patterns of net primary production on tallgrass prairie (Chapter 12, this volume) are primarily functions of the demography of grass populations.

Forb species show variable responses to spring burning. Most increase in abundance with decreasing fire frequency (Gibson and Hulbert 1987), a pattern that appears to be general in tallgrass prairies (Kucera and Koelling 1964; Zimmerman and Kucera 1977). Many annual and perennial forbs that initiate growth early in the growing season (e.g., *Senecio plattensis* Nutt, *Viola pedatifida* G. Don, *Lomatium foeniculaceum* [Nutt.] Coult. & Rose, *Zigadenus nuttallii* A. Gray) are killed by fire or severely damaged by direct effects on the fate of buds and subsequent vegetative and floral development, such that they may be unable to set seed and complete their life cycle (Gibson 1988). Annuals and short-lived perennials that experience direct mortality from fire may show patterns of abundance that track the fire cycle closely, as do *Erigeron strigosus* Muhl. ex Willd. and *Viola rafinesquii* Greene (Fig. 6.6). Over longer time scales, gradual reduc-

Table 6.1. Growth, flowering, and vegetative reproduction responses to fire in several common Konza Prairie forbs (means ± SE)

	Frequently Burned Sites	Long-Term Unburned Sites	Probability
Solidago canadensis			
Total plant mass (g)	4.1 ± 0.4	9.0 ± 0.9	0.0001
Flower heads/plant	155 ± 31	412 ± 77	0.003
Rhizomes/plant	1.9 ± 0.2	3.4 ± 0.3	0.0005
Echinacea angustifolia			
Total plant mass (g)	5.3 ± 0.5	9.0 ± 0.7	0.001
Flower heads/plant	1.4 ± 0.2	2.2 ± 0.2	0.01
Vernonia baldwinii			
Total plant mass (g)	10.0 ± 0.9	11.4 ± 1.6	NS
Flower heads/plant	50.8 ± 6.2	78.2 ± 7.4	0.01
Salvia azurea			
Total plant mass (g)	10.8 ± 1.9	5.42 ± 1.9	0.09
Flowers/plant	44.4 ± 10.2	43.0 ± 10.2	NS
Baptisia bracteata			
Total plant mass (g)	8.5 ± 0.4	12.8 ± 0.4	0.003
Flower heads/plant	17.2 ± 2.9	15.0 ± 2.9	NS
Oenothera speciosa			
Total plant mass (g)	0.70 ± 0.06	0.78 ± 0.06	NS
Flowers/plant	4.0 ± 0.3	4.9 ± 0.3	0.07

Note: Probability values indicate significant differences between fire frequencies, based on data from Knapp (1984a), Hartnett (1990, 1991) and Damhoureyeh (1996).

tions in soil seed bank populations with annual burning (Abrams 1988) also may contribute to the gradual declines in populations of annuals and other short-lived species. Forbs that are active later and flower in mid- to late summer (e.g., *Ratibida columnifera* [Nutt.] Woot. & Standl., *Solidago canadensis* L., *Vernonia baldwinii* Torr.) show reductions in growth, flowering, and vegetative reproduction in response to frequent burning (Table 6.1; Knapp 1984a; Hartnett 1990, 1991). However, the effects of fire on these species are more likely indirect, i.e., enhanced growth and competitive ability of the matrix grasses in the postfire environment (Hartnett 1991). Thus, increases in abundances of many forbs with increasing time since fire may be attributed to competitive release. A few forb species such as *Salvia azurea* Lam. and *Kuhnia eupatorioides* L. increase in frequency and/or cover with increasing fire frequency. Towne and Knapp (1996) showed that the densities of most legumes were two to three times higher on annually burned than on infrequently burned watersheds on Konza Prairie. This study supports the hypothesis that potential nitrogen fixers increase in relative abundance because of greater nitrogen limitation in frequently burned sites (Chapter 13, this volume).

A counterintuitive result of long-term studies on prairie forb populations is that fire effects on plant reproduction often do not result in parallel effects on abundances. For example, *Ratibida columnifera* populations show significantly

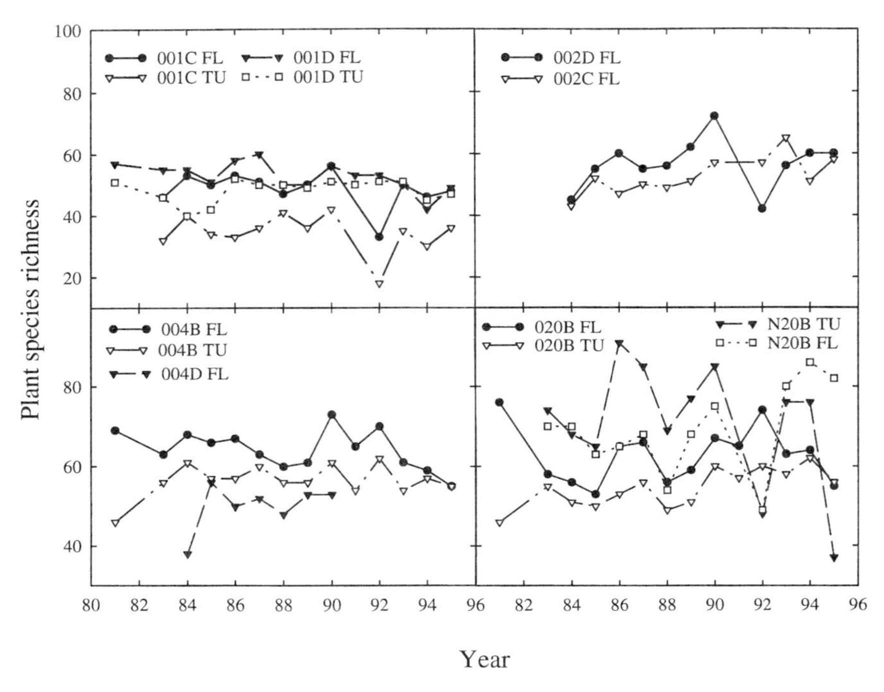

Figure 6.8. Year-to-year variation in plant species richness on watersheds undergoing different burn regimes, 1981–1995. 001C, 001D = annually burned; 002C, 002D = biannually burned; 004B, 004D = burned every 4 years; 020B = long-term unburned; N20B = long-term unburned + grazed by *Bos bison*. TU = lowland Tully soils; FL = upland Florence soils.

lower growth, flowering, and seed production in frequently burned than in unburned prairie, yet they show no corresponding reduction in abundance in frequently burned sites and generally no correlation between seed production and abundance among sites (Hartnett 1991). This may be due in part to the high temporal and spatial variations in plant species abundances driven stochastically by climatic variability, which are often of much greater magnitude than the variation associated with fire regime (Fig. 6.8). The lack of correlation between seed production and population abundance also reflects the general unimportance of sexual reproduction in regeneration and maintenance of local populations. Although many species flower profusely and produce large seed crops, successful regeneration via seed is nonetheless extremely rare in time and space, usually confined to disturbed microsites (see later discussion) coupled with a significant period of favorable moisture and microsite conditions after germination (Hartnett and Keeler 1995; Glenn-Lewin et al. 1990). Populations of most tallgrass prairie plants regenerate primarily via vegetative reproduction (Hartnett and Keeler 1995). Much of the literature on fire effects on natural plant populations focuses on responses of flowering and seed production, when, in fact, these responses may have little impact on plant population dynamics. Studies that attempt to understand the regulation of plant population dynamics on tallgrass prairie and the influence of population processes on patterns of aboveground net primary production (ANPP) should focus on the demography and controls on vegetative reproduction and shoot/tiller dynamics. This dominant influence of vegetative reproduction on local population dynamics may be characteristic of other highly productive subhumid grasslands but not of semiarid grasslands, such as the shortgrass steppe, where seed production and seedling establishment are quite important (e.g., Coffin and Lauenroth 1989).

The net result of the different responses to fire among taxa is that increasing fire frequency results in an increase in the relative abundance of the dominant C_4 grasses, a decrease in the abundance of C_3 grasses, forbs, and woody species, and a decrease in plant species richness and diversity (Fig. 6.8). The strong dominance of the matrix grasses and the rarity of satellite forbs under annual burning also result in less community heterogeneity in frequently burned prairie (Nellis and Briggs 1989; Collins and Gibson 1990), which is consistent with reduced patch-scale variability in ANPP in annually burned prairie (Chapter 12, this volume). In general, fire alters plant community structure in tallgrass prairie by modulating the intensity and patterns of plant competition (Chapter 9, this volume). Moreover, there may be important fire-topographic interactions because species richness is typically higher in upland than lowland sites (Fig. 6.8).

The Role of Animals

Plant Species Responses to Herbivory

Studies of important herbivores on Konza Prairie have focused on "bottom-up" controls on their community structure (Chapters 8, 9, this volume). However, herbivores in tallgrass prairie have important reciprocal effects on plants. LTER studies have examined plant responses to a variety of terrestrial invertebrate consumers, including leaf-feeding, seed-feeding, and gall-forming insects. With the reintroduction of *B. bison* to Konza Prairie in 1987 and the addition of cattle to several watersheds in 1992, long-term effects of large mammalian grazers on plant populations now are being assessed.

The tolerance of prairie plants to herbivory (their capacity for compensatory regrowth and reproduction following damage) is determined by a variety of factors such as the timing and intensity of defoliation, resource availability, and the intensity of competition from neighboring plants. For example, the wasp *Antistrophus silphii* (Hymenoptera) induces galls on shoot apical meristems of the perennial composite *Silphium integrifolium* Michx. Under natural levels of competition from neighboring plants in the field, *Silphium* poorly tolerated gall damage; shoot growth and reproduction were reduced, and flowering was delayed (Fig. 6.9; Fay and Hartnett 1991). However, when grown without competitors and with abundant nutrients and water, *Silphium* was much more tolerant of gall damage; extensive compensatory growth minimized lost biomass, leaf area, and fecundity (Fay et al. 1996). Thus, this forb possesses the necessary morphologi-

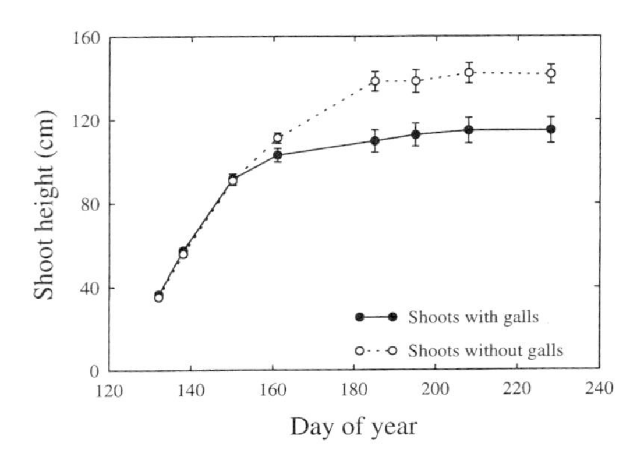

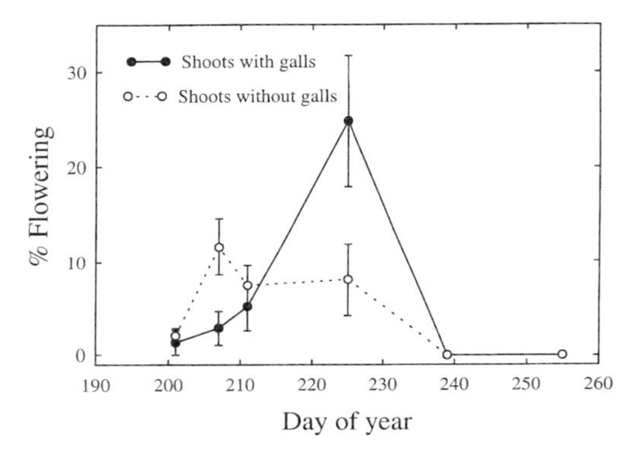

Figure 6.9. Impact of gall insect damage on seasonal patterns of shoot growth and flowering (means ± SE) in *Silphium integrifolium* shoots galled by *Antistrophus silphii* and ungalled control shoots. From Fay and Hartnett (1991), with permission of Springer-Verlag.

cal and physiological mechanisms to compensate for insect damage, but competition and resource limitation on tallgrass prairie constrain these responses. Poor tolerance of insect damage may typify forbs in the highly competitive tallgrass prairie plant communities (Fay et al. 1996).

Reproductive tissues of tallgrass prairie plants frequently are heavily attacked by insects, which can destroy large fractions of a plant's seed crop. Temporal and spatial characteristics of tallgrass prairie plants are often of key importance in such interactions, and to various degrees the host plant may succeed in escaping in either time or space from its insect consumers. For example, most seeds produced by the prairie legume *Baptisia australis* (L.) R. Br. var *minor* (Lehm) S. Wats. are destroyed by combined feeding of blister beetles, moths, and weevils (Fig. 6.10). Weevils attack pods early in the flowering period, whereas blister beetles attack later during flowering, leaving *Baptisia* with a narrow window when it can flower successfully (Evans et al. 1989b). The conflicting selection pressures imposed by these herbivores may select for diversity of flowering phenologies among individuals in local populations (Evans et al. 1989b). In addition, blister beetle feeding may indirectly lessen damage caused by moths and weevils. When groups of blister beetles aggregate on *Baptisia*, only 20% of the flowers and fruits escape consumption (Evans 1990). As pods are consumed by beetles, the internal-feeding moths and weevils are killed. The neighborhood

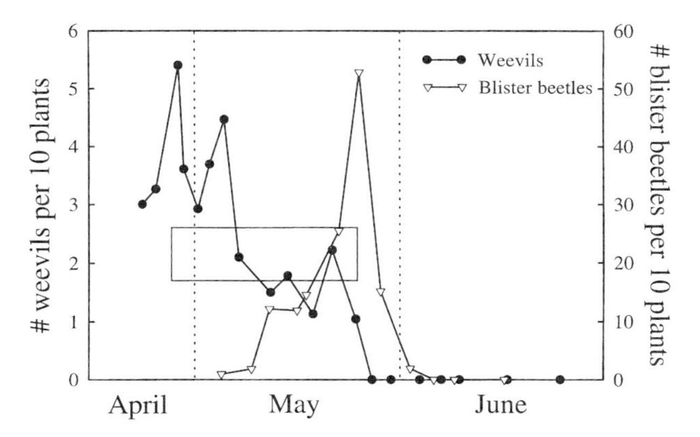

Figure 6.10. Timing of reproduction in *Baptista australis* (box) compared with the seasonal trends in densities (means ± SE) of weevils and blister beetles feeding on *Baptista* seeds and pods. Vertical dashed lines separate months. Redrawn from Evans et al. (1988b).

context influences not only plant tolerance to herbivory, as in *Silphium*, but also escape or defense from herbivores. For example, the abundance of milkweed bugs (*Lygaeus kalmii* [Hemiptera: Lygaeidae]) on prairie milkweeds (*Asclepias viridis* Walt.) on Konza Prairie and their effects on fecundity depend strongly on the abundance of neighboring conspecifics (Evans 1983).

Although impacts of particular insects on their host plant species are relatively easy to demonstrate, the collective effects of insects on plant community structure remain elusive. Gibson et al. (1990) summarized a 4-year study on Konza Prairie in which insects were removed from annually burned sites by frequent applications of carbaryl. Insect removals caused minor changes in plant species richness and diversity. For example, annual forb species richness was reduced and perennial forb richness increased in some years but not in others. Frequently burned tallgrass prairies, strongly dominated by perennial grasses and forbs, may be relatively tolerant of the presence of herbivorous insects. In addition, high year-to-year variation in plant performance may mask the impacts of herbivores, necessitating longer-term experiments to resolve the role of these consumers.

Plant responses to large mammalian herbivores also vary in time and space depending upon resource availability, plant competition, and the frequency and intensity of defoliation, all of which are influenced strongly by fire regime. An initial study of effects of *B. bison* grazing on growth and tillering of two of the dominant grasses, *A. gerardii* and *Panicum virgatum* L., suggests that compensatory growth responses of grazed tallgrass prairie grasses vary considerably, even within species. *Bos bison* grazing caused a short-term increase in shoot relative growth rates (Fig. 6.11) but no effect on end-of-season plant biomass and reduced tiller growth and survivorship during the following season (Vinton and Hartnett 1992). These responses to defoliation show additional variation depend-

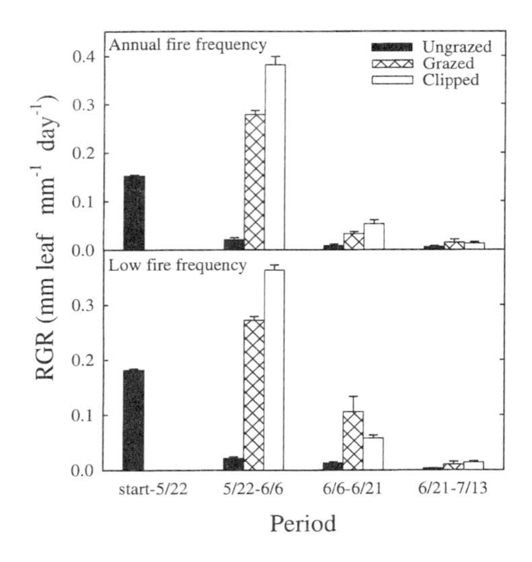

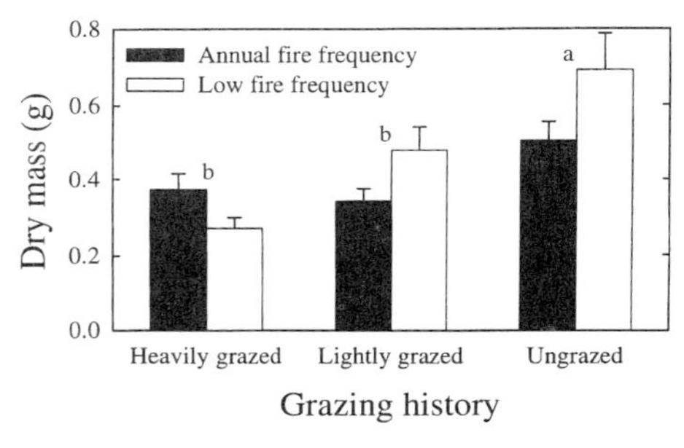

Figure 6.11. Relative growth rates during April to mid-July for burned and unburned *Andropogon gerardii* and end-of-season dry weights under various grazing histories (means ± SE). Tiller growth began in mid-April, burning occurred on 10 April, and defoliation occurred from 21 to 24 May 1989. For dry weights, means with different letters were significantly different ($P < 0.05$) according to Duncan's multiple range test. From Vinton and Hartnett (1992), with permission of Springer-Verlag.

ing upon fire, topographic position, local plant densities, and the growth stage at which the plants are defoliated (Hartnett 1989). As observed for prairie forbs such as *Silphium*, the short-term compensatory growth response of these grasses also is constrained by plant competition; the amount of regrowth following defoliation decreases markedly with increasing plant (tiller) density (Hartnett 1989). Just as fire modulates the intensity of plant competition, it alters a suite of other environmental factors that have direct consequences for plant responses to herbivory.

Fire and topography also influence patterns of ungulate grazing at scales ranging from diet selection within small patches up to the landscape level (Hartnett et al. 1997; Pfeiffer and Hartnett 1995; Vinton et al. 1993). Fire effects on *B. bison* diet selection at small scales alter plant population structure and dynamics. For example, in unburned prairie, *B. bison* show a strong preference for the dominant rhizomatous *A. gerardii* and avoid grazing neighboring individuals of *A. scoparius*. The latter is a bunchgrass that accumulates a clump of standing dead tillers that act as a physical deterrent to grazing. When the prairie is burned, this physical barrier is removed and *B. bison* graze both species in direct proportion to their abundances (Pfeiffer and Hartnett 1995). Furthermore, grazing alters the demography of tillers in *A. scoparius* and shifts its population size structure toward a greater frequency of small individuals (Pfeiffer and Hartnett 1995). Because drought tolerance in *A. scoparius* is size-dependent, this grazing-induced shift in population size structure may result in greater mortality during drought years, accounting for the grass's decline in abundance over time in grazed prairie. Thus, the interactions of *B. bison* grazing, topography, fire, and climate can influence the composition, relative abundances, and structure of tallgrass prairie plant populations.

Within the first few years following reintroduction of *B. bison* onto Konza Prairie, these large grazers significantly altered plant species abundances and diversity. The frequency and cover of cool-season graminoids (e.g., *Poa pratensis* L., *Agropyron smithii* Rydb., and *Carex* spp.) and some forbs (e.g., *Aster ericoides* L. and *Oxalis stricta* L.) consistently are higher in prairie grazed by *B. bison*, whereas the dominant warm-season matrix grasses and other forbs (e.g., *Solidago missouriensis* Nutt.) decreased in response to *B. bison* grazing (Hartnett et al. 1996). Responses in several other species vary depending on fire frequency and topographic position.

Responses of forbs to grazing on Konza Prairie are complex and vary significantly among plant species, grazer species, fire regimes, and plant life history stages (Damhoureyeh 1996). Some forbs (e.g., *Baptisia bracteata* Muhl. ex Ell. var *glabrescens* [Larisey] Isley, *Oenothera speciosa* Nutt., and *Vernonia baldwinii*) show increased growth and reproduction in grazed sites, indicating competitive release in response to selective grazing of the dominant matrix grasses (Fahnestock and Knapp 1993; Damhoureyeh 1996). Forbs that show reduced performance in grazed sites likely are affected negatively by disturbances generated by ungulate nongrazing activities because none of the forbs studied were consumed directly by *B. bison* or cattle (Damhoureyeh 1996). Furthermore, similar to patterns of forb responses to fire, no correlation occurred between grazer effects on seed production and on population densities, further indicating that variation in sexual reproduction plays a minor role in regulating local population abundances.

At several spatial scales, plant species diversity is significantly higher in prairie grazed by *B. bison* (Table 6.2). This and other plant responses to *B. bison* are generally larger in annually burned than in infrequently burned prairie, consistent with the greater use of burned watersheds by *B. bison* during most of the growing season (Vinton et al. 1993). Greater species richness in grazed prairie is likely a result of greater microsite diversity generated by the grazing and nongrazing activities of *B. bison*, whereas increases in evenness of species abundances on grazed sites is likely due to the preferential grazing of the competitively dominant matrix grasses and concomitant increases in abundances of the subordinate species. Like fire, grazing also modulates the intensity of plant com-

Table 6.2. Plant species diversity components in Konza Prairie sites subjected to various combinations of burning and grazing

	S		E		H'	
Site	Ungrazed	Grazed	Ungrazed	Grazed	Ungrazed	Grazed
Annually burned uplands	27	35	0.42	0.53*	1.40	1.87*
Annually burned lowlands	28	43*	0.51	0.49*	1.70	1.82*
Periodically burned uplands	27	32*	0.53	0.59*	1.76	2.03*
Periodically burned lowlands	29	41*	0.38	0.50	1.29	1.85

Source: From Hartnett et al. (1996).

Note: Annually burned = sites subjected to annual spring prescribed burns. Periodically burned = sites burned at 4-year intervals. Grazed = sites grazed year-round by *B. bison*. S = species richness (number of species per 0.01 ha). H' = Shannon species diversity index ($H' = \Sigma\ p_i * \ln p_i$, where p_i = the relative cover of species *i*). E = evenness index = $H'/H'_{max} = H'/\ln S$).

*Significantly different from ungrazed sites at $P < 0.05$ by Mann-Whitney U test.

petition. However, unlike frequent fire, grazing suppresses rather than enhances the competitive dominants, resulting in very different effects on plant species diversity and community heterogeneity (Chapter 9, this volume).

Changes in plant species abundances in response to *B. bison* on Konza Prairie generally are consistent with responses of "increaser," "decreaser," and "invader" species classified by J. E. Weaver and coworkers in their early long-term studies of effects of livestock grazing in tallgrass prairie (Dyksterhuis 1958; Voigt and Weaver 1951; Weaver and Hansen 1941; Weaver 1968). There are some notable exceptions, however. *A. scoparius* is classified as a "decreaser" under livestock grazing, but its abundance increased in response to *B. bison* grazing on infrequently burned sites on Konza Prairie (Hartnett et al. 1996) and showed no significant response on burned prairie. In addition, several C_3 grasses classified as "increasers" under cattle grazing show no significant increase in response to *B. bison*. Many forbs also show markedly different demographic responses to *B. bison* versus cattle grazing (Damhoureyeh 1996). These plant population responses and other differences in vegetation responses to *B. bison* versus cattle may be explained partly by differences in management (e.g., seasonal versus year-round grazing) and partly by differences in both the grazing and nongrazing behaviors of the herbivores per se (Hartnett et al. 1996, 1997).

Effects of Small-Animal Disturbances

Small mammals also affect tallgrass prairie plant populations through herbivory and by the generation of various small-scale disturbances. Animal disturbances provide unique microsites for plant colonization and establishment and support an assemblage of plant species that is distinct from surrounding undisturbed vegetation. Local composition of plant species varies among different disturbance types such as badger or pocket gopher mounds, prairie mole burrow systems, and *B. bison* wallows (Gibson 1989). The combination of the various activities and disturbances associated with large grazers, the numerous effects of small animals, and plant responses to these patches illustrates the important contribution of small-scale soil disturbance in maintaining species richness and heterogeneity in tallgrass prairie.

In a 4-year Konza Prairie study involving reductions of small-mammal densities in fenced plots, few significant effects on the biomass of most plant species were observed (Gibson et al. 1990), yet a clear pattern of response occurred among life-form classes. Reduction of animal densities resulted in an increased abundance of C_4 grasses and a decreased abundance of annual forbs. Although this study did not separate effects of herbivory from responses to animal disturbances, the general results suggest that herbivory has small effects on tallgrass prairie vegetation relative to disturbance.

The plains pocket gopher (*Geomys bursarius*) is an important burrowing animal in Great Plains grasslands, and Konza Prairie field studies have examined the soil disturbances associated with its burrowing and mound-building activities. These two types of disturbances differentially alter essential plant resources; root herbivory occurs belowground, and the deposition of subsurface soil in mounds aboveground buries plants and alters the availability of light, water, and nutrients. The spatial patterns of burrows are highly uniform, whereas mounds occur in strongly clumped distributions, producing a spatially explicit pattern of influence on the adjacent plant community. Plant growth and reproduction is inhibited over the disturbance, and increased resources and plant growth adjacent to the disturbance result in a competition-induced wave of biomass emanating outward at least 50 cm from the disturbed site (Reichman et al. 1993). Current studies on Konza Prairie are examining the relationship between the spatial patterns of these disturbances, spatial patterns of resources at several scales, and their consequences for pattern in the prairie vegetation. In addition, ongoing studies of the processes and sources of plant colonization and establishment on mounds and burrows (e.g., seed rain, seed banks, vegetative reproduction, and regrowth of buried plants) will help elucidate the mechanisms driving small-scale vegetation dynamics on these disturbances.

Complex Biotic Interactions: The Role of Mycorrhizae

In addition to the influence of fire, competitors, and herbivores, symbiotic associations with mycorrhizal fungi are ubiquitous and play a key role in tallgrass prairie plant populations (Hetrick and Bloom 1983). Arbuscular mycorrhizae are mutualistic associations between fungi (Endogonaceae) and plant roots. The plant translocates carbohydrate to the fungus, and the fungus benefits the host plant through increased nutrient acquisition, growth, water relations, and disease resistance (Mosse et al. 1981; Allen 1991; Read 1984). Plant species of tallgrass prairie vary significantly in their dependency on the symbiosis and their growth responses to fungal infection. The warm-season grasses are obligate mycotrophs and show large positive growth responses to mycorrhizae. They will not grow in native prairie soil without the symbiosis (Hetrick et al. 1988). Cool-season grasses generally are facultative mycotrophs and show smaller growth responses to mycorrhizal infection. Forbs vary in their responses to mycorrhizae, from large positive to negative growth responses (Hetrick et al. 1992).

LTER studies on mycorrhizae in tallgrass prairie have focused on their roles in belowground processes; in plant growth and demography; and in modulating plant responses to competitors, fire, and grazers. Mycorrhizae exert a strong influence on plant-plant interactions in tallgrass prairie (Hetrick et al. 1989, 1994; Hartnett et al. 1993). The C_4 matrix grasses show strong competitive dominance under mycorrhizal conditions, whereas less mycorrhizal-responsive subordinate grass and forb species do better in competition under conditions where mycorrhizal fungal activity is reduced.

Earlier studies in other grasslands (e.g., Chiariello et al. 1982) indicated that hyphae of mycorrhizal fungi can form functional interconnections among neighboring plants, allowing interplant translocation of nutrients. Studies on Konza Prairie showed nonrandom patterns of transfer of phosphorus (P) between neighboring plant species (Fischer-Walter et al. 1996), and a complementary greenhouse study showed that interplant P transfer by mycorrhizae may contribute a significant proportion of the total P acquired by certain prairie grasses (Chua et al. unpublished data). Interplant nutrient transfer via mycorrhizal hyphae could have a significant influence on plant competition and community structure.

The effects of mycorrhizae on the population dynamics of co-occurring prairie plants vary among species and among different life history stages (seedling establishment, flowering, tillering) within species (Hartnett et al. 1994). Flowering and densities of the cool-season grasses and some forbs increase when mycorrhizal fungi are suppressed, whereas mycorrhizal infection significantly enhances flowering and tillering of the dominant warm-season grasses (Hartnett et al. 1994). Burning stimulates active mycorrhizal colonization of warm-season grasses in tallgrass prairie (Bentivenga and Hetrick 1991), which suggests that enhanced ANPP following fire (Chapter 12, this volume) and enhanced competitive ability of the C_4 grasses may result, at least partly, from enhancement of mycorrhizal symbiosis. These patterns further suggest that mycorrhizal effects on the dynamics of the subordinate cool-season graminoid and forb populations likely are mediated indirectly through effects of the symbiosis on the competitive dominance of their neighbors. Effects of mycorrhizae on seedling emergence and establishment vary among grass and forb species, and responses at the seedling stage often differ considerably from growth responses of established plants. For example, flowering and tiller densities increased among the cool-season grasses, but seedling emergence rates decreased when mycorrhizal fungi were suppressed (Hartnett et al. 1994).

Mycorrhizal fungi also may have important effects on plant-grazer interactions in tallgrass prairie. Mycorrhizal symbiosis significantly enhances the grazing tolerance (compensatory regrowth capacity following defoliation) of *A. gerardii*, an effect that diminishes with increasing clipping frequency (Hetrick et al. 1990). Studies in other grasslands indicate that compensatory growth capacity of grasses is limited by P availability (Chapin and McNaughton 1989), which further suggests that mycorrhizal associations significantly increase the grazing tolerance of grasses. With repeated intensive defoliation, however, mycorrhizal colonization of roots decreased and host plant growth benefits were lost (Hetrick et al. 1990). Mycorrhizae may enhance the grazing tolerance of dominant prairie grasses under low to moderate grazing intensities by increasing the availability of nutrients for regrowth, but these fungi may decrease grazing tolerance under intense grazing regimes because regrowth of foliage and mycorrhizal fungi may compete for limited carbohydrate resources. These hypotheses currently are being explored through additional long-term studies on Konza Prairie involving manipulation of mycorrhizal symbiosis under grazing regimes of various intensities.

Konza Prairie studies suggest that complex interactions exist among plant competition, grazing, fire, and mycorrhizal symbiosis. Environmental variation associated with season, fire, grazing, microclimate, and edaphic factors has both direct effects on plant communities and indirect effects mediated through changes in mycorrhizal activity. Given that plant species show differential responses to mycorrhizal infection, any shifts in mycorrhizal activity caused by these factors likely alter the composition and relative abundances of plant species. In addition, significant interactions among disturbances, fungal community structure, and soil invertebrate grazers of mycorrhizal hyphae also are predicted to influence mycorrhizal effects on plant communities, but these interactions have not yet been explored.

Summary

Significant insights have been gained into the ecology of tallgrass prairie plant populations through long-term research. Research on Konza Prairie has greatly increased our understanding of patterns and mechanisms of fire effects on plant populations and species interactions, and our knowledge of the roles of grazing and climate variation is increasing. Numerous general patterns and conclusions have emerged.

Plant species abundance and distribution patterns on tallgrass prairie are bimodal, conforming to predictions of the "core-satellite" hypothesis. The matrix grasses are abundant, widely distributed, and temporally stable and predictable, whereas the more numerous interstitial species are localized, dynamic, and temporally unpredictable. Evidence indicates that competition is important in regulating populations of the dominant matrix grasses but not the satellite species, whose nonequilibrium patch dynamics are influenced by disturbances operating at different scales.

Patterns of variation in species richness are driven primarily by the nonequilibrium dynamics of the satellite species. The high temporal and spatial variability of the plant populations is related to the vagaries of climate and mosaic patterns of grazing, fire, topography, and soil disturbances operating at different frequencies and spatial scales. For many populations, variation among years associated with annual weather variability is much greater than spatial or temporal variation associated with fire or grazing treatment. Temporal dynamics also differ considerably among plant species, some showing significant interannual changes in abundance in response to fire cycles or weather, and others characterized by much greater population stability.

Prairie plant species show differential responses to fire and grazing, resulting in significant effects on species composition and relative abundances. Plant demographic responses to fire or grazing vary depending on life history and phenology and the timing of fire. Annual spring burning on tallgrass prairie decreases woody plant density and the frequency of many interstitial C_3 grasses and forbs, but it increases the relative abundances of the dominant warm-season grasses. Fire also influences patterns of large ungulate herbivory, plant responses to herbivory, and important plant symbionts such as mycorrhizal fungi. These all affect the population size structure, competition, and demography of grasses and forbs. Fire, grazing, and drought all interact in complex ways to shape patterns of plant species abundances and distribution.

The general picture to emerge from long-term plant population studies on Konza Prairie is one of complex interactions among species and among the major processes that influence them (Fig. 6.1). This is consistent with the central theme that the biota and ecosystem processes in tallgrass prairie are regulated by multiple limiting factors that change in importance in time and space (Chapter 1, this volume). It is also consistent with the views of Belovsky and Joern (1995) that population limitations constantly change, making single-limitation explanations for population dynamics unrealistic.

An inevitable result of our synthesis of the long-term ecological studies on tallgrass prairie is that the gaps in the story have become more clearly evident. Assessing what we now know about prairie plant populations also has revealed what we still do not know and has raised a number of unanswered questions. What are the demographic consequences of variation in the season of fire? What are the long-term effects of fire and grazing on genetic diversity within plant populations? Are native and domestic grazers significantly different in their effects on plant populations? To what extent are plant demographic responses to grazers direct or indirect effects of grazer-induced changes in other biotic interactions of plants? How do grazer effects vary across the heterogeneous tallgrass prairie landscape? What specific climatic factors or other processes control between-year fluctuations in plant populations? These questions represent some of the challenges facing researchers in the years to come.

Acknowledgments This is contribution no. 97-191-B from the Kansas Agricultural Experiment Station, Kansas State University, Manhattan.

7

Diversity of Terrestrial Macrofauna

Donald W. Kaufman
Philip A. Fay
Glennis A. Kaufman
John L. Zimmerman

Thousands of vertebrate and invertebrate species occur in grassland and forest habitats on Konza Prairie. However, LTER population and community studies have focused only on selected groups of ecologically important species, including small mammals, large mammalian grazers (bison and cattle), insectivorous birds, and grasshoppers. We also have gathered general information on several less studied groups, such as large mammals, reptiles, amphibians, and several orders of insects. In this chapter, we present background information on the diversity of terrestrial vertebrates and invertebrates studied to date on Konza Prairie.

Mammals

Mammals constitute a functionally diverse and important array of consumers on Konza Prairie (Fig. 7.1a; Table 7.1). Of the 7 orders, Rodentia and Carnivora are the most diverse, with 17 species representing 6 families and 9 species representing 4 families, respectively (Table 7.1). The 10 remaining species are members of Artiodactyla (2 families), Insectivora (2 families), Chiroptera (1 family), Lagomorpha (1 family), and Didelphimorphia (1 family).

The mammalian fauna of the northern Flint Hills is much less diverse today than it was before Euro-American settlement (14 species of carnivores, artiodactyls, rodents, and lagomorphs were purportedly extirpated; Bee et al. 1981; Kaufman and Kaufman 1997). Further, the impact of human activities on the diversity of animals in the region of Konza Prairie, in the Flint Hills, and in the Great Plains was greater for mammals than for birds, reptiles, amphibians, and

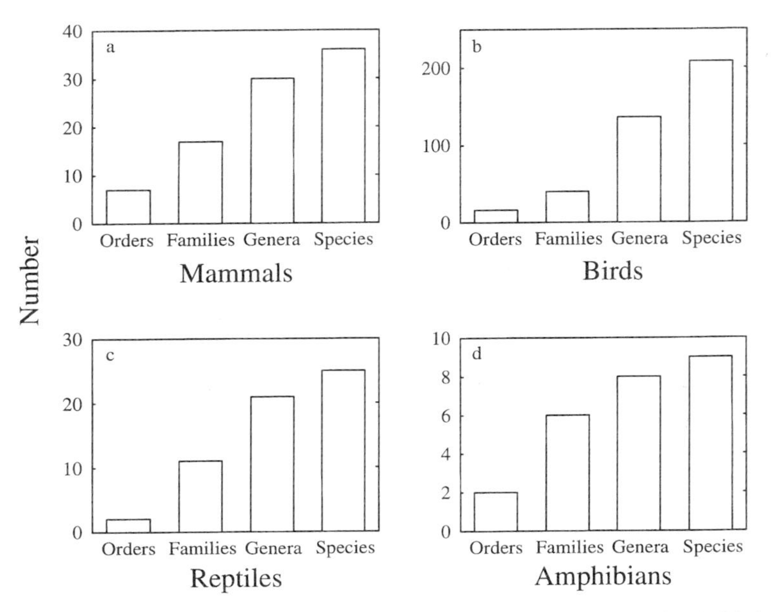

Figure 7.1. Numbers of orders, families, genera, and species of mammals (a), birds (b), reptiles (c), and amphibians (d) documented for Konza Prairie.

insects. Most extirpations occurred in the late 1800s and early 1900s; however, the spotted skunk (*Spilogale putorius*) disappeared from many areas of Kansas during the last 25 years.

Herbivorous mammals that remain absent from the region of Konza Prairie are the white-tailed jackrabbit (*Lepus townsendii*), Franklin's ground squirrel (*Spermophilus franklini*), elk (*Cervus elaphus*), and pronghorn (*Antilocapra americana*). Extirpated carnivores are the gray wolf (*Canis lupus*), black bear (*Ursus americanus*), grizzly bear (*U. arctos*), eastern spotted skunk, northern river otter (*Lutra canadensis*), and mountain lion (*Felis concolor*). In contrast, beaver, bison (a captive herd), white-tailed deer, and mule deer, which were once absent, now occur on Konza Prairie (Table 7.1).

Although a general loss of mammalian diversity occurred, three species that were not present before the late 1800s now occur in native habitats on Konza Prairie (Table 7.1). Of these, the hispid cotton rat immigrated north and the least weasel south into the Konza Prairie region following environmental changes caused by human activities (Finck et al. 1986). Further, the house mouse was introduced by humans and now occurs infrequently in native prairie and woodland habitats on Konza Prairie and throughout the Great Plains (Kaufman and Kaufman 1990a).

In addition to the 36 species listed in Table 7.1, 7 species of Chiroptera, 3 of Rodentia, 2 of Carnivora, and 1 of Lagomorpha occur in the northern Flint Hills region but have not been recorded on Konza Prairie (Bee et al. 1981; Finck et al. 1986). The black-tailed jackrabbit (*Lepus californicus*), gray fox (*Urocyon cin-*

Table 7.1. Species and status of mammals observed at Konza Prairie, 1981–1996

Common and Scientific Names	Status[a]
Order Didelphimorphia	
Virginia oppossum (*Didelphis virginiana*)	Common
Order Insectivora	
Elliot's short-tailed shrew (*Blarina hylophaga*)	Common
Least shrew (*Cryptotis parva*)	Present
Eastern mole (*Scalopus aquaticus*)	Present
Order Chiroptera	
Hoary bat (*Lasiurus cinereus*)	Present
Big brown bat (*Eptesicus fuscus*)	Present
Order Lagomorpha	
Eastern cottontail (*Sylvilagus floridanus*)	Common
Order Rodentia	
Woodchuck (*Marmota monax*)	Occasional
Thirteen-lined ground squirrel (*Spermophilus tridecemlineatus*)	Common
Fox squirrel (*Sciurus niger*)	Common
Plains pocket gopher (*Geomys bursarius*)	Common
Hispid pocket mouse (*Chaetodipus hispidus*)	Uncommon
American beaver (*Castor canadensis*)	Occasional
Plains harvest mouse (*Reithrodontomys montanus*)	Occasional
Western harvest mouse (*Reithrodontomys megalotis*)	Common
Deer mouse (*Peromyscus maniculatus*)	Common
White-footed mouse (*Peromyscus leucopus*)	Common
Hispid cotton rat (*Sigmodon hispidus*; immigrant)	Local
Eastern woodrat (*Neotoma floridana*)	Common
Prairie vole (*Microtus ochrogaster*)	Common
Woodland vole (*Microtus pinetorum*)	Present
Southern bog lemming (*Synaptomys cooperi*)	Local
House mouse (*Mus musculus*; introduced)	Rare
Meadow jumping mouse (*Zapus hudsonius*)	Present
Order Carnivora	
Coyote (*Canis latrans*)	Common
Red fox (*Vulpes vulpes*)	Present
Raccoon (*Procyon lotor*)	Common
Long-tailed weasel (*Mustela frenata*)	Present
Mink (*Mustela vison*)	Present
Least weasel (*Mustela nivalis*; immigrant)	Present
American badger (*Taxidea taxus*)	Common
Striped skunk (*Mephitis mephitis*)	Common
Bobcat (*Lynx rufus*)	Common
Order Artiodactyla	
Mule deer (*Odocoileus hemionus*)	Occasional
White-tailed deer (*Odocoileus virginianus*)	Common
American bison (*Bos bison*; reintroduced herd)	Captive

Sources: From Finck et al. (1986) and McMillan et al. (1997).

Note: Data will be updated as necessary and are available at http://climate.konza.ksu.edu.

[a]Categories, defined relative to abundances typical of species, are: common = abundance is high in appropriate habitat(s) in most years; local = abundance is high in local sites within appropriate habitat(s) in most years; uncommon = abundance is low in most years even in best habitats; rare = abundance is very low across the site, but species is present in most years; occasional = abundance is very low across site in best years and zero in many years; present = species recorded for site, but not enough information is available to assign other status; and captive = maintained in captive herds.

eroargenteus), and other unverified species likely are present only infrequently if at all (Finck et al. 1986).

Of the small, terrestrial mammals found on Konza Prairie, deer mice, western harvest mice, short-tailed shrews, and prairie voles are the common species found in grassland habitats (Finck et al. 1986; Kaufman et al. 1989; Kaufman et al. 1990; Bixler and Kaufman 1995b; Clark et al. 1995; Kaufman et al. 1995b). Thirteen-lined ground squirrels and southern bog lemmings also are grassland species but at relatively low densities. Of these six species, deer mice and thirteen-lined ground squirrels generally respond favorably to conditions created by prairie fires, whereas voles, harvest mice, short-tailed shrews, and bog lemmings respond negatively to these same conditions (Kaufman et al. 1983; Kaufman et al. 1988b; Clark and Kaufman 1990; Kaufman et al. 1990; Chapter 8, this volume). The hispid cotton rat also occurs in prairie habitats but is most common in habitats dominated by dense herbaceous cover (Peterson et al. 1985; Brillhart et al. 1995) and typically is not common in native prairie (Finck et al. 1986; Kaufman et al. 1990).

Of the small mammals present in woodland habitats in the Flint Hills, white-footed mice are the most abundant, whereas eastern woodrats are common only locally (Kaufman et al. 1993a, 1995a). White-footed mice do spill into ungrazed tallgrass prairie but never reach the high densities found in wooded habitats (Clark et al. 1987; McMillan and Kaufman 1994; Kaufman et al. 1995a).

Birds

Of the vertebrates, birds have the greatest diversity, with 208 species representing 16 orders recorded during 1981–1992 (Fig. 7.1b; Zimmerman 1985, 1993). Passeriformes includes 126 species, or over 50% of the species recorded (Table 7.2). The four most diverse families of this order are Emberizidae, Muscicapidae, Tyrannidae, and Troglodytidae. Additionally, over 80% of all species are included in the four most diverse orders (Passeriformes, Falconiformes, Charadriiformes, and Anseriformes), whereas only 10% of all species are found in the nine least diverse orders (Table 7.2).

Although more than 200 species of birds have been recorded for Konza Prairie, nearly half of these are rare to occasional visitors, and many are present only during spring and autumn migrations. Only 13% of the 208 species (n = 27) are summer-winter residents, with about 70% of these occurring only in forested habitats. Additionally, 31% of the avian species are present only during summer and 10% only during winter. Based on LTER research, abundance of even the common summer and winter residents varies greatly among species and among years and habitats for individual species (Finck 1986; Zimmerman 1993).

Because of migratory patterns, avian species richness varies greatly between summer and winter in both burned and unburned prairie (Zimmerman 1993). For example, the avian community in tallgrass prairie during summer consists of 11 and 5 regularly occurring species in unburned and annually burned sites, respectively (Table 7.3). The five species in annual burns also are common in unburned

Table 7.2. Orders and most diverse families (5 or more species) of birds recorded at Konza Prairie, 1971–1992

Order	Families	Genera	Species
Podicipediformes	1	1	1
Pelecaniformes	1	1	1
Ciconiiformes (Ardeidae)	1	7	7
Anseriformes (Anatidae)	1	6	11
Falconiformes	3	9	17
Accipitridae		7	12
Galliformes	2	4	4
Gruiformes	1	3	3
Charadriiformes	3	11	14
Scolopacidae		8	10
Columbiformes	1	2	2
Cuculiformes	1	1	2
Strigiformes	2	6	7
Strigidae		5	6
Caprimulgiformes	1	3	4
Apodiformes	2	2	2
Coraciiformes	1	1	1
Piciformes (Picidae)	1	4	6
Passeriformes	18	75	126
Emberizidae		37	67
Tyrannidae		5	9
Muscicapidae		6	11
Troglodytidae		6	7
Hirundidae		4	5
Vireonidae		1	5
All orders	40	136	208

Source: From Zimmerman (1993).

sites. Differences between annually burned and unburned prairie are due to the presence in unburned prairie of Henslow's sparrow, a species that is dependent on litter and standing dead vegetation (Zimmerman 1988), and the similarly dependent common yellowthroat and several species dependent on woody vegetation. In contrast, winter bird communities are dominated by American tree sparrows in both unburned and annually burned prairie.

As in prairie habitats, richness of common species in gallery forest (continuously distributed forest associated with streams) and attenuated gallery forest (disjunct patches of trees associated with ravines and stream channels) is higher in summer than in winter (Table 7.3). Six of the common summer species occur in both gallery forest and attenuated forest, whereas four of the common winter species occur in both forest types. Five common species are present in gallery forest during both summer and winter, whereas three occur in attenuated forest during both summer and winter.

Avian diversity in tallgrass prairie in the Konza Prairie region has been influenced less by human activities since Euro-American settlement than has mam-

Table 7.3. Common summer and winter birds recorded in grassland and forest habitats of Konza Prairie, 1981–1990

Habitat	Summer	Winter
Tallgrass prairie		
Unburned prairie	Dickcissel	American tree sparrow
	Brown-headed cowbird	
	Grasshopper sparrow	
	Eastern meadowlark	
	Mourning dove	
	Common yellowthroat	
	Northern bobwhite	
	Upland sandpiper	
	Henslow's sparrow	
	Brown thrasher	
	Field sparrow	
Annually burned prairie	Dickcissel	American tree sparrow
	Grasshopper sparrow	
	Eastern meadowlark	
	Brown-headed cowbird	
	Upland sandpiper	
Deciduous forest		
Gallery forest	Great crested flycatcher	Black-capped chickadee
	Red-headed woodpecker	Blue jay
	Blue jay	Red-headed woodpecker
	Black-capped chickadee	Northern bobwhite
	Northern bobwhite	White-breasted nuthatch
	Yellow-billed cuckoo	Red-bellied woodpecker
	Eastern phoebe	American crow
	White-breasted nuthatch	
	Tufted titmouse	
	Mourning dove	
	House wren	
Attenuated gallery forest	Mourning dove	Red-headed woodpecker
	Field sparrow	Black-capped chickadee
	Northern cardinal	Blue jay
	Great crested flycatcher	American robin
	Black-capped chickadee	White-breasted nuthatch
	Brown-headed cowbird	Brown creeper
	Blue jay	Golden-crowned kinglet
	House wren	
	Dickcissel	
	Yellow-billed cuckoo	
	White-breasted nuthatch	
	Brown thrasher	

Source: From Zimmerman (1993).

Note: Within each group, species are listed in decreasing order of abundance. Gallery forest consists of continuous forest that is associated with streams, whereas attenuated forest consists of disjunct patches of trees that are associated with ravines and stream channels. A periodically updated listing of all bird species observed on Konza Prairie is available at http://climate.konza.ksu.edu.

malian diversity. For example, greater prairie chickens and upland sandpipers are common on Konza Prairie and elsewhere in the Flint Hills tallgrass prairie (Thompson and Ely 1989; Zimmerman 1993). Further, no grassland birds have been extirpated from the Flint Hills tallgrass prairie. Konza Prairie also offers important habitat for Henslow's sparrow, which for nesting requires the standing dead vegetation that develops in ungrazed or lightly grazed, unburned prairie sites (Zimmerman 1988).

Of woodland species that likely occurred in wooded habitats on Konza Prairie, the passenger pigeon and Carolina parakeet have become extinct through hunting and habitat destruction, respectively (Thompson and Ely 1989). Additionally, the wild turkey was extirpated from Kansas but has been reintroduced (Thompson and Ely 1989); wild turkeys on Konza Prairie now use mostly lowland sites consisting of a mix of wooded and prairie sites.

Introduced species in the Flint Hills region include the ring-necked pheasant, rock dove, European starling, and house sparrow (Thompson and Ely 1989, 1992). The three smaller species occur occasionally in native tallgrass habitats on Konza Prairie, but none are abundant there (Zimmerman 1993). Ring-necked pheasants occur in cultivated fields, weedy lowland prairie, and, less commonly, in upland prairie. Recent immigrants to the Flint Hills region such as house finches, great-tailed grackles, and cattle egrets occur only rarely on Konza Prairie.

Reptiles

Reptiles are less diverse on Konza Prairie than either mammals or birds (Heinrich and Kaufman 1985; Kazmaier 1993a). The 25 species known for Konza Prairie represent 11 families but only 2 orders (Fig. 7.1C). Of those orders, Chelonia contains only 3 species of turtles, whereas Squamata includes 7 species of lizards (suborder Sauria) and 15 of snakes (suborder Serpentes) (Table 7.4).

In addition to these 25 species, 4 aquatic turtles, 2 lizards, and 9 snakes occur in the general region (Collins 1993). Of these, species likely to occur on Konza Prairie include the five-lined skink (*Eumeces fasciatus*), western ribbon snake (*Thamnophis proximus*), plains garter snake (*T. radix*), western and eastern hognose snakes (*Heterodon nasicus* and *H. platirhinos*), plains blackhead snake (*Tantilla nigriceps*), and massasauga (*Sistrurus catenatus*) (Heinrich and Kaufman 1985; Collins 1993). However, both hognose snakes are species in need of conservation in Kansas (Collins 1993), and the lack of sightings at Konza Prairie may result from a general decline in their abundance.

The ornate box turtle, the only terrestrial turtle on Konza Prairie, is common on the site (Table 7.4; Heinrich and Kaufman 1985). Of the lizards and snakes, western slender glass lizards, Great Plains skinks, and eastern yellow-bellied racers are the common species in prairie habitats, especially in unburned prairie (Heinrich and Kaufman 1985; J. Cavitt personal communication). Many of the other lizards and snakes occur in prairie associated with rocky and limestone outcrop areas, but their abundance and habitat distribution are not well known.

Table 7.4. Species and status of reptiles and amphibians observed at Konza Prairie, 1981–1996

Common and Scientific Names	Status[a]
Class Reptilia	
Order Chelonia	
Snapping turtle (*Chelydra serpentina*)	Uncommon
Painted turtle (*Chrysemys picta*)	Present
Ornate box turtle (*Terrapene ornata*)	Common
Order Squamata	
Suborder Sauria	
Collared lizard (*Crotaphytus collaris*)	Local
Texas horned lizard (*Phrynosoma cornutum*)	Uncommon
Great plains skink (*Eumeces obsoletus*)	Common
Northern prairie skink (*Eumeces septentrionalis*)	Present
Ground skink (*Scincella lateralis*)	Present
Six-lined racerunner (*Cnemidophorus sexlineatus*)	Local
Western slender glass lizard (*Ophisaurus attenuatus*)	Common
Suborder Serpentes	
Western worm snake (*Carphophis vermis*)	Present
Ringneck snake (*Diadophis punctatus*)	Common
Flathead snake (*Tantilla gracilis*)	Present
Yellow-bellied racer (*Coluber constrictor*)	Common
Great Plains rat snake (*Elaphe emoryi*)	Common
Black rat snake (*Elaphe obsoleta*)	Common
Prairie kingsnake (*Lampropeltis calligaster*)	Present
Common kingsnake (*Lampropeltis getula*)	Common
Milk snake (*Lampropeltis triangulum*)	Common
Gopher snake (*Pituophis catenifer*)	Present
Northern water snake (*Nerodia sipedon*)	Present
Brown snake (*Storeria dekayi*)	Present
Common garter snake (*Thamnophis sirtalis*)	Common
Lined snake (*Tropidoclonion lineatum*)	Present
Copperhead (*Agkistrodon contortrix*)	Occasional
	(continued)

Further, western worm snakes, ringneck snakes, and rat snakes are associated with shrub and forested habitats on Konza Prairie; northern water snakes are associated with flowing streams; and common garter snakes are found in an array of moist habitats.

Reptiles generally are at risk of injury or death from prairie fires. Their vulnerability probably is heightened when snakes are active aboveground but are less mobile because of cool temperatures. For example, 12 racers, 11 Great Plains rat snakes, 1 lined snake, and 1 slender glass lizard were killed by a prescribed prairie fire on 26 April 1983 (Heinrich and Kaufman 1985). In this specific case, individuals that apparently were emigrating from a winter hibernaculum were killed within a small area when it was encircled by fire.

Table 7.4. continued

Common and Scientific Names	Status[a]
Class Amphibia	
Order Caudata	
Tiger salamander (*Ambystoma tigrinum*)	Present
Order Anura	
Plains spadefoot (*Spea bombifrons*)	Present
Woodhouse's toad (*Bufo woodhousii*)	Common
Northern cricket frog (*Acris crepitans*)	Common
Cope's gray treefrog (*Hyla chrysocelis*)	Common
Striped chorus frog (*Pseudacris triseriata*)	Common
Plains leopard frog (*Rana blairi*)	Common
Bullfrog (*Rana catesbeiana*)	Common
Plains narrowmouth toad (*Gastrophryne olivacea*)	Present

Sources: From Heinrich and Kaufman (1985) and Kazmaier (1993a), with information on status from Heinrich and Kaufman (1985), J. Cavitt (pers. comm.), and R. S. Matlack (pers. comm.).

Note: A periodically updated listing of all reptile and amphibian species observed on Konza Prairie is available at http://climate.konza.ksu.edu.

[a]Categories, defined relative to abundances typical of species, are: common = abundance is high in appropriate habitat(s) in most years; local = abundance is high in local sites within appropriate habitat(s) in most years; uncommon = abundance is low in most years even in best habitats; rare = abundance is very low across the site but species is present in most years; occasional = abundance is very low across site in best years and zero in many years; and present = species recorded for site, but not enough information is available to assign other status.

Amphibians

Amphibians are the least diverse of the four classes of terrestrial vertebrates on Konza Prairie, with only nine species documented (Heinrich and Kaufman 1985). These amphibians represent two orders, with Caudata (one salamander species) being less diverse than Anura (two species of toads and six of frogs; Fig. 7.1d; Table 7.4). These nine species are the common amphibians of the region, and the American toad (*Bufo americanus*) and Great Plains toad (*B. cognatus*) are the only undocumented amphibians that are likely to occur on Konza Prairie (Collins 1993).

Adult amphibians use terrestrial habitats on Konza Prairie (Heinrich and Kaufman 1985) because they are less dependent on standing and flowing water than are larval amphibians. Adult amphibians undoubtedly are minor consumer components of tallgrass prairie habitats, although leopard frogs were observed in prairie in wet periods (Heinrich and Kaufman 1985) and the plains spadefoot, Woodhouse's toad, and the plains narrowmouth toad probably occur in localized areas of the prairie (Collins 1993). Woodhouse's toad and the plains narrowmouthed toad were collected in limestone outcrop sites dominated by woody vegetation (Heinrich and Kaufman 1985). Prairie sites dominated by limestone outcrops probably offer better habitat for these two species than does nonrocky prairie because the limestone rocks provide refuge sites.

Larval amphibians are restricted to aquatic habitats, including streams, ponds, seeps, and temporary pools. Deep buffalo wallows (relict wallows created before the 1860s and current wallows made by the captive herd) that occur in upland prairie likely are used for breeding sites by some species of frogs and toads in wet years. Field observations indicate that striped chorus frogs commonly breed in temporary pools that develop in wallows during wet springs and summers (B. R. McMillan personal communication). However, a preliminary study suggests that tadpoles of chorus frogs from wallows metamorphose into adults only in wet years (N. M. Gerlanc unpublished data).

Insects

Insects are conspicuous and diverse elements of the terrestrial macrofauna of Konza Prairie. This diversity is typical of most terrestrial ecosystems, in which insects often account for over 50% of the species (Strong et al. 1984). Previous studies in tallgrass prairie (Risser et al. 1981) and general observations suggest that the insect fauna of Konza Prairie may include over 3,000 species from 200 families in 25 orders. Although highly diverse, few of these many species are likely to be endemic. Rather, the majority of insect species present in tallgrass prairie probably evolved elsewhere and subsequently spread to this habitat (Risser et al. 1981).

A total of 307 insect species from 62 families in 6 orders, which accounts for only about 10% of the species likely to occur, has been documented from Konza since 1977 (Table 7.5). Insects are widely distributed in all prairie habitats and occupy prominent roles as herbivores, pollinators, predators, parasites, and decomposers. Herbivores are probably the most conspicuous functional group of insects found in tallgrass prairie. Many members of the well-documented insect orders (Table 7.5) are herbivorous for all or part of their life cycles (Strong et al. 1984). Because of their abundance and potential impact on vegetation, herbivorous insects constituted a logical focal group for LTER studies of the dynamics and distributions of animal populations. Grasshoppers (family Acrididae) were studied in detail because they are abundant and easily sampled and identified (Evans et al. 1983).

Acridid grasshoppers on Konza Prairie are represented by 36 species (Table 7.5). Twenty-four of these species were the focus of a multiyear study of community dynamics (Evans 1988a) in which more than 25,000 individual grasshoppers were collected. Grasshoppers with diets specialized on grasses were most abundant among the 24 species. These grass-feeding acridids included two species, *Phoetaliotes nebrascensis* and *Orphulella speciosa*, that accounted for over 65% of the sampled grasshoppers. Of the 22 remaining species, 13 feed largely on forbs or on both grasses and forbs. Three of these forb- and mixed-feeding species, *Melanoplus scudderi*, *Hypochlora alba*, and *M. keeleri* accounted for over 20% of the samples (Evans 1988b). Although these values demonstrate average patterns, the distribution and local abundance of individual grasshopper species depend on the abundance and distribution of their host plants, which, in turn, depend on fire and drought (Evans 1984, 1988a).

Table 7.5. Orders and common families (represented by 4 or more species) for 307 species of insects documented for Konza Prairie, 1977–1996

Order/Families	Families	Genera	Species
Orthoptera (Acrididae)	1	23	36
Hemiptera	20	52	58
Pentatomidae		9	11
Miridae		10	10
Lygaeidae		7	8
Homoptera	9	56	82
Cicadellidae		38	56
Delphacidae		5	7
Dictyopharidae		2	4
Coleoptera	24	83	112
Chrysomelidae		28	48
Hydrophilidae		10	10
Carabidae		5	7
Coccinelidae		5	6
Dytiscidae		3	5
Cerambycidae		4	4
Mordellidae		1	4
Hymenoptera	2	3	6
Halictidae		2	5
Lepidoptera	6	13	13
Hesperidae		5	5
Nymphalidae		4	4
All orders	62	230	307

Note: Numbers of families, genera, and species is given for each order. A periodically updated listing of all insect species collected on Konza Prairie is available at http://climate.konza.ksu.edu.

Pollinators, predators, parasites, and detritivores are also important functional groups of the insect fauna of Konza Prairie. Hymenoptera and Diptera are the major pollinators of prairie forbs (Parrish and Bazzaz 1979; Bertin 1989). Predaceous and parasitic species are important natural enemies of many prairie insects and also of some vertebrates. Predaceous species are found in most insect orders; some familiar predaceous insects on Konza Prairie include robber flies (Asilidae), assassin bugs (Reduviidae), and the praying mantis (Mantidae). Among noninsects, spiders may be the most abundant but understudied of the invertebrate predators of the tallgrass prairie (Risser et al. 1981). Insects that parasitize insects are mostly in the orders Hymenoptera and Diptera, whereas insects that parasitize vertebrates are in Diptera, Phthiraptera (lice), and Siphonaptera (fleas). Detritivorous insects participate in organic matter breakdown. Although most of these species live near the soil surface (Chapters 4, 14, this volume), some members of Coleoptera, Hymenoptera, Orthoptera, and Diptera (e.g., dung beetles, ants, crickets, and sarcophagid flies) feed on plant litter, dung, and animal carcasses, thus speeding the decomposition of these plant and animal materials (Risser et al. 1981).

Finally, two species of insects of conservation importance may occur at Konza Prairie. The regal fritillary butterfly (*Speyeria idalia*) occurs in virgin tallgrass prairie. This butterfly has declined in many areas, with only fragmented populations remaining. This fragmentation has prompted its consideration for endangered species status (Nagel et al. 1991). However, populations of the regal fritillary may be relatively stable in the Flint Hills because of the extensive tracts of relatively undisturbed prairie. The second insect, the American burying beetle (*Nicrophorus americanus*), is a federally listed endangered species found in tallgrass prairies. Recent evidence suggests that these beetles now occupy only 10% of their former range (Lomolino et al. 1995). Populations of the American burying beetle occur in prairies in Oklahoma and Nebraska (Ratcliffe and Jameson 1992; Quisenberry and Purcell 1993), but surveys on Konza Prairie have failed to find this endangered species (D. Mulhern personal communication)

Summary

In this chapter, we have outlined the broad patterns of diversity for terrestrial vertebrates and insects on Konza Prairie. With the exception of some large mammals, the diversity of the terrestrial macrofauna remains generally high. Against this background of high diversity, the macrofauna of Konza Prairie exhibits complex population and community dynamics. Populations and communities are in constant flux, driven by interactions among fire, native and domestic grazers, and topographic and climatic variations (Chapters 6, 8, 9, this volume). These interactions create the mosaic of habitats, vegetation, and physical features necessary to maintain the array of native animals associated with Flint Hills tallgrass prairie. In other words, it is not possible to remove fire and grazing by large herbivores from tallgrass prairie and maintain the diversity of terrestrial animals.

The maintenance of the natural diversity of mammals, birds, reptiles, amphibians, and insects in tallgrass prairie depends on the presence of both fire and grazing. Too frequent use of fire or too high intensity of grazing could lead to lowered species diversity and, ultimately, lowered overall productivity. The same is true for very infrequent use of fire and the absence of large grazers from tallgrass prairie. The impacts of these two extremes are of great practical and conservation concern. A challenge for grassland ecologists and conservationists is to develop ecologically based grazing management strategies that maintain prairie productivity, as well as high biodiversity.

Acknowledgments We thank J. Cavitt, N. M. Gerlanc, R. S. Matlack, and B. R. McMillan for comments about distribution and abundance of reptiles and amphibians. This is contribution no. 97-182-B from the Kansas Agricultural Experiment Station, Kansas State University, Manhattan.

8

Animal Populations and Communities

Donald W. Kaufman
Glennis A. Kaufman
Philip A. Fay
John L. Zimmerman
Edward W. Evans

Important consumers in tallgrass prairie include mammals (large ungulate grazers and a suite of small mammals), invertebrates (several guilds of aboveground insects and soil invertebrates), and birds (both grass- and woody-dependent species; see Chapter 7, this volume). These consumers represent abundant and ecologically important groups that vary considerably in space and time in their relative importance to the trophic structure and ecosystem dynamics of the prairie. A primary goal of the LTER Program is to understand the ecology of these populations and communities, their spatial distributions and temporal patterns of abundances, and the processes controlling their dynamics and structure. As is true for the flora, the population dynamics and evolution of animals in tallgrass prairie have been molded by the selective pressures associated with fire, large ungulate grazers, and an annually and seasonally variable climate. These factors directly, and indirectly through their effects on vegetation structure and composition, influence the spatial and temporal distributions of birds, mammals, and invertebrate consumers in tallgrass prairie.

Risser et al. (1981) predicted that high spatial and temporal variations would be characteristic of consumer populations in tallgrass prairie, compared with those of forest systems, where vegetation structure itself reduces the amplitude of environmental fluctuations.

> Responses of populations of the grasslands to features of the physical environment such as weather should be more obvious than in many other terrestrial vegetation types. Little buffering of environmental conditions occurs in the grasslands, since most organisms are exposed directly to a wide range of variation and extremes of temperature, humidity, wind, precipitation, and a host of other factors. Because the environment is so variable, and because the organisms are directly exposed, we

> can expect the relationships between biota and environment to be close and apparent. (Risser et al. 1981, p. 2)

Much of the research on animal populations on Konza Prairie has attempted to infer process from pattern and to understand controls on population and community structure and dynamics in light of current ecological theory. Spatial and temporal distributions of animal populations and the responses of animals to experimental manipulations have shed important light on the roles of equilibrium and nonequilibrium processes influencing community structure of small mammals, birds, and grasshoppers. For example, studies of the responses of insects to fire regimes and drought provide insight into controls on their population dynamics. Long-term patterns displayed by prairie bird populations similarly provide important evidence relevant to alternative hypotheses on the regulation of grassland avifaunas. For some taxa, especially small mammals, it is important to recognize that significant alterations in their communities during the past 150 years have limited our ability to fully understand their ecological characteristics and dynamics as they occurred in the prairie landscape before pre-European settlement. Nonetheless, LTER studies have provided a wealth of information on population patterns and processes and the roles of fire, grazing, and climate in the tallgrass prairie.

In this chapter, we summarize our LTER and associated studies of the population-community patterns and dynamics of small mammals, birds, and insects. Some of these studies, particularly focal LTER projects, are ongoing, whereas other short-term studies of processes and patterns have been concluded.

Small Mammals

The distribution and abundance of mammals in the Great Plains were largely altered both directly and through environmental change by hunting, ranching, farming, and development practices (Jones et al. 1983; Benedict et al. 1996; Kaufman and Kaufman 1997). In most if not all cases, these changes occurred before detailed study of ecological characteristics of prairie mammals was done. As a result, little is known about abundance, spatial distribution, and habitat relationships of most species before anthropogenic changes took place. However, distributional ranges and general patterns of habitat association of rodents and shrews were known within the anthropogenically altered landscape of eastern Kansas (Bee et al. 1981) when long-term studies were initiated in 1981. Short-term studies of population dynamics and habitat relationships of rodents and shrews also have been conducted in various grassland, woodland, and agricultural habitats in the Great Plains. Responses of small mammals to fire were studied with various degrees of rigor in a number of grassland sites before 1980 (Kaufman et al. 1990), but little work was directed toward the effects of grazing. However, none of these earlier investigations, whether focused on abundance, spatial distribution, effects of prairie fires or grazing, or habitat requirements of small mammals in native tallgrass prairie, considered long-term temporal variation in the factors studied.

Our continuing study of small mammals is focused on rodents and shrews in prairie sites that are subjected to different burning regimes with some sites now grazed by *Bos bison*. Permanent transects (traplines of 20 stations with 15-m interstation intervals) also sample across the major topographic gradient, upland-slope-lowland. Our specific objectives are directed toward understanding the temporal patterns of abundance of small mammals, the influence of fire and climate on this abundance, the influence of fire and topoedaphic conditions on the spatial distribution of small mammals, and possible causes of observed temporal and spatial patterns.

Community Composition

During our LTER studies, 12 species of rodents and 2 species of shrews were recorded along sampling transects placed in prairie sites. The relative proportions of individual species in the small mammal community were strikingly different from those expected from geographic distributions and habitat associations of small mammals in eastern Kansas (Bee et al. 1981). During 1981–1995, the community of small mammals was dominated numerically by deer mice (47%), followed by western harvest mice (14%), white-footed mice (12%), short-tailed shrews (11%), prairie voles (7%), hispid cotton rats (3%), thirteen-lined ground squirrels (3%), and southern bog lemmings (1%). Collectively, the abundances of hispid pocket mice, eastern woodrats, house mice, plains harvest mice, least shrews, and meadow jumping mice represented only 1 to 2% of the total abundance of all small mammals in prairie habitats. These six species will not be considered further in any detail.

Deer mice numerically dominated the community of eight common small mammals in spring, summer, and autumn, although their percent contribution was lower in autumn (42% of all individuals of the eight common species) than in spring and summer (54–61%). Percent compositions were similar among seasons for western harvest mice (14–16%), white-footed mice (9–12%), and thirteen-lined ground squirrels (2–5%) but less so for prairie voles (3–10%) and southern bog lemmings (0.2–2%). Seasonal differences in percent composition were high for short-tailed shrews (autumn: 18%; summer: 6%; spring: less than 1%) and cotton rats (autumn: 5%; spring: 1%; summer: less than 1%) because of greater abundances in autumn than in the other two seasons. Additionally, some ground squirrels likely were hibernating when autumn samples were taken.

Species-specific patterns demonstrated large differences between the expected and observed contributions of individual species to the community of small mammals in tallgrass prairie. First, the prairie vole ranked fifth, rather than first, in its numerical contribution to the small mammal community, which was not expected from its known association with productive grasslands (Birney et al. 1976; Klatt and Getz 1987). The low abundance of voles also has been documented elsewhere on Konza Prairie by Bixler and Kaufman (1995b). Further, deer mice ranked first by a very large margin and were followed by western harvest mice, short-tailed shrews, and white-footed mice. Especially surprising was

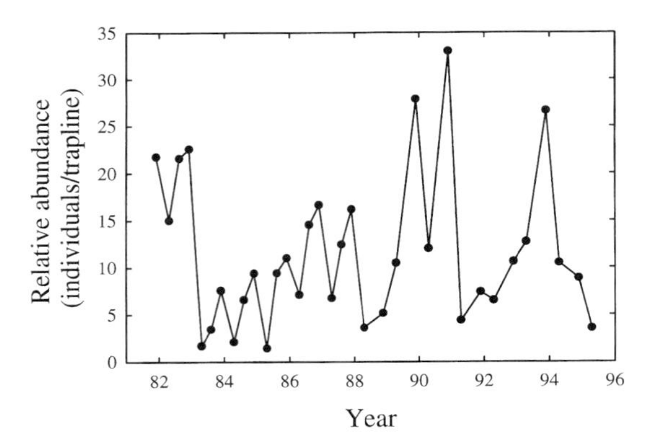

Figure 8.1. Relative abundance (individuals/trapline) of all small mammals during spring, summer, and autumn from autumn 1981 through spring 1995 on Konza Prairie. Small mammals were sampled in summer only from 1982 to 1987. Relative abundance is the average number of individuals/trapline in burned and unburned prairie during each sampling period.

the observation that short-tailed shrews and white-footed mice, a woodland species, were more common than were prairie voles.

Of the 14 species of small mammals observed in prairie habitats, 8 were categorized as omnivores (deer mice, western harvest mice, white-footed mice, thirteen-lined ground squirrel, eastern wood rat, house mouse, and plains harvest mouse); 2 as carnivores (short-tailed shrews and least shrews); 3 as herbivores (prairie voles, cotton rats, and southern bog lemmings); and 1 as a granivore (hispid pocket mouse). Because of the unexpected patterns of abundances, the percent contributions to the small mammal community by functional groups did not show the dominance by herbivores that we expected. Rather, the community was composed of 78% omnivores, 11% carnivores, 11% herbivores, and less than 1% granivores.

Temporal Variation in Abundance

Relative abundance varied widely among species, with average values of 5.5 individuals per trapline for deer mice, 1.7 for western harvest mice, 1.4 for white-footed mice, 1.3 for short-tailed shrews, 0.8 for prairie voles, 0.4 for cotton rats, 0.3 for thirteen-lined ground squirrels, and 0.1 for southern bog lemmings from autumn 1981 to spring 1995. Further, relative abundance varied greatly among years for each of the eight common species; the general pattern of variation is illustrated for all small mammals in Figure 8.1. To quantify the magnitude of temporal variation, the standard deviation (s) of the common log of relative abundance was calculated for each species (Ostfeld 1988) during autumn, when populations of small mammals were typically at the highest point of their annual cycle of abundance. Relative abundances were fairly constant in autumn during 1981–1994 for white-footed mice ($s = 0.27$) and deer mice ($s = 0.28$). In contrast,

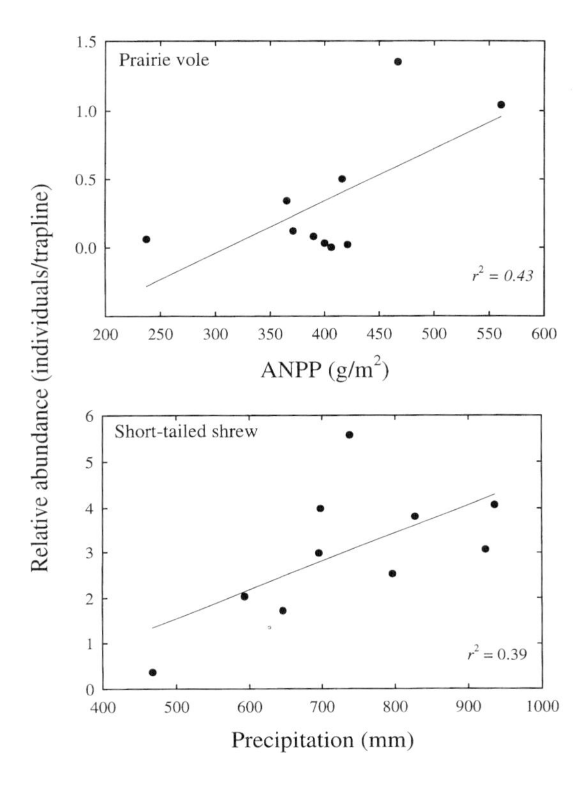

Figure 8.2. Relationships between relative abundance (individuals/trapline) of prairie voles (*Microtus ochrogaster*) during spring (1982–1991) and above-ground net primary production (ANPP) during the previous growing season on Konza Prairie and relative abundance of short-tailed shrews during autumn (1981–1990) on Konza Prairie and precipitation at Manhattan, Kansas, during the previous January–September.

autumn abundances were highly variable for thirteen-lined ground squirrels ($s = 0.71$), southern bog lemmings ($s = 0.76$), western harvest mice ($s = 0.86$), prairie voles ($s = 0.90$), and cotton rats ($s = 1.03$). Short-tailed shrews ($s = 0.52$) were intermediate between these groups in the magnitude of variability.

For voles and bog lemmings, temporal patterns of abundance were typical of arvicolid rodents that cycle with a 3- to 5-year periodicity (Taitt and Krebs 1985). Further, these cyclic patterns of voles and bog lemmings were synchronized highly (i.e., autumn abundances of these two species were correlated highly, $r = 0.90$, $d.f. = 12$, $P < 0.01$). In contrast, temporal patterns of abundance for deer mice, western harvest mice, short-tailed shrews, white-footed mice, cotton rats, and ground squirrels were typical of noncyclic fluctuations.

Spring abundances of deer mice, western harvest mice, ground squirrels, and short-tailed shrews were related neither to winter temperature or precipitation nor to grass stem or primary production during the previous growing season (Kaufman and Kaufman 1997). However, spring abundances of prairie voles, cotton rats, and bog lemmings apparently were influenced by vegetative conditions on Konza Prairie during the previous growing season but not by climatic factors during winter. Significant positive correlations were recorded between relative abundance of prairie voles and both primary production (Fig. 8.2) and grass production, between relative abundance of cotton rats and both primary production and grass production, and between relative abundance of southern bog lemmings and grass production (Kaufman and Kaufman 1997). These vegetative conditions provided both food and cover for these herbivorous rodents. In addition, spring abundance of white-footed mice was correlated positively with both primary production and grass production.

In contrast to relationships between measures of plant productivity and spring abundances of herbivorous rodents, autumn abundances of small mammals during a 10-year period did not exhibit consistent responses to changes in climate or vegetation. This suggests that no single factor is acting as a primary determinant of the observed temporal patterns of abundance (Chapter 1, this volume). However, significant correlations were found between some factors (summer temperature, summer and current year precipitation, and grass and primary production) and autumn abundance of individual species (Kaufman and Kaufman 1997). Cotton rats were the only species to respond to vegetation, increasing in abundance with an increase in grass production. Southern bog lemmings and short-tailed shrews responded to increasing summer temperatures with a decrease in abundance. Both short-tailed shrews (Fig. 8.2) and white-footed mice increased in abundance in response to increased current-year precipitation and summer precipitation, respectively, whereas thirteen-lined ground squirrels decreased in abundance with increasing summer precipitation. Only two general patterns between autumn abundance and environmental conditions emerged: decreases in autumn abundance of short-tailed shrews were related to water stress associated with low precipitation and high summer temperature, and increases in autumn abundance of cotton rats were related to food and cover associated with high primary production.

Patterns of Spatial Distribution

Based on autumn sampling, the 14 species of small mammals observed could be placed into four groups related to spatial-temporal distribution considering all transects sampled from 1981 to 1990. First, the deer mouse and short-tailed shrew were distributed widely through space and time, with captures on 89% and 82% of transects, respectively. Second, the western harvest mouse and white-footed mouse were captured on about one-half of the transects (57% and 46%, respectively). Third, the prairie vole and thirteen-lined ground squirrel were captured on about one-fourth of the transects (30% and 21%, respectively). Finally, eight species were highly localized in space and/or time, with captures on 12% or less of the transects: the cotton rat (12% of transects), southern bog lemming (11%), hispid pocket mouse (5%), house mouse (4%), plains harvest mouse (4%), least shrew (3%), eastern woodrat (3%), and meadow jumping mouse (<1%).

The number or percent of transects on which a species was trapped was correlated highly with average density during autumns from 1981 to 1990 ($r = 0.89$, *d.f.* = 12, $P < 0.001$). Further, no species was relatively common and highly localized spatially, and no species was rare and widely distributed. The intermediate distribution of prairie voles and western harvest mice was due to the averaging of years with high abundance and a broad spatial distribution and those with low abundance and a narrow spatial distribution. For example, western harvest mice were caught on 98% of the transects when abundant in 1981 and 1982 but on only 18% of transects when uncommon in 1983 and 1984. In contrast, white-footed mice and ground squirrels consistently were uncommon, and their intermediate distributional patterns were related to habitat features that differed among transects. For example, the distance between transects and nearby ravines and woody vegetation likely influenced the capture of white-footed mice (Clark et al. 1987) and, therefore, the spatial distribution of this woodland species on prairie transects. Similarly, differences in breaks sites among transects probably influenced the observed spatial distribution of ground squirrels (Clark et al. 1992; Kaufman et al. 1995b). Finally, fire history likely influenced observed spatial distributions because some small mammals respond negatively to fire and others respond positively to fire (Kaufman et al. 1990).

The major topoedaphic conditions on Konza Prairie (flat, shallow-soil upland; rocky limestone breaks; and gently sloped, deep-soil lowlands; Chapter 3, this volume) influence the spatial distribution of small mammals. For example, deer mice and thirteen-lined ground squirrels selected breaks over upland and lowland (Kaufman et al. 1988b; Clark et al. 1992; Kaufman et al. 1995b). Further, deer mice and ground squirrels were associated with upland over lowland in a site with less physical relief and only upland and lowland conditions (Peterson et al. 1985; Brillhart et al. 1995). In contrast, western harvest mice, white-footed mice, and cotton rats selected lowland over breaks and upland (Clark 1989; Kaufman et al. 1988b; Kaufman et al. 1995b). In sites with only upland and lowland and low physical relief, cotton rats were associated with lowland but not upland (Brillhart et al. 1995).

Distribution of captures along LTER transects suggested a positive association with lowland by short-tailed shrews and southern bog lemmings, but little pattern for prairie voles (Kaufman et al. 1995b). In contrast, detailed study of small mammals on a large trapping grid (13 ha) during 1983–1987 demonstrated the selection of lowland by short-tailed shrews (Clark et al. 1995) and southern bog lemmings (Clark 1989) but selection of upland by prairie voles (Bixler and Kaufman 1995a) and least shrews (Clark et al. 1995).

Responses to Fire

Based on earlier studies (Kaufman et al. 1990), we expected population responses to fire by small mammals to be either negative (decreases in numbers of prairie voles, western harvest mice, and short-tailed shrews) or positive (increases in numbers of deer mice). Indeed, a strong negative response by western harvest mice and a strong positive response by deer mice became evident very quickly (Kaufman et al. 1983, 1988a, 1988b). However, long-term data sets were needed to demonstrate and understand fire responses by other species.

Continued study of fire responses along LTER transects demonstrated a strong fire-positive response by deer mice and a weak fire-positive response by thirteen-lined ground squirrels (Table 8.1; Kaufman et al. 1990). In contrast, fire-negative responses were observed for southern bog lemmings, short-tailed

Table 8.1. Fire effects on common small mammals in tallgrass prairie habitats based on captures of individuals along LTER sampling transects at Konza Prairie

Species	Pattern Observed	Percent Censuses with Pattern	Total Number of Censuses
Fire-positive species			
Deer mouse	B > U	95	19
Thirteen-lined ground squirrel	B > U	67	18
Fire-negative species			
Southern bog lemming	U > B	92	12
Short-tailed shrew	U > B	88	17
Western harvest mouse	U > B	76	17
Prairie vole	U > B	65	17

Source: From Kaufman et al. (1990).

Note: All patterns were significantly different at $P < 0.05$ except for ground squirrels, where $P < 0.10$. The pattern for prairie voles approached the critical probability ($P < 0.15$). B > U indicates where abundance was greater in burned than in unburned prairie, whereas U > B indicates the opposite pattern. Total number of censuses is the number of census periods where abundance was greater than zero in either burned or unburned prairie.

shrews, western harvest mice, and prairie voles. Intensive sampling elsewhere on Konza Prairie also demonstrated the fire-positive responses for deer mice and thirteen-lined ground squirrels (Clark and Kaufman 1990; Clark et al. 1992) and fire-negative responses for western harvest mice, prairie voles, short-tailed shrews, and southern bog lemmings (Clark 1989; Clark and Kaufman 1990; Clark et al. 1995). In contrast to these six species, white-footed mice exhibited no differential response to fire. Preliminary analyses of data collected from autumn 1981 to spring 1991 suggested that none of the species exhibiting fire-positive or fire-negative responses differed in their responses between annually and periodically burned prairie. However, the abundance of white-footed mice was greater in periodically burned prairie. This difference likely was caused by a lack of woody vegetation in annually burned sites (Chapter 6, this volume).

Fire also shifted the community composition of small mammals as a result of decreases in fire-negative species and increases in fire-positive species (Table 8.2). The sum of fire-positive responses was numerically greater than that of fire-negative responses, so that the total abundance of small mammals was greater in burned prairie than in unburned prairie. Because of these numerical changes, the postfire community was dominated strongly by deer mice, which resulted in lower species richness and diversity in burned prairie than in unburned prairie.

What are the causes of the observed fire effects? An occasional small mammal was found dead after prairie fires in spring, but no major fire-induced mortality was apparent for rodents or shrews on Konza Prairie. This contrasts with the observations that fire mortality for some small mammals in prairie may be high under unusual conditions, e.g., young rodents in aboveground nests (Erwin and Stasiak 1979). On Konza Prairie, emigration of individuals likely is respon-

Table 8.2. Relative abundance (individuals/trapline) and percent abundance of small mammals for the community of 14 species captured in unburned and burned prairie at Konza Prairie

	Abundance		Percent	
Species	Unburned	Burned	Unburned	Burned
Deer mouse	3.53	8.44	35.19	63.40
Western harvest mouse	2.51	1.14	25.02	8.56
Short-tailed shrew	1.74	0.92	17.35	6.91
White-footed mouse	0.90	0.97	8.97	7.29
Prairie vole	0.66	0.55	6.58	4.13
Hispid cotton rat	0.12	0.73	1.20	5.48
Thirteen-lined ground squirrel	0.29	0.42	2.89	3.15
Southern bog lemming	0.16	0.03	1.60	0.23
Hispid pocket mouse	0.04	0.05	0.40	0.38
Eastern woodrat	0.03	0.01	0.30	0.08
House mouse	0.02	0.02	0.20	0.15
Plains harvest mouse	0.02	0.02	0.20	0.15
Least shrew	0.01	0.01	0.10	0.08
Meadow jumping mouse	0.004	0.003	0.04	0.02

Note: Average values are based on 26 sampling periods during autumn 1981 to spring 1991 with autumn samples in 1981–1990, spring samples in 1982–1991, and summer samples in 1982–1987.

sible for decreases in abundance of fire-negative species. For example, western harvest mice, prairie voles, and southern bog lemmings moved from burned sites to nearby unburned sites within hours or days of a prairie fire (Clark and Kaufman 1990). Abundance of short-tailed shrews also decreased, but neither movements to nearby unburned prairie nor direct mortality to fire or heat was observed. We assume that shrews succumbed to starvation or predation in the burned site, but some movement to more distant unburned sites was possible. In contrast to decreases in fire-negative rodents and shrews, the abundance of deer mice was not depressed following prairie fire. Rather, numbers of deer mice increased within weeks because of immigration (Clark and Kaufman 1990; Kaufman et al. 1988a) and subsequently reproduction in burned sites. Immigration has not been demonstrated as the cause of the fire-positive response of thirteen-lined ground squirrels, although a short-term increase in abundance suggested that it does occur.

Removal of litter and standing dead vegetation by prairie fire appears to be the proximate cause of the fire-negative and fire-positive responses of small mammals. Further, species-specific patterns of recovery to new highs or lows during the first few years after a fire in ungrazed prairie probably are dependent on the accumulation of a relatively deep litter layer. For example, abundance of deer mice decreased, whereas abundances of western harvest mice and short-tailed shrews increased with increasing depth of plant litter (Kaufman et al. 1989). Results of laboratory experiments also support the notion that litter depth is an important habitat feature that influences habitat selection and spatial distribution of small mammals in grassland habitats. Deer mice and western harvest mice ate more seeds from under sparse than under deep litter (Kaufman and Kaufman 1990b), which was consistent with a low cost of searching for and/or handling of seeds. However, a greater proportion of seeds was eaten by harvest mice (43%) than by deer mice (15%) from under deep litter. This may reflect a difference in ease of movement through the litter by the smaller species. In a more complex experiment, rodent behavior also was consistent with fire responses because western harvest mice and prairie voles preferred patches with moderate litter for foraging and nesting and avoided patches with sparse litter (Clark and Kaufman 1991). Further, both harvest mice and voles preferred to nest in litter rather than in wooden nest boxes. In contrast, deer mice and white-footed mice chose sparse litter over moderate litter for foraging and also preferred to use nest boxes in areas with sparse litter rather than to build their own nests in the plant litter layer.

What other factors might influence fire responses of small mammals? Fire creates ash by burning prairie vegetation, and rodents living in burned prairie ingest the ash during the first few days after fire. Does the ash differentially influence small mammals? A field experiment demonstrated that ash-coated millet seeds were neither preferred nor avoided by either deer mice or cotton rats (Rustiati and Kaufman 1993). This lack of a differential response suggests that ash has little impact on fire responses of small mammals.

Responses to Grazing

Interspecific differences in the responses of small mammals to plant form, depth of the litter layer, and relative openness of the soil surface suggest that grazing by large mammalian herbivores (*B. bison* or cattle) should alter the distribution and abundance of rodents and shrews. For example, grazing reduces the height and density of the plant canopy and litter layer and increases the proportion of bare soil. Grazing also alters plant species composition (Chapters 6 and 9, this volume), but how this change may affect small mammals is not known.

Captures of small mammals in ungrazed sites on Konza Prairie and nearby grazed prairie indicate that grazing increased the abundance of deer mice and decreased the abundance of prairie voles and western harvest mice on grazed sites compared with ungrazed sites (Clark et al. 1989). This numerical increase of deer mice on grazed prairie was consistent with earlier studies in grasslands (Phillips 1936; Reynolds 1980; Reynolds and Trost 1980; Grant et al. 1982). Also, numbers of prairie voles were decreased greatly in northern Oklahoma tallgrass prairie subjected to cattle grazing. Not only did grazing affect the abundance of grassland rodents, but removal of the potential herbaceous fuel load by grazing likely changed the impact of fire on woody vegetation. As a result, white-footed mice were more abundant in grazed rangeland that had scattered shrub patches than in topographically similar burned sites on Konza Prairie that lacked shrubs.

In addition to grazing-induced increases in deer mice, Grant et al. (1982) also found that cattle grazing had a strong positive influence on thirteen-lined ground squirrels and hispid pocket mice in tallgrass prairie in northern Oklahoma. Further, strong negative effects of grazing were observed not only for prairie voles but also for the short-tailed shrew and hispid cotton rat. The responses of these six species of small mammals were consistent with their general habitat requirements and their responses to fire; we would expect similar responses to moderate or heavier grazing in the general region of Konza Prairie.

Biotic Interactions

Small mammals are potential prey for an array of mammalian carnivores, raptors, and snakes, but use of small mammals as prey has been studied only for coyotes on Konza Prairie. Coyotes are the common mammalian carnivores on Konza Prairie, where they often travel on roads and fireguards. Analyses of teeth, bone, and hair from 1989 to 1991 revealed that small mammals were important components of the diet of coyotes on Konza Prairie (Brillhart and Kaufman 1994). The most frequently eaten small mammals were the prairie vole, cotton rat, southern bog lemming, and harvest mouse, which were recorded in 26%, 22%, 13%, and 6%, respectively, of scats analyzed. Other rodents and shrews were consumed much less frequently. The importance of rodents in general and voles and cotton rats in particular was consistent with studies in mixed-grass prairie (Kaufman et al. 1993b; Brillhart and Kaufman 1995).

Although rodents are eaten commonly by individual coyotes, the low abundance of coyotes on Konza Prairie and large numbers of rodents probably result in coyotes having little influence on the distribution and abundances of most small mammals within a prairie landscape. This is particularly true for abundant species such as the deer mouse and short-tailed shrew, which often are common but only infrequently are preyed upon by coyotes. We found evidence of deer mice in less than 5% of the scats analyzed from 1989 to 1991, even though the deer mouse is the most common small mammal on Konza Prairie (about 60% of the total numbers of small mammals captured from 1989 to 1991; Brillhart and Kaufman 1994). Infrequent use of deer mice by coyotes also is consistent with observations in mixed-grass prairie (Kaufman et al. 1993b; Brillhart and Kaufman 1995).

In addition to serving as prey for a variety of predators, different species of small mammals prey on a variety of invertebrates and an occasional vertebrate; graze or browse on plant foliage, buds, or rhizomes; and eat or disperse plant seeds. Studies of small mammals as predators, herbivores, and dispersers of seeds have been few in tallgrass prairie; however, the general effects of prairie voles on vegetation composition were examined experimentally on Konza Prairie. In this case, prairie voles maintained at high densities did shift the composition of prairie vegetation (Gibson et al. 1990; Kaufman and Bixler 1995). However, extrapolation of the effects of prairie voles to the overall landscape is difficult because the densities of voles on Konza Prairie seldom reach levels similar to the experimental densities and then only on a very local scale.

In addition to effects on foliage, prairie voles also cut down the culms of several species of grasses and forbs to access and subsequently eat the seeds (Kaufman and Bixler 1995). Consistent with vole behavior under experimental conditions, field observations on Konza Prairie and in mixed-grass prairie in central Kansas suggest that the seed heads of prairie grasses and forbs are harvested by voles, cotton rats, and deer mice. Laboratory experiments demonstrated that cotton rats also will cut down stems and thereby obtain seeds from the herbaceous canopy (Jekanoski and Kaufman 1993). In contrast, field observations on deer mice (Kaufman and Kaufman 1997) and laboratory observations on deer mice, white-footed mice, and western harvest mice (Jekanoski and Kaufman 1995) demonstrated that these seed-eating rodents will climb into the herbaceous canopy to harvest seeds directly from plants. Additionally, free-living deer mice not only harvest seed heads of prairie forbs but also disperse a small proportion of seeds in the process of handling and transporting the seed heads from a few centimeters to over 20 m before consuming the seeds. Although we know that rodents eat and disperse plant seeds, we do not understand how such predation and dispersal of seeds affects the plant community.

Birds

Like the flora, which is not unique to the tallgrass prairie (Wells 1970a; Chapter 5, this volume), no endemic birds occur in the tallgrass prairie (Mengel 1970). Rather, the assemblage of bird species presently found in the tallgrass prairie is derived from species inhabiting grassland habitats in other geographic regions. Further, few bird species are characteristic of the tallgrass prairie (Cody 1966; Risser et al. 1981), with two opposing conceptual views suggesting why this is so. One view (Conrad 1939) is that the bird community is in equilibrium at a species richness determined by competition for limited resources. Thus, the community of grassland birds is saturated, and all available resources have been partitioned (Cody 1968). The alternative view (Gleason 1939) is that the resources are not limiting, and the species that occur in tallgrass prairie are those adapted to the spatial, temporal, and climatic characteristics of the prairie environment (Wiens 1974).

Bioenergetic studies suggest that food resources are not limiting during the breeding season (Wiens 1973). The maximum energy demand during the breeding season of the entire avian community was estimated to be 3 kcal/m^2. This is less than 0.1% of the energy available in the standing crop of herbivorous insects in tallgrass prairie (Wiens and Dyer 1975). An average of nine medium-sized grasshoppers/m^2 can support an avian community in grasslands throughout a breeding season (Zimmerman 1993). Thus, the available evidence does not support the saturated, resource-limited hypothesis as the determinant of the low species richness of grassland birds. Instead, tolerance of drought appears to be the important determinant of membership in the bird community of the tallgrass prairie (Zimmerman 1992).

The primary objective for the study of birds within the LTER Program has been to assess species richness and relative abundances in avian communities during midwinter (January) and during the breeding season (June) in grassland and gallery forest habitats. Study areas were chosen so that a variety of combinations of fire frequency and grazing could be assessed for their effects on the abundance and richness of birds through time.

Factors Regulating Populations

Because most species of grassland birds migrate and food appears abundant during the breeding season, the most stringent regulation of these populations probably occurs on the wintering areas (Fretwell 1972). Generally, the physical structure of the vegetation, and not food resources, sets the carrying capacity for the bird species in tallgrass prairie. Although not examined for all species, this appears to be a reasonable hypothesis for species of grassland birds studied to date (Wiens 1969; Kahl et al. 1985; Zimmerman 1988). Furthermore, because vegetative structure is affected directly by unpredictable perturbations such as fire (Collins and Gibson 1990), drought (Briggs et al. 1989), and grazing (Herbel and Anderson 1959; Collins and Gibson 1990), grassland birds have been confronted with a habitat that exists in a nonequilibrium state (Chapter 1, this volume).

Numerous mechanisms may be involved in the regulation of avian populations around a carrying capacity determined by the unpredictable effects of drought, fire, and grazing. For example, the structure of the vegetation is correlated highly with male abundance and degree of polygyny in dickcissels (Zimmerman 1971). Vegetative structure affects the nest microclimate, which in turn affects the fledging weights of the young via effects on the behavior of the female (Blankespoor 1970). Territorial behavior provides the mechanism that allows some males to obtain good habitat patches, whereas other males remain mateless in poor patches or obtain no territory at all. However, some unsuccessful males may seek out suitable habitat elsewhere (Schartz and Zimmerman 1971; Zimmerman and Finck 1989), which permits exploitation of newly available habitat and increases their probability for reproductive success. Social parasitism by the brown-headed cowbird severely impacts the productivity of dickcissels in the prairie (Zimmerman 1983), but predation remains the major cause of nest loss in dickcissels (Zimmerman 1984). Less is known about the regulation of other bird populations on tallgrass prairie, but some data exist (e.g., Lanyon 1957).

Avian Communities in Grassland and Forest Habitats

As expected, the most abundant breeding birds found in prairie habitats on Konza Praire were largely those that depend on grass/forb substrates to meet their nesting requirements (Zimmerman 1992, 1993). Common species in this group include the dickcissel, grasshopper sparrow, and eastern meadowlark (Table 8.3). However, a characteristic suite of birds also occurs that depend upon the woody vegetation within the tallgrass prairie for nesting and/or foraging. Common woody-dependent species include the eastern kingbird and brown thrasher (Table 8.3). Because fire reduces woody cover in the prairie, this latter set of species is represented poorly in frequently burned grassland.

Furthermore, the tallgrass prairie in the Flint Hills of Kansas is characterized

Table 8.3. Relative abundance (mean ± 1 SE) of birds (individuals/km) in unburned and burned prairie at Konza Prairie 1981–1990

Species	Unburned	Burned
Common species		
Grass-dependent species		
Dickcissel	13.9 ± 0.90	12.3 ± 1.86
Grasshopper sparrow	8.3 ± 0.67	6.8 ± 0.90
Eastern meadowlark	8.0 ± 0.41	6.8 ± 0.80
Mourning dove	4.0 ± 0.39	2.2 ± 0.45
Common yellowthroat	3.1 ± 0.39	0.2 ± 0.17[a]
Upland sandpiper	2.9 ± 0.46	5.2 ± 0.70
Henslow's sparrow	2.8 ± 0.59	0.0
Red-winged blackbird	1.6 ± 0.22	3.8 ± 0.37
Ring-necked pheasant	1.0 ± 0.16	0.2 ± 0.10[a]
Woody-dependent species		
Northern bobwhite	3.0 ± 0.47	1.2 ± 0.44[a]
Brown thrasher	2.8 ± 0.23	0.7 ± 0.18[a]
Field sparrow	2.8 ± 0.45	0.0
Bell's vireo	2.3 ± 0.52	0.0
American goldfinch	2.3 ± 0.55	0.0
Eastern kingbird	1.4 ± 0.18	1.5 ± 0.19
House wren	1.3 ± 0.29	0.0
Baltimore oriole	0.4 ± 0.09	0.3 ± 0.10[a]
Habitat-independent species		
Brown-headed cowbird	8.4 ± 0.51	6.1 ± 1.24
Uncommon species		
Grass-dependent species		
Killdeer	0.4 ± 0.03	0.0
Common nighthawk	0.3 ± 0.10	1.0 ± 0.36
Lark sparrow	0.1 ± 0.03	0.0
Woody-dependent species		
Northern flicker	0.7 ± 0.15	0.2 ± 0.10
Yellow-billed cuckoo	0.6 ± 0.13	0.0
Eastern towhee	0.6 ± 0.56	0.0
Eastern bluebird	0.4 ± 0.24	0.0
Red-headed woodpecker	0.3 ± 0.06	0.0
Blue jay	0.3 ± 0.08	0.0
Northern cardinal	0.3 ± 0.10	0.0
European starling	0.3 ± 0.12	0.0
American robin	0.2 ± 0.05	0.1 ± 0.08
Great crested flycatcher	0.2 ± 0.06	0.0
Gray catbird	0.2 ± 0.08	0.0
Loggerhead shrike	0.2 ± 0.15	0.0
Common grackle	0.1 ± 0.02	0.8 ± 0.28
Warbling vireo	0.1 ± 0.03	0.0
Yellow warbler	0.1 ± 0.03	0.0
Tufted titmouse	0.1 ± 0.04	0.0
Black-capped chickadee	0.1 ± 0.05	0.0
Blue grosbeak	0.1 ± 0.05	0.0

Table 8.3. continued

Species	Unburned	Burned
Red-bellied woodpecker	0.1 ± 0.05	0.0
Eastern phoebe	0.1 ± 0.07	0.0
Orchard oriole	0.1 ± 0.09	0.0

Source: From Zimmerman (1993).

Note: Bird species not included were raptors, prairie chickens, and swallows. Common species were present every year, whereas uncommon species were present in 20 to 90% of the years.

[a]Present less than every year in burned prairie.

by the development of deciduous (gallery) forest along the lower reaches of the prairie watersheds. This habitat has its own characteristic bird community that constitutes about 50% of all the species known for Konza Prairie (Zimmerman 1993), even though the gallery forest constitutes only 7% of the area of the site (Knight et al. 1994).

More than two-thirds of the species that are permanent residents, such as blue jays and American robins, occur in the gallery forest. This likely is due to gallery forest being a highly structured habitat that provides protection from the extremes of winter temperatures. For many of the permanent resident species found on Konza Prairie, however, the individuals present in the winter may not be the same individuals that were present during the summer.

Four structurally different avian habitats containing unique avian communities are found on Konza Prairie. In increasing complexity of vegetative structure, these habitats are burned prairie, unburned prairie, attenuated forest, and gallery forest (Zimmerman 1993). Both the total abundance and species richness of the avian communities in winter were correlated with the complexity of the habitat (Fig. 8.3; Zimmerman 1993). During the breeding season, however, this pattern is obscured. Species richness was similar in the two forested communities and unburned grassland but distinctly lower in annually burned grassland (Fig. 8.3). In contrast, total abundance was highest in the gallery forest but lower and similar in the attenuated forest and the two grassland habitats.

Not only does community composition vary among the four major habitats, but the annual variability in total abundance and richness of avian communities shows distinct seasonal differences (Fig. 8.3). In June, the coefficients of variation for both total abundance and richness of species were low and similar among the four community types, reflecting little change in the suitability of the habitats from year to year. In winter, however, annual variation for both variables was higher and increased with decreasing vegetative complexity (Zimmerman 1993). These data suggest that avian communities are less subject to the interannual variation in climate during summer than during winter. The avian community in the open, less protected grassland sites is affected more during winter than that in the more protected forested sites.

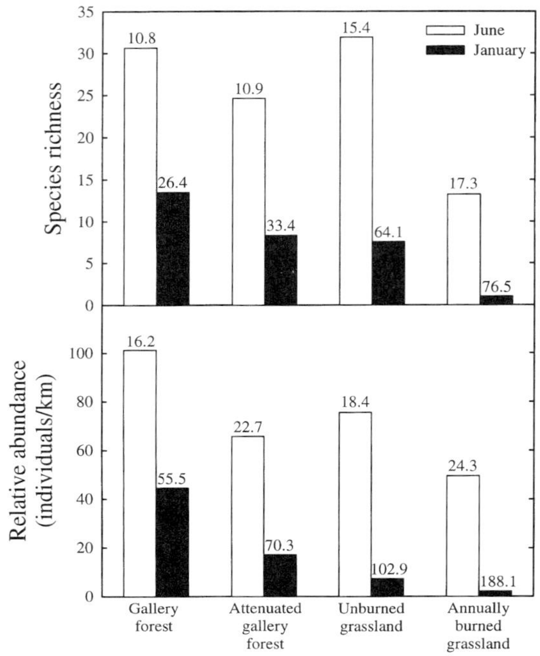

Figure 8.3. Species richness and relative abundance (individuals/km) for birds during summer and winter in major habitats on Konza Prairie. Values are means for 1981–1990; coefficient of variation is shown above each bar. From Zimmerman (1993), with permission of the University Press of Kansas.

Seasonal and Temporal Variation in Prairie

The avian community of the tallgrass prairie displays dramatic seasonal shifts in species composition (Zimmerman 1993). Of the species recorded for Konza Prairie, 46% are migratory and present only during spring and autumn, whereas only 13% are present throughout the year. Almost all species breeding in grassland habitats are migratory. About one-third of these species winter in the neotropics (e.g., dickcissel) or in the temperate zone of South America (e.g., upland sandpiper), whereas the remainder winter in the temperate zone of North America. A similar pattern has been described for grazed prairie in Oklahoma (Grzybowski 1982).

During the breeding season, the avian communities of both unburned and burned prairie include both common grass/forb-dependent and woody-dependent species that are present every year (Table 8.3; Zimmerman 1992). Additionally, a second set of species, the uncommon species, in both communities occur in only 20 to 90% of the years. The proportion of common to uncommon species is similar in both annually burned grassland (47%) and unburned grassland (42%). However, both values are about twice those of the proportions of common species in the gallery forest and attenuated gallery forest communities (22% and 18%, respectively). The relatively high proportion of common species in grassland habitats as compared with forested habitats probably reflects the relatively low species richness of avian communities in grasslands.

The magnitude of temporal variation in relative abundance of common and uncommon species during June varied widely (Table 8.3). In unburned prairie, common grass-dependent species were less variable (average CV = 38% for 9 species) than were common woody-dependent species (average CV = 57% for 8 species), uncommon grass-dependent species (average CV = 75% for 3 species), and uncommon woody-dependent species (average CV = 137% for 21 species). In burned prairie, the average CV was 80 to 90% for both grass-dependent and woody-dependent common species, whereas the average CV was over 100% for both groups of uncommon species. In unburned prairie, the common grass-dependent species with low temporal variation (CV < 30%) were the eastern meadowlark, dickcissel, and grasshopper sparrow, whereas the brown thrasher was the only woody-dependent species with low temporal variation. In burned prairie, the least variable common grass-dependent species was the red-winged blackbird (CV = 31%), and the least variable common woody-dependent species was the eastern kingbird (CV = 40%).

The abundance of birds in grassland habitats on Konza Prairie during winter is extremely low. The only consistent species is the American tree sparrow, which has been recorded in every year in unburned grassland and in half of the years in annually burned grassland (Zimmerman 1993). Because greater seed set occurs in annually burned prairie (Knapp and Hulbert 1986), tree sparrow abundance was expected to be greater in burned grassland than in unburned prairie. However, the relative abundance of this species was lower on burned prairie, but the difference was not statistically significant (Zimmerman 1993). The apparent preference for unburned prairie likely is a response to the greater availability of woody cover, which provides a refuge for escape from predators, as well as amelioration of nighttime temperature extremes (Pulliam and Mills 1977; Grubb and Greenwald 1982).

The decline of neotropical migrants, especially those inhabiting grasslands (Knopf 1994), has been of national concern in North America (Robbins et al. 1989). In Kansas, significant declines of the mourning dove, eastern kingbird, red-winged blackbird, eastern meadowlark, dickcissel, and grasshopper sparrow have been recorded (S. Droege unpublished data). All of these species are common in grasslands on Konza Prairie, but none of them show declining trends in abundance over the decade 1981–1990. In fact, the abundance of common yellowthroats on Konza Prairie increased during that period, in contrast with patterns for Kansas as a whole (Zimmerman 1993). Differences between temporal patterns for these birds on Konza Prairie (largely ungrazed) and elsewhere in Kansas suggest that habitat deterioration may have occurred in areas grazed by livestock.

In contrast to the relative stability of the passerine species of Konza, greater prairie chickens (Fig. 8.4) have undergone population cycles, with peak populations in 1981 and 1988, low populations in both 1983 and 1984, and a continually declining population from 1991 through 1996 (Fig. 8.5; Zimmerman 1993). From 1981 to 1996, 20 different leks were identified on Konza Prairie, although no more than 13 were active in any given year. Even during years of low population numbers, usually at least eight leks remained active. When prairie chicken populations are high, competition may be intense, and nonterritorial birds may

Figure 8.4. Male greater prairie chicken displaying on a lek in tallgrass prairie. (R. Robel)

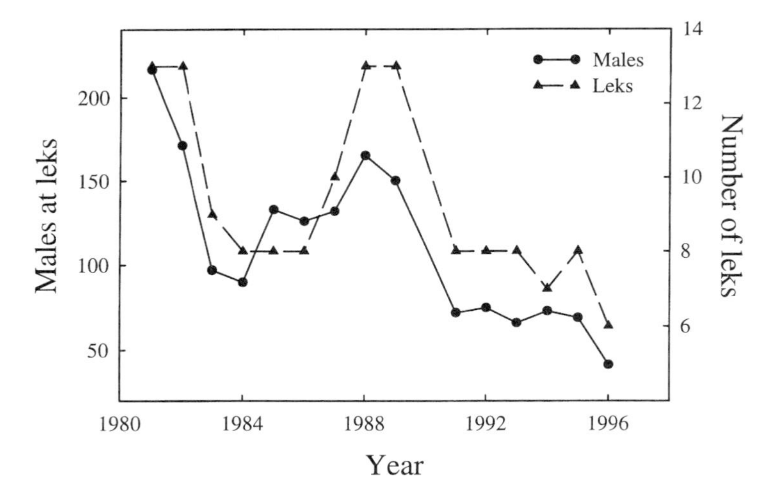

Figure 8.5. Number of male prairie chickens at leks and number of leks found during spring surveys on Konza Prairie from 1981 to 1996.

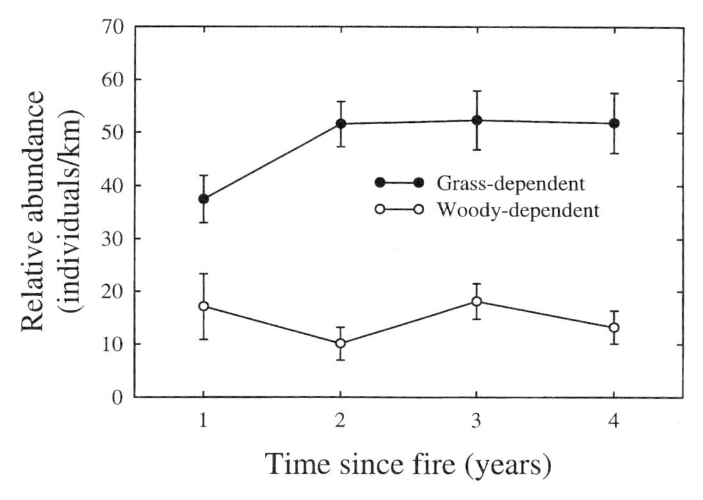

Figure 8.6. Relative abundances of birds (mean ± SE for indiviuals/km) for common grass-dependent species and both common and uncommon woody-dependent species in summer over a 4-year cycle of fire on Konza Prairie during 1981–1990. Mean relative abundances were 37.6 ± 2.8 and 5.9 ± 0.8 for grass- and woody-dependent species, respectively, in annually burned prairie. Mean relative abundances were 45.6 ± 2.2 and 21.7 ± 2.0 8 for grass- and woody-dependent species, respectively, in unburned prairie. From Zimmerman (1993), with permission of the University Press of Kansas.

be unable to enter existing leks (Horak 1985). Consequently, these males set up small, satellite booming grounds. The number of satellite leks recorded each year on Konza Prairie is related positively to the total number of males observed (Zimmerman 1993), which supports this notion.

Effects of Fire

Annual and periodic fires during nondrought years have no significant effects on the total relative abundance of most grass/forb-dependent birds (Fig. 8.6; Zimmerman 1992, 1993). Henslow's sparrow is an exception to this pattern. This species requires habitats with dense standing dead biomass and litter (Zimmerman 1988). However, adult male Henslow's sparrows move into and begin singing in annually burned prairie in July. It is possible that these males are selecting territories for the next year because they do not nest at this time. However, banding studies from Konza Prairie indicate that Henslow's sparrows are not philopatric (unpublished data). Therefore, the presence of Henslow's sparrows in annually burned prairie remains unexplained.

Relative abundances of common grass/forb-dependent species in burned prairie decrease significantly in drought years as evidenced by significant correlation between abundance and March soil moisture (Zimmerman 1992). However, this relationship was not evident in unburned prairie. Dissimilarity of these patterns may be due to relative differences in vegetative structure in burned and unburned prairie in drought and normal/wet years. For example, aboveground plant biomass generally is greater in burned prairie than in unburned prairie in both normal and wet years (Hulbert 1988; Chapter 12, this volume). However, plant biomass is reduced in burned prairie in drought years (Briggs et al. 1989; Briggs and Knapp 1995). Finally, the relative abundances of common woody-dependent species in both burned and unburned prairie are correlated positively with March soil moisture (Zimmerman 1992).

Because fire reduces the amount of woody vegetation (Collins and Gibson 1990), woody-dependent species in burned sites are decreased in abundance or eliminated (Table 8.3; Zimmerman 1992). The exceptions to this generalization are species such as the eastern kingbird and Baltimore oriole that nest in mature trees, which largely are unaffected by fire. Woody-dependent species of birds are less affected by periodic prairie fires (Fig. 8.6) than by annual occurrence of fire. But populations of woody-dependent species in infrequently burned sites still remain lower than those in long-term unburned tallgrass prairie (Zimmerman 1993).

Effects of Grazing

The species richness of prairie bird communities is related inversely to the intensity of grazing. Further, evenness of the proportion of species is decreased because grazing increases the carrying capacity for a few species but decreases the carrying capacity for a majority of species (Ryder 1980; Kantrud 1981). Data from bird transects in prairie grazed by *B. bison* currently demonstrate no significant correlation between avian richness and grazing. This lack of pattern probably is due to a moderate intensity of grazing by *B. bison*.

However, intensive study of the grass/forb-dependent breeding birds in tallgrass prairie suggests that they are affected by cattle grazing (Zimmerman 1997). A significant reduction occurred in the relative abundances of these species on grazed/burned treatments during the breeding season. However, similar to the responses to *B. bison* grazing, avian richness did not differ among fire/grazing treatments.

Although no differences exist among the probabilities that nests will survive in grazed/unburned prairie, ungrazed/burned prairie, and ungrazed/unburned prairie, the probability that nests of grass/forb-dependent species will be successful is significantly lower in grazed/burned prairie than in grazed/unburned prairie (Zimmerman 1997). Additionally, nesting begins about 2 weeks later for all grass/forb-dependent species in grazed/burned prairie than in the other three treatments. Despite the later initiation of nests, no differences in weights of fledged young occurred among the four treatments, which suggests that food resources are adequate in all treatments. However, for the dickcissel, the production of young was significantly lower in grazed/burned prairie than in grazed/unburned prairie.

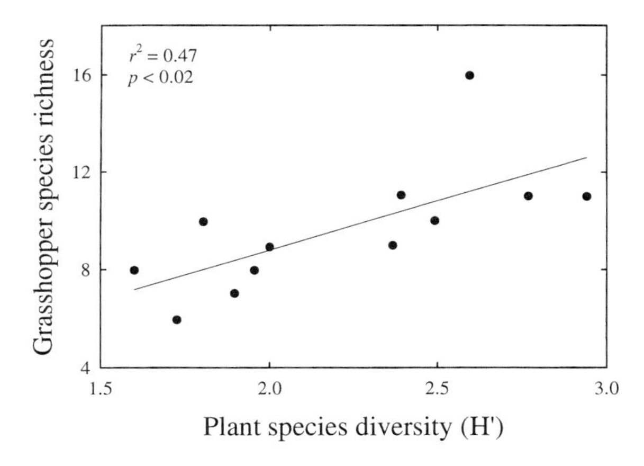

Figure 8.7. Relationship between species richness of grasshoppers and diversity of plant species (H') on Konza Prairie during 1981. From Evans (1988b), with permission of the National Research Council of Canada.

Grasshoppers

Before initiation of LTER studies at Konza Prairie, little was understood about the relationships among populations and communities of terrestrial invertebrates in tallgrass prairie and fire, climate, and grazers. This lack of information and understanding reflected the tremendous taxonomic and ecological diversity of these organisms rather than a low level of research effort focused on invertebrates. Grasshoppers have been the focal invertebrate taxa for LTER studies at Konza Prairie because of their abundance and potential impacts on the prairie ecosystem. Before LTER, two major studies presented different and sometimes conflicting views of the role of grasshoppers in tallgrass prairie ecosystems. For example, Knutson and Campbell (1976) concluded that grasshopper assemblages were in equilibrium, were not limited by food supply, and caused little damage to tallgrass prairie pastures. In contrast, French et al. (1979) and Scott et al. (1979) concluded that herbivorous invertebrates in grassland ecosystems used all available food and played an energetic role in grassland ecosystems second only to belowground invertebrates.

LTER studies of population and community dynamics of terrestrial invertebrates have been focused on two general topics: the inherent nonequilibrium condition of communities of acridid grasshopper caused by fire and the underlying biotic interactions among grasshopper species.

Effects of Fire on Community Structure

The composition of the grasshopper community at a given location depends on the species composition of the plant community. Therefore, both the periodicity and the seasonality of prairie fires are important influences on grasshoppers. For example, species diversity of grasshoppers and plants increases with increasing time since fire (Evans 1984, 1988b; Chapter 6, this volume). Thus, a positive correlation exists between the species richness of grasshoppers and the species diversity of plants (Fig. 8.7).

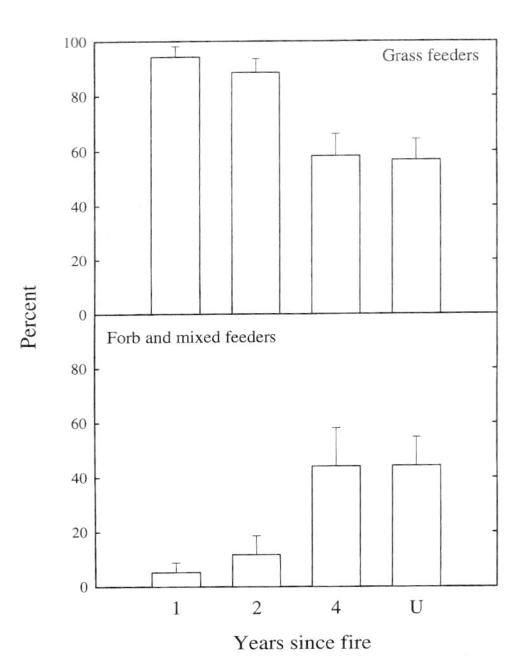

Figure 8.8. Percent of grass feeder and forb/mixed feeders in grasshopper communities in tallgrass prairie on Konza Prairie over a 4-year period. Sites included prairie burned at 1-, 2-, and 4-year intervals and prairie left unburned (U). From Evans (1988b), with permission of the National Research Council of Canada.

Forb-feeding species were the most dynamic component of the grasshopper community in response to fire (Evans 1988b). The abundance of forb-feeding grasshoppers strongly increased as the frequency of fire decreased (Fig. 8.8). In contrast, relative abundance of grass-feeding grasshoppers decreased with less frequent fires, although their abundance remained much higher and more stable than that of forb-feeding species. Therefore, the grasshopper community appears to fit the core-satellite model (Hanski 1982; Collins and Glenn 1997), with a few grass-feeding core species and a number of forb-feeding satellite species.

The nonequilibrium nature of grasshopper communities in tallgrass prairie emerged when community dynamics were examined in successive growing seasons after a fire (Evans 1988a). A correlated random-walk model of community dynamics suggested that the structure of local grasshopper communities displaced by fire tended to return to a characteristic species composition. However, the interannual changes in local grasshopper communities in the years following fire were subtle. Species richness and diversity varied irregularly during a 4-year cycle of fire (Evans 1988a), but the ratio of grass-feeding grasshoppers to forb-feeding species generally increased after a fire and then declined in subsequent years without fire (Fig. 8.9). Thus, fire places broad limits on the structure of the grasshopper community. As a result, species composition remains relatively stable at individual sites, but relative abundances exhibit moderate variability.

The mechanism of fire effects on grasshopper populations are not well understood. Fire compresses the seasonal cycle of abundance of grasshoppers (Fig. 8.10). Grasshopper abundance peaks earlier and declines to lower levels in annually burned prairie than in infrequently burned prairie (Evans et al. 1983). Be-

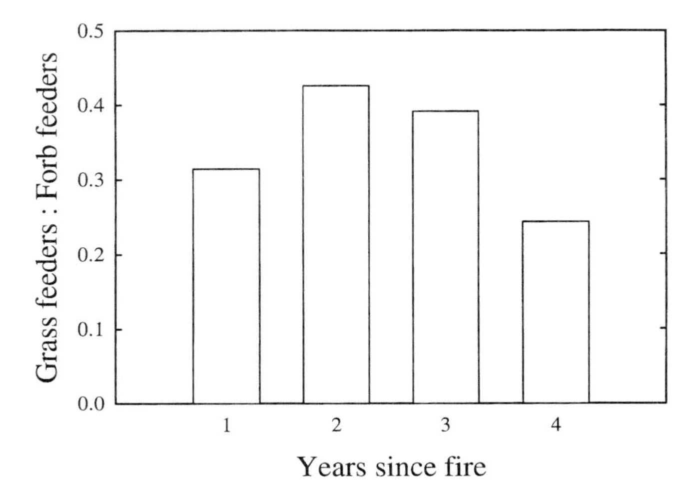

Figure 8.9. Temporal changes in the ratio of number of grass-feeding grasshoppers to forb- and mixed-feeding grasshoppers during each year of a 4-year cycle of fire on Konza Prairie. From Evans (1988a), with permission of Oikos.

cause grasshopper eggs overwinter in the soil, the early peak of abundance in annually burned prairie may arise from enhanced egg development caused by warm soil temperatures prevailing after fire (Knapp 1984b). The greater decline in the abundance of grasshoppers may have several causes, including fire-induced changes in host plant quality and phenology and increased rates of predation. Forb-feeding grasshoppers in particular appear to be affected adversely in the growing season immediately following a spring fire (Evans 1984).

Biotic Interactions

The unpredictability of the responses by grasshoppers to fire may arise partly from biotic interactions that occur between sequential fires. The grasshopper community of the tallgrass prairie contains species that vary in dietary overlap. As a result, competition may be an important biotic factor that structures local communities. Competitive interactions among grass- and forb-feeding grasshoppers were examined in years of moderate productivity and during a year of drought. During the drought, primary productivity at the study sites was 65% of that during an average year. The dominant grass-feeding species, *Phoetaliotes nebrascensis*, was expected to have strong competitive interactions with other grass-feeding species. However, experimental additions and removals of *P. nebrascensis* (Evans 1989) caused no changes in grass or forb biomass, no effect on other orders of phytophagous insects, and at most a slight effect on the abundance of other Orthoptera. Therefore, little evidence exists to indicate that *P. nebrascensis* competes with co-occurring grasshoppers or other phytophagous insects. Similar results were found after additions and removals of *P. nebrascensis* during a drought year (Evans 1992). At typical grasshopper densities, competi-

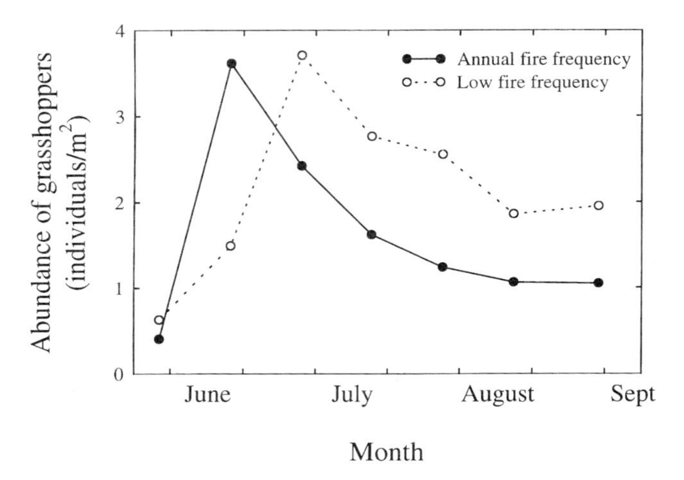

Figure 8.10. Abundance (individuals/m²) of grasshoppers trapped in burned and unburned prairie on Konza Prairie during 1982. Modified from Evans et al. (1983).

tive interactions appear to play a small role in structuring grasshopper communities in the tallgrass prairie. This supports the hypothesis that grasshoppers in tallgrass prairie are not food-limited (Knutson and Campbell 1976).

Summary

Research during the past 15 years has greatly enhanced our understanding of spatial and temporal patterns of populations and communities of small mammals, birds, and grasshoppers and the potential mechanisms by which fire and grazers affect these taxa. Our knowledge of the interacting roles of grazing and climatic variation on animal populations within topographically diverse landscape of the Flint Hills tallgrass prairie is less developed but is increasing. Overall, several general patterns have emerged from our analyses of both the primary LTER data sets and associated short-term research efforts. These patterns are summarized in the following.

1. *Population abundances of small mammals, birds, and grasshoppers are highly variable in both space and time.* Abundances of small mammals, birds, and grasshoppers in tallgrass prairie are characterized by high temporal and spatial variability related to the vagaries of climate and mosaic patterns of grazing, fire, and topographic relief. Further, our LTER animal studies demonstrate that these factors operate at many different temporal scales and periodicities and at many different spatial scales. Differences in scale effects among small mammals, birds, and grasshoppers and among species within each taxon likely are due to differences in food and/or habitat requirements (e.g., generalist versus specialist), differences in vagility (e.g., flying versus walking); differences in life span (annual versus perennial); or some combination of these. Further, temporal

variation in abundance of many species, driven in part by weather, often is much greater than spatial variation in abundance associated with fire and grazing.

2. *Population-level mechanisms underlying fire-induced dynamics of animal communities are taxon- and species-specific.* Vertebrates generally experience little direct mortality during spring fires. Instead, most species of birds and small mammals respond to fire by either immigrating into or emigrating away from burned prairie based on the effects of fire on vegetation structure. For example, species present after a fire are those that require little standing vegetation and no litter layer. In contrast, insects may exhibit more species-specific patterns than birds and small mammals because of the diversity, complexity, and specificity of their life histories and habitat requirements. For example, many species of grasshoppers overwinter as eggs in the soil and are mobile as adults. Therefore, spring fire has little direct effect on grasshopper abundances. However, sedentary insect species that are aboveground during fires suffer catastrophic mortality (Fay and Samenus 1993), which negates any possible indirect influence on these insects of fire-induced changes in plant species composition or structure.

3. *Population-level mechanisms underlying grazing-induced responses of animal community assemblages are also taxon-specific and similar to fire responses.* Like fire, grazing by cattle or *B. bison* modifies plant species composition and reduces vegetative cover and the accumulation of plant litter. Many animals respond to these grazing-induced changes in habitat features in largely the same manner as they react to fire-induced changes. For small mammals, fire-negative and fire-positive species also appear to be "grazing-negative" and "grazing-positive" species, respectively. For birds, species richness is related negatively to grazing intensity. This response occurs because the carrying capacity for most, but not all, species of birds is decreased through reduced nesting success caused by the absence of suitable vegetative structure. The effects of grazing on grasshoppers have not yet been examined. Because grazing by large mammals increases the abundance and diversity of forbs, we expect grasshopper diversity to increase because of increased abundance of forb-feeding species. Also, grasshopper abundance should peak earlier in the growing season on grazed than on ungrazed prairie because grazing reduces cover and litter accumulation. Finally, grazing may reduce overall insect abundance on Konza Prairie. Total numbers of insects were lower in grazed sites than in ungrazed sites in tallgrass prairies in Nebraska and Oklahoma (P. Fay unpublished data).

4. *Species assemblages of small mammals, birds, and grasshoppers conform to the predictions of Hanski's (1982) core-satellite hypothesis.* The communities of each of the LTER focal taxa are characterized by the presence of several abundant, widely distributed, and relatively stable species, accompanied by a larger number of rare, localized, and dynamically variable species. The temporally variable pattern within each of these communities is driven primarily by changes in the distribution and abundance of the rare species. The occurrence of common and rare species and the consistency with which species remain in one category or the other suggests that both equilibrium and nonequilibrium factors influence the coexistence of animal species in tallgrass prairie. Furthermore, the general fit of the core-satellite concept to three communities of markedly differ-

ent organisms offers broad support for that hypothesis and suggests that this pattern will occur in most taxa of grassland animals (Chapters 6, 9, this volume).

5. *Fire indirectly influences the composition and dynamics of animal communities by altering the physical structure and species composition of prairie vegetation.* Fire-induced changes in the physical structure and species composition of vegetation clearly are associated with changes in population abundance and community composition of small mammals, birds, and grasshoppers. Changes in the physical structure of the vegetation are particularly important in modifying the structure and composition of small mammal and bird communities. For example, the presence or absence of litter and standing dead vegetation influences the population abundance of many species of small mammals and some species of birds and, therefore, the community structure of both groups. The physical structure created by shrubby plants also determines the presence or absence of a characteristic group of bird species that required shrubs for nesting and foraging. In contrast to birds and mammals, grasshopper assemblages depend on plant species composition per se rather than the physical structure associated with those species. Further, forb-feeding grasshoppers are the most dynamic members of the community because they increase or decrease in concert with changes in their preferred host plants.

6. *Climate and topography are sources of spatial and temporal heterogeneity in animal populations and communities.* Climate has long been recognized as an important influence on animal populations, although the mechanisms underlying population responses to varying climate are understood poorly (Price et al. 1989; Belovsky and Joern 1995). However, the temporal dynamics of tallgrass prairie animal communities are mediated at least partly by climate. For example, temporal variations in relative abundances of some species of birds and small mammals are related to interannual differences in temperature and precipitation. Spatial heterogeneity also is introduced into animal communities by species-specific topographic preferences. Most small-mammal species are specialists that preferentially select either uplands, lowlands, or rocky outcrops, whereas a few species are generalists that show no preference among topographic positions in the prairie landscape. Topographic preferences in birds probably relate more to preferences for the life-form of the vegetation that grows in certain topographic positions. For example, avian species that prefer shrubs are associated with wooded rocky outcrops, whereas avian species that prefer trees occur in gallery forest, which usually is associated with lowland. Of the LTER focal taxa, grasshoppers may be the least linked to topography because of their size, mobility, and host plant affinities.

7. *Spatial and temporal variations in population abundance and community structure require multifactoral explanations.* Although climate, grazing, fire, and topography have important effects on vegetative features, interactions among these factors create a complex, patchy, and highly dynamic landscape for animals in the tallgrass prairie. For example, fire influences grazing patterns, grazing patterns influence vegetative structure, and these together with climate affect future fire intensity. As a result, fire, grazing, and climate are inseparably linked in their direct and indirect influences on the composition and structure of prairie

vegetation and, in turn, the whole array of prairie animals. Because the relative importance of grazing, fire, and climate varies at any one place and time, population limitations constantly change, making single-limitation explanations for population and community dynamics unrealistic (Belovsky and Joern 1995). Thus, Risser et al. (1981) were quite correct in their observation that "the relationships between biota and environment [are] close and apparent," but the complexity of these relationships is only beginning to emerge. Further, many of the important factors influencing abundance and distribution of animal populations are "bottom-up" rather than "top-down" regulators (Hunter and Price 1992) that act via the composition and structure of prairie vegetation.

Acknowledgments Partial funding was provided by the National Science Foundation (BSR-8307571, BSR-8305436, and BSR-8505861) for studies of mammals and insects and by the National Science Foundation (Undergraduate Research Programs), Kansas Department of Wildlife and Parks, and Chapman Fund of the American Museum of Natural History for studies of birds. This is contribution no. 97-186-B from the Kansas Agricultural Experiment Station, Kansas State University, Manhattan.

9

Disturbance, Diversity, and Species Interactions in Tallgrass Prairie

Scott L. Collins
Ernest M. Steinauer

The previous four chapters have described the kinds of plants and animals found on Konza Prairie (Chapters 5, 7) and discussed the basic structure of plant and animal communities in this grassland (Chapters 6, 8). In this chapter, we will focus explicitly on how community structure responds to disturbances such as fire and grazing. Over the past 15 years, research at the Konza Prairie LTER site has greatly increased our understanding of the factors that control community and ecosystem dynamics in tallgrass prairies (Chapters 12, 13, this volume). In general, fire and grazing are the overriding, large-scale factors that drive dynamics in this tallgrass ecosystem. A third important factor that overlays fire and grazing is climate variability (Anderson 1982; Bragg 1995). The direct effects of fire, in particular, on biotic and abiotic environments have been documented extensively (Vogl 1974; Wright and Bailey 1982; Collins and Wallace 1990; Whelan 1995), yet fire interacts with climate, grazing, and topography to create a complex set of indirect effects in prairies (Seastedt and Knapp 1993), the consequences of which are not well understood. Although much remains to be learned regarding basic community and ecosystem processes in tallgrass prairie, the history of research on fire at Konza provides a solid foundation for future studies of the interactive effects of multiple driving variables on pattern and process at all hierarchical levels.

In this chapter, we will begin by summarizing the effects of fire frequency on plant community structure and dynamics and the mechanisms that control plant community response to fire. We then will ask whether or not the effects of fire on plant community structure translate to similar or different patterns in abundance and diversity at higher trophic levels. Finally, we briefly will describe some of the consequences that the reintroduction of large grazers have had on community and ecosystem processes in this tallgrass prairie.

We have chosen to develop this chapter along these lines for three reasons. First, disturbance theory has served as the primary framework for much of the plant community research at Konza Prairie. This occurs as a logical consequence of the landscape-level burning regime that is the cornerstone of the Konza Prairie LTER Program (Hulbert 1985; Chapter 1, this volume). Second, plant community structure acts as the primary conduit between belowground processes and resource availability for higher trophic levels through pathways such as plant species composition and diversity, resource quality, habitat structure, and primary production. Third, the background work on fire sets the stage for the reintroduction of *Bos bison*, thus allowing an integrative assessment of fire and grazing effects on grassland structure and function.

Fire Effects on Community Dynamics at Konza Prairie

The following discussion assumes that fire is, in fact, a disturbance in grasslands (Fig. 9.1). Some have questioned this notion because fire is necessary to prevent invasion and establishment of woody species in mesic prairie (Bragg and Hulbert 1976; Evans et al. 1989a; Chapter 1, this volume). By definition, a disturbance is a relatively discrete event that destroys biomass and alters resource availability (Pickett and White 1985). Based on this definition, fire is indeed a disturbance in grasslands because, as shown throughout this volume, fire destroys living and standing dead biomass and alters the availability of critical resources such as N and light. The debate regarding whether or not fire is a disturbance arises partly from confusing the effects of a single fire with the long-term disturbance regime in grasslands (Collins 1990). Although the impacts of each fire on community structure and ecosystem processes can be quantified, a perspective that focuses on the disturbance regime obfuscates the effects of individual fires with processes that occur between fire events (e.g., dispersal, competitive interactions). Therefore, we consider each fire event to be a disturbance in tallgrass prairie, even though the complex disturbance regime in prairies, which includes frequent burning, is essential in maintaining grassland structure and function.

Plant Community Structure

Research at Konza Prairie and elsewhere clearly indicates that community structure and ecosystem functioning in tallgrass prairie are impacted strongly by fire frequency (Daubenmire 1968; Rice and Parenti 1978; Abrams et al. 1986; Gibson and Hulbert 1987; Seastedt 1988; Ojima et al. 1990; Collins 1992; Seastedt 1995; Collins et al. 1995; Turner et al. 1997). Plant species composition, in particular, differs dramatically between annually burned and less frequently burned sites on Konza Prairie (Gibson and Hulbert 1987; Collins et al. 1995). An ordination of vegetation dynamics on sites burned at 1-, 4-, and 20-yr intervals reveals this distinction (Fig. 9.2). Annually burned sites are dominated strongly by

Figure 9.1. Historically, fires occurred frequently in tallgrass prairie. On Konza Prairie, prescribed fires are used to evaluate the effects of this disturbance on individual-, population-, community-, and ecosystem-level processes. (A. Knapp)

C_4 perennial grasses. Although C_4 grasses retain dominance in infrequently burned sites, C_3 grasses, forbs, and woody species are considerably more abundant (Fig. 9.3), resulting in greater diversity and heterogeneity in unburned prairie. In fact, the flora on annually burned sites is a nested subset of that found on less frequently burned areas (Collins et al. 1995). Thus, the differences revealed in the ordination reflect shifts in dominance between frequently and infrequently burned sites, rather than difference in composition per se.

In addition, the ordination shows that the annually burned sites are less variable from year to year than infrequently burned treatments (Fig. 9.2). These differences again result from dominance by long-lived, highly competitive C_4 grasses on annually burned sites (Fig. 9.3). The implication from the ordination analysis is that community stability is related inversely to disturbance frequency, a pattern that contradicts general predictions from disturbance-diversity theory (Horn 1974; Pickett and White 1985; Wu and Loucks 1995). Anderson and Brown (1986) suggested that fire was a stabilizing force in grasslands in comparison with fire effects on forest vegetation. However, the observation that compositional stability increases within a community type as disturbance frequency increases may be unique to tallgrass prairie vegetation (Chapter 17, this volume).

The conceptual framework we have used to focus our analyses of plant community structure and dynamics at Konza is the intermediate disturbance hypothesis (IDH; Connell 1978). This hypothesis is based on three primary assumptions

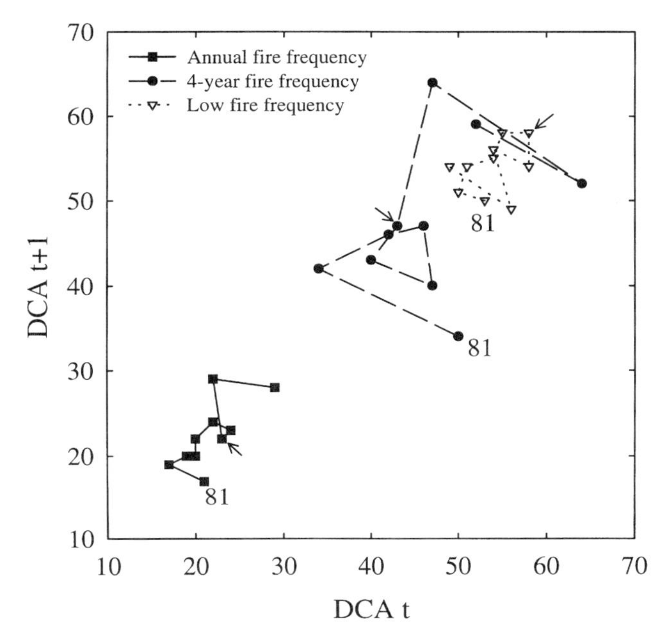

Figure 9.2. Phase-space plots of a detrended correspondence analysis ordination of vegetation changes since 1981 on three burning treatments at Konza Prairie. Axis scores are sums of axes 1 and 2 scores weighted by their respective eigenvalues. Arrows indicate the drought year of 1988. Each site contains four transects along which five 10-m^2 permanent quadrats are evenly spaced, for a total of 20 plots per site. Aerial cover of plants in these plots is estimated in spring and late summer. The ordination indicates that vegetation on 1-year burn cycles differs from vegetation on less frequently burned prairie. Also, directional change is occurring on 1-year and 4-year burn cycles but not in unburned vegetation.

(Collins and Glenn 1997). The first assumption is that competition is an important regulator of species diversity. Second, the IDH model assumes that disturbance reduces the abundance and competitive ability of the dominant species. The third assumption states that these changes in the competitive environment increase resource availability for less competitive species. Given these three assumptions, the IDH predicts that species diversity will be maximized at intermediate frequencies of disturbance because the community contains a mixture of tolerant and intolerant species, and weak and strong competitors. If disturbances are too frequent in space or time, all but the most tolerant species are eliminated, whereas if disturbances are too rare, competitive exclusion reduces species diversity.

Konza Prairie contains a series of replicated landscape-scale management units in which spring burns have been applied at different frequencies, in some cases for over 20 years (Chapter 1, this volume). This provides an appropriate forum to test predictions derived from the IDH. Although the fire regime currently employed at Konza Prairie spans likely historic fire return intervals for the

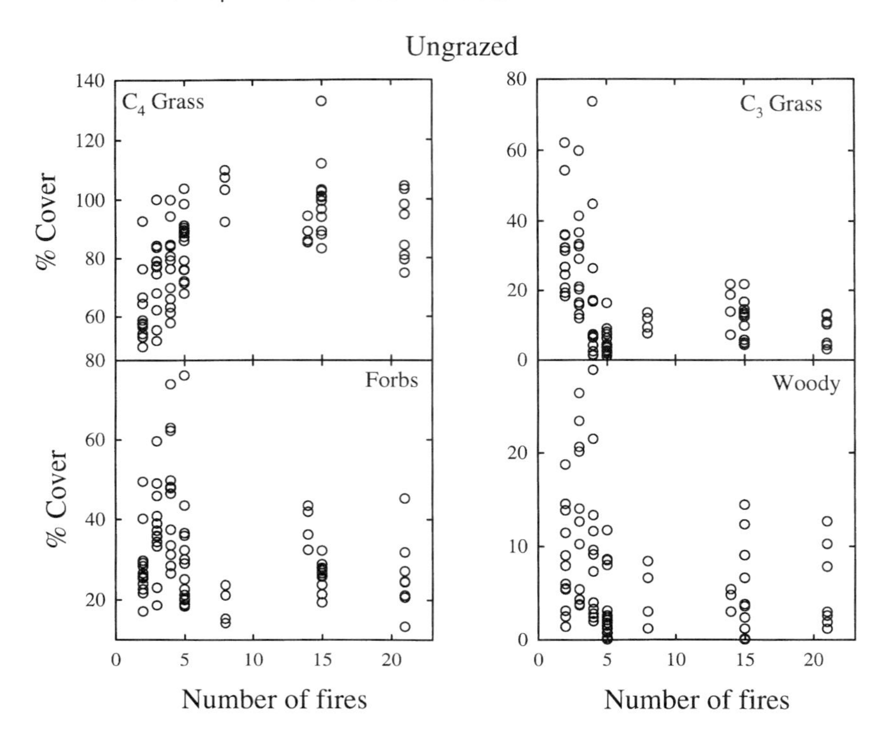

Figure 9.3. The relationship between percent cover of C_4 grasses, forbs, C_3 grasses, and woody species and the number of times a site has been burned during the interval from 1972 to 1993 in ungrazed prairie. Data are summarized by transect rather than the entire site as in Figure 9.2. From Steinauer and Collins (unpublished data).

area (Bragg 1995), our results demonstrate that no intermediate fire frequency exists at which diversity is maximized. Instead, increasing frequency of disturbance by fire increases the dominance of C_4 grasses, thereby reducing the number of species and decreasing heterogeneity (Fig. 9.4). Thus, the community-level response of vegetation to fire on Konza Prairie does not conform to the predictions of the IDH.

We now can turn our attention to the mechanisms that cause the decreases in plant species richness and heterogeneity with increasing fire frequency. One obvious mechanism is direct mortality from fire. However, little evidence indicates that significant mortality occurs as a direct result of burning (Gibson and Hulbert 1987; Hartnett 1991). Direct mortality is likely important in eliminating seedlings of woody species and may affect some annuals and C_3 grasses. Cover and richness of these functional groups have been shown to decrease as fire frequency increases (Gibson 1988; Steinauer and Collins unpublished data; Figs. 9.3, 9.4). This decrease in species that cannot tolerate frequent burning conforms to one of the predictions of the IDH. That is, diversity is low when disturbances are frequent because only a few species can tolerate frequent distur-

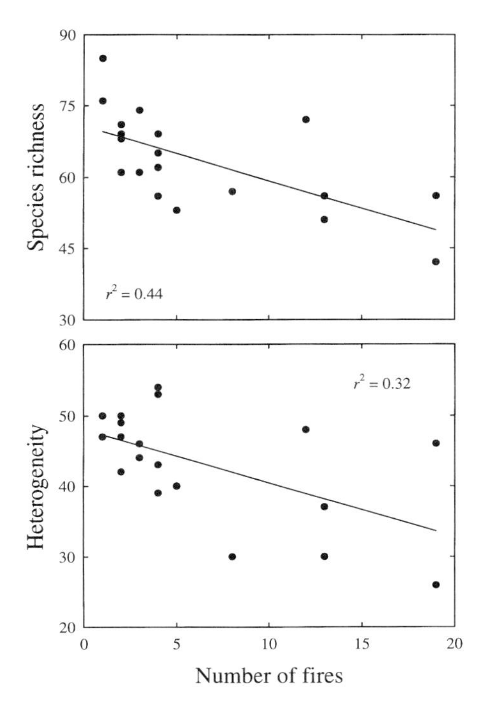

Figure 9.4. Relationship between number of times a site has been burned during the interval from 1972 to 1990 and plant species richness and community heterogeneity. Species richness is measured as the total number of species, and heterogeneity is the percent dissimilarity in species abundance from one point to another within a site. Redrawn from Collins et al. (1995) and Collins (1992), with permission of the Ecological Society of America.

bances. However, the difference in richness between annually burned and infrequently burned sites is not that large (Collins et al. 1995). So direct mortality from fire is, at best, a minor mechanism affecting plant species diversity in this tallgrass prairie.

A second mechanism leading to a decline in species diversity and heterogeneity with frequent fire is competition. As noted earlier, the IDH assumes that diversity will be maximized at intermediate levels of disturbance if the disturbance negatively impacts the competitively superior species. In grasslands, fire directly and indirectly alters N, moisture, and light availability (Knapp 1984b; Knapp and Seastedt 1986; Turner et al. 1997). These changes in bottom-up controlling factors *increase* the competitive ability of the C_4 grasses, leading to a decrease in species diversity. Because dominance by C_4 grasses increases across the burned landscape, heterogeneity decreases as well (Gibson et al. 1993; Collins 1992; Collins et al. 1995).

Data from a long-term experiment with burning, mowing, and fertilizer addition, commonly referred to as the "belowground plots," (Chapters 13, 14, this volume) support this contention. On the belowground plots, species richness is somewhat lower with annual burning compared with no burning (Fig. 9.5), a pattern consistent with diversity on the larger management units on Konza Prairie. Diversity is even lower on fertilized plots, suggesting that belowground resources are not limiting species diversity. If they were, diversity should increase when more resources are available. However, species richness on treatments

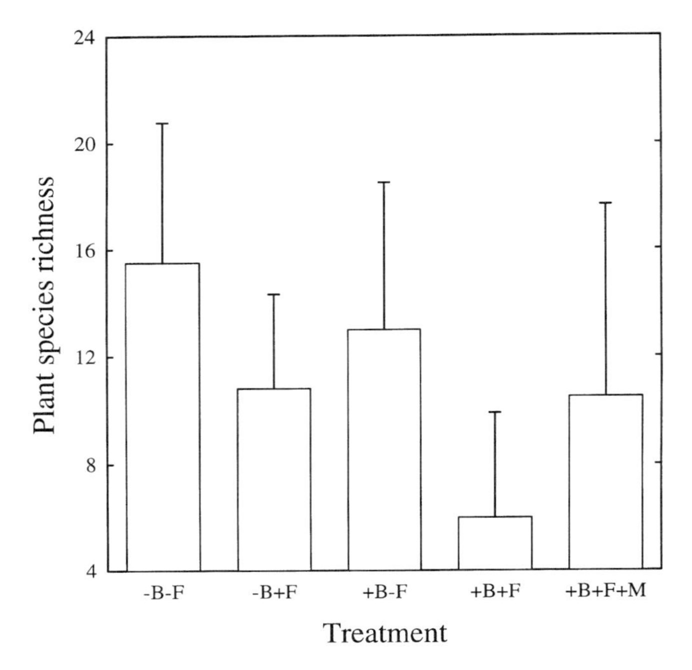

Figure 9.5. Plant species richness in the belowground plots (Chapter 14, this volume). The treatments include annual burning (+B), fertilizer addition (+F), and mowing (+M). Species richness is measured in one 10-m^2 circular plot in each of four replicates per treatment. Error bars are 95% confidence intervals. From Collins et al. (unpublished data).

that were burned, fertilized, and mowed annually was significantly higher than on treatments that were burned and fertilized but not mowed (Fig. 9.5; Collins et al. unpublished data). In this case, competition for light is the primary mechanism causing the decrease in plant species diversity in response to frequent burning. When belowground resources were increased in the absence of mowing, species diversity continued to decline, but when the aboveground canopy was reduced by mowing, diversity remained similar to that found on control plots. This illustrates that richness can be maintained under an annual burning treatment if another disturbance, in this case mowing, negatively affects the competitive dominants that otherwise would thrive under frequent burning.

Results from the belowground plots and other studies (Chapter 6, this volume) at Konza Prairie do not agree with those of Wilson and Shay (1990), who conducted a similar, but short-term, experiment in Saskatchewan prairies. They reported that competition for soil resources following burning was not a significant mechanism affecting community structure in these northern prairies. However, Canadian grasslands are dominated by C_3 grasses, which tend to decrease with burning, unlike the C_4 grasses that dominate prairies in warmer climates (Knapp 1984b; Svejcar and Browning 1988). Thus, the role of competition and fire in structuring grassland plant communities may increase along a latitudinal gradient throughout the Great Plains.

Primary production in tallgrass prairie is limited by a variety of factors depending upon time of year and length of time since burning. In general, the most important factors are light, N, and soil moisture. Tilman and Wedin (1991a, b) have shown that prairie plants differ in their ability to survive as resource availability decreases, and, in general, the dominant species can persist at the lowest resource levels. Moreover, they have shown that the C_4 grasses can reduce soil N below the level at which many other prairie plants can survive. This ability to decrease soil N helps to create a competitive environment that favors dominance by the C_4 grasses and results in low diversity (Tilman 1982; Tilman and Wedin 1991a, b; Wilson and Tilman 1991).

Fire effects on the N cycle have received intensive investigation at Konza Prairie (Seastedt et al. 1991; Seastedt 1995). Both modeling (Ojima et al. 1990) and field data (Turner et al. 1997; Chapter 13, this volume) demonstrate that extractable soil N is generally lower on annually burned compared with unburned prairie. Despite high soil N levels, primary production is limited in unburned prairie because of low light levels at the soil surface caused by litter accumulation (Knapp and Seastedt 1986; Ojima et al. 1990; Rice and Garcia 1994). By removing litter, spring burning increases light and temperature levels at the soil surface and creates a microclimate favorable to C_4 grasses, which rapidly incorporate available N. Therefore, production peaks in tallgrass prairie often occur following fire in infrequently burned areas. Because fire volatilizes N in aboveground pools, frequent burning eventually depletes soil N. Also, the belowground tissue of C_4 grasses, which dominate in frequently burned areas, has a high C:N ratio and decomposes slowly, creating a feedback system that leads to further decreases in available soil N and further increases in the intensity of competition after fire.

The effect of fire on soil moisture is an additional bottom-up factor favoring dominance by C_4 grasses. Moisture levels are high in unburned prairie because litter accumulation limits evaporation. High light levels at the soil surface in burned prairie increase evapotranspiration and result in warm, relatively dry conditions. However, under conditions of normal precipitation, aboveground production is greater on annually burned sites than on unburned sites, despite reduced moisture levels (Briggs and Knapp 1995). The warm, high-light, and relatively dry conditions following fire increase the competitive advantage of C_4 grasses over C_3 grasses and forbs. In summary, aboveground net primary production is typically higher on burned compared with unburned sites on comparable topography (Briggs and Knapp 1995), despite reductions in soil resources (N and moisture) on burned sites. This results from increased productivity and dominance of C_4 grasses following fire. Indeed, production is distributed more evenly across burned than unburned prairie (Briggs and Knapp, 1995; Chapter 12, this volume). This parallels the decrease in species diversity and heterogeneity in burned prairie landscapes.

Overall, the community and ecosystem responses to fire in tallgrass prairie do not conform to the primary prediction of the IDH. The data from Konza Prairie support the assumption of the IDH that competition is an important mechanism regulating plant populations and controlling species diversity in tallgrass prairie. However, the second assumption that disturbance, in this case fire, negatively impacts the competitive dominants is not supported. Instead, fire increases the competitive ability of the C_4 grasses. The third assumption regarding resource availability also does not hold in grasslands. Fire decreases moisture and N availability and increases available light early in the growing season, all of which favor the growth of the dominant species and lead to a decrease in species diversity and community heterogeneity.

Thus, a clear picture of fire effects on plant community structure has emerged from the long-term empirical and experimental research at Konza Prairie. In the absence of large herbivores, the system is strongly driven by bottom-up forces associated with light, soil resource availability, and differential ability to compete under low resource conditions. Although light availability increases with burning, the abundance of other critical limiting resources, N and water, declines as fire frequency increases. This is especially true in upland areas where net primary production (NPP) is likely limited by water. These changes in resource availability favor the growth and dominance of a small cadre of perennial C_4 grasses and forbs. As dominance by these competitive species increases, general declines in plant species diversity and community heterogeneity occur.

Animal Community Structure

Given the distinct effects of fire frequency on plant community structure and dynamics within and among burning treatments, it seems plausible that similar responses will occur among consumer species that depend on the primary producers for food and habitat structure. Indeed, Swengel (1996) and Anderson et al. (1989) found that the abundance of several insect groups in prairies declined significantly following a single burn. In addition, Zimmerman (1992) noted that bird species diversity was low on frequently burned grasslands at Konza Prairie. This suggests a potential parallel relationship between responses of producers and consumers in grasslands. Detailed mechanistic experiments on fire effects on community structure of most consumer groups have yet to be conducted at Konza Prairie. Nevertheless, we can use the long-term data sets on small mammal, grasshopper, and bird populations at Konza to test hypotheses derived from the IDH.

The response of bird species richness to fire frequency is similar to that of plants. General declines in bird species richness and total number of individuals occur as fire frequency increases (Fig. 9.6). Grasshoppers seem to show the opposite trend. In general, more species of grasshoppers occur on frequently burned sites (Fig. 9.6). Number of individuals of grasshoppers is slightly more variable but also appears to increase as fire frequency increases (Fig. 9.6). Small mammal abundance and species richness are unrelated to fire frequency (Fig. 9.6). Thus, the responses of consumers to fire frequency do not necessarily mirror those of plants. Given that fire homogenizes the plant community, consumer diversity would be expected to decrease as well. Although this appears to be true for birds, it is not so for grasshoppers or small mammals.

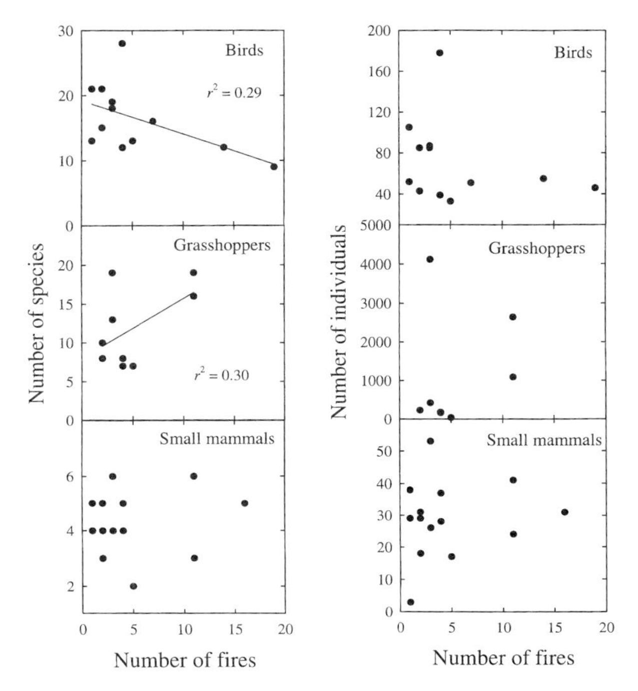

Figure 9.6. Relationships between the number of times a site has been burned and species richness and number of individuals of birds, grasshoppers, and small mammals at Konza Prairie.

The variable responses among consumers could reflect differences in the way individuals in these groups perceive vegetation structure. Given that birds are responding primarily to the broadly visual, three-dimensional structure of habitat (James 1971), it is not surprising that bird diversity decreases with burning because woody cover is reduced and the grassland habitat becomes more structurally homogeneous (Zimmerman 1993). Small mammals and grasshoppers, on the other hand, perceive the habitat from within the vegetation. Members of these assemblages are responding to vegetation as cover and a source of food (Evans 1988a; Kaufman et al. 1990). Food quality varies between grasses and forbs, and within these plant functional groups, further variation in quality occurs among burn treatments. Indeed, a clear positive relationship exists between ANPP and the average number of small mammals found on Konza Prairie from one year to the next (Fig. 9.7). Thus, the small mammals appear to respond more to the distribution of primary production, which is related to burning, than to the direct effects of fire per se.

The question of what factors regulate the distribution and abundance of grasshoppers in grasslands is particularly intriguing (Fowler et al. 1991; Joern 1992; Belovsky and Joern 1995). At broad spatial scales (e.g., 100 km^2), grasshopper community structure is controlled by differences in habitat structure (Kemp et al. 1990). At intermediate scales within a habitat type (e.g., 100 ha),

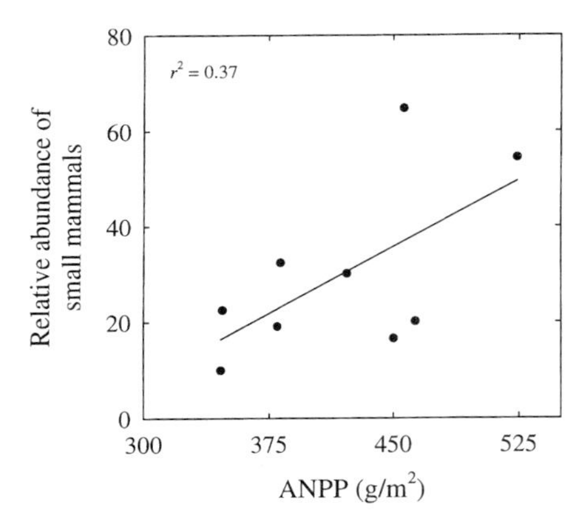

Figure 9.7. Relationship between average aboveground net primary production (ANPP) and relative abundance of small mammals on Konza Prairie over a 4-year period.

such as grassland at Konza Prairie, grasshopper species are widely distributed spatially and show little relationship to fire frequency and habitat differences within a plant community type (Collins and Glenn 1997). Thus, habitat differences are less important to grasshoppers than to plants, birds, and small mammals. At local scales within burning treatments (e.g., 100 m^2), grasshopper abundance fluctuates irregularly over time (Fig. 9.8). At this spatial scale, field experiments reveal little or no interspecific competition among grasshoppers even under severe drought that drastically limits primary production (Evans 1988a, 1992).

As noted earlier, frequent burning reduces forb biomass and greatly enhances the cover of C_4 grasses (Fig. 9.4). Given that grasshoppers can be classified as either forb or grass specialists, shifting dominance to grasses should reduce grasshopper diversity by eliminating those species that forage on forbs. The opposite occurs, however. More grasshopper species are found as fire frequency increases, although the relationship of fire to the number of individuals of grasshoppers is less clear (Fig. 9.6).

Trophic interactions may play an important role in regulating grasshopper abundance on Konza. Grasshopper abundance in many grassland habitats appears to be sufficient to support higher trophic levels under all but the most extreme conditions (Evans 1992; Wiens and Rotenberry 1979). On average, no relationship exists between the number of individuals of grasshoppers and the number of individuals of birds from one year to the next at Konza Prairie ($r^2 = 0.10$, $P = 0.30$). This may imply that numbers of birds are positively but only marginally related to total grasshopper biomass. Within burning treatments, however, the increase in grasshopper richness and abundance on frequently burned sites may reflect lower predation rates, because the number of birds decreases on those sites (Fig. 9.6). Fowler et al. (1991) demonstrated experimentally that bird predation reduced grasshopper abundance in a North Dakota prairie. This suggests that top-down control may have some influence on regulation of grasshoppers at Konza Prairie. Our inferences are based on weak, correlative data, but the patterns ob-

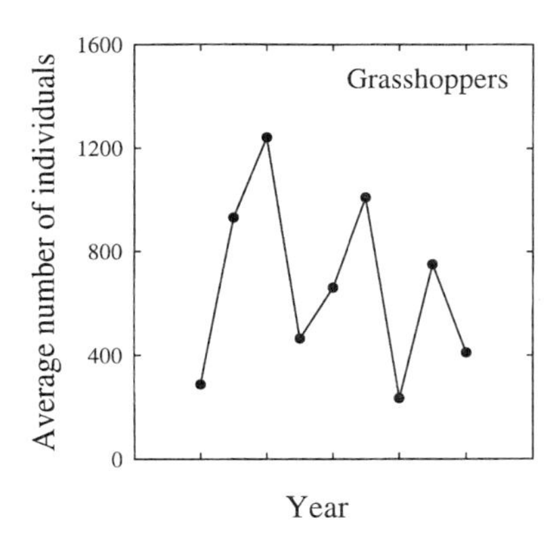

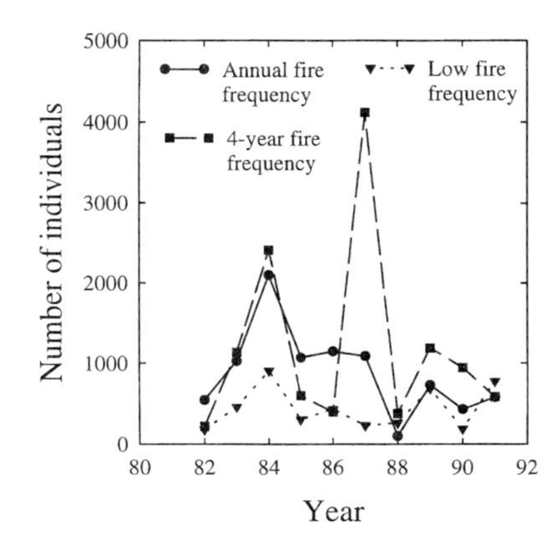

Figure 9.8. Average number of individuals of grasshoppers and total number of individuals of grasshoppers on sites burned at 1 and 4-year intervals or not burned on Konza Prairie from 1982 to 1991. The data from Konza Prairie as a whole suggest that the abundance of grasshoppers tends to cycle at 2- to 3-year intervals. Within a management unit, population abundances fluctuate somewhat independently.

served are consistent with the generalization by Belovsky and Joern (1995) that top-down controls predominate in C_4 grassland systems.

Given the weak relationships between birds, which are primary consumers of grasshoppers, and grasshopper abundance, population regulation at Konza Prairie also may be related to bottom-up forces (Schmitz 1993; Belovsky and Joern 1995). Using experimental food chains that included grasshoppers, Schmitz (1993) demonstrated the predominance of bottom-up controls on food chain structure in a C_3-dominated palouse prairie in western Montana. Results from this experiment suggested that rate of resource supply drives the system more so than the top-down effects of predation, which in this case was by spiders. Similar patterns were observed by Belovsky and Slade (1993). Clearly, conditions affecting grasshopper distribution and abundance, as well as the impacts of these important consumers on plant community structure, deserve further attention at Konza Prairie.

Herbivore Effects on Community Structure

Obviously, the simple patterns of community structure that result solely from burning are highly unrepresentative of the conditions found in grasslands with complex disturbance regimes typical of presettlement landscapes. However, the long-term management plan for Konza Prairie was designed to provide baseline information on fire effects prior to the introduction of large herbivores. In October 1987, *B. bison* were released into a 469-ha portion of Konza Prairie; as the size of the herd increased, the total area grazed was expanded to 949 ha in 1992. Because we have considerable data on fire effects in the absence of large herbivores, current studies can concentrate on the interactive effects of fire and grazing on grassland structure and function. These studies already are showing that the myriad activities of *B. bison* will dramatically alter the relatively simple patterns produced by fire alone.

Collins and Benning (1996) suggested that *B. bison* serve as "keystone engineers" in grassland ecosystems. Keystone engineers reflect a combination of two somewhat independent concepts: keystone species (Bond 1993) and ecological engineers (Jones et al. 1994). *Keystone species* are those species that impact the integrity of ecological systems to an extent that is far greater than their size or number in that system (Paine 1969; Bond 1993). For example, *B. bison* biomass at Konza Prairie is estimated to be 11 to 12 g m^{-2}, whereas the C_4 grass *Andropogon gerardii* averages at least five times that amount. Despite their relatively small biomass, *B. bison* have a tremendous impact on prairie structure and function that qualifies them as keystone species in tallgrass prairie ecosystems.

Ecological engineers are species that modulate available resources by causing physical state changes in biotic and abiotic systems (Jones et al. 1994). *Bos bison* fits this definition quite well. For example, these large ungulates create wallows, i.e., small depressions in the soil devoid of vegetation, during dust-bathing behavior in which an animal rolls back and forth on its side. As these wallows revegetate, they may provide establishment sites for ruderals or fill with rainwater to create habitat islands that support flora and fauna dramatically different from that of the surrounding upland prairie (Uno 1989). Because *B. bison* can regulate biotic and abiotic conditions through behaviors such as wallowing, as well as impact interspecific interactions in plant communities by grazing, these ungulates serve as keystone engineers in tallgrass prairie ecosystems.

Bos bison behavior has important effects on ecosystem processes in tallgrass prairie. One subtle consequence of ecological engineering by *B. bison* is the retention of N in grazed prairies following burning. As noted earlier, frequently burned grasslands continuously lose N through volatilization by fire. Because grazers move N from aboveground plant biomass to the soil via defecation, fire volatilizes less N in grazed compared with ungrazed prairie. Therefore, grazing by large herbivores reduces N losses on frequently burned sites (Hobbs et al. 1991).

The redeposition of N in urine and feces has other small-scale engineering consequences. In shortgrass prairie, plants covered by the feces of large herbivores are killed, which initiates small-scale patch dynamics in this system (Coffin and Lauenroth 1988). In contrast, urine deposition in mixed and tallgrass prairies creates resource patches within which aboveground production and leaf N concentrations are increased (Day and Detling 1990; Jaramillo and Detling 1992a; Steinauer and Collins 1995). These sites then serve as focal areas for future grazing events by vertebrate and invertebrate herbivores (Jaramillo and Detling 1992b; Steinauer 1994). Also, experiments at Konza using simulated bovine urine have shown that urine patches increase the heterogeneity of the grassland plant community (Steinauer 1994). In the absence of grazing, patchiness is reflected in increased aboveground biomass where urine has been deposited. In areas grazed by *B. bison*, heterogeneity is expressed as zones of low

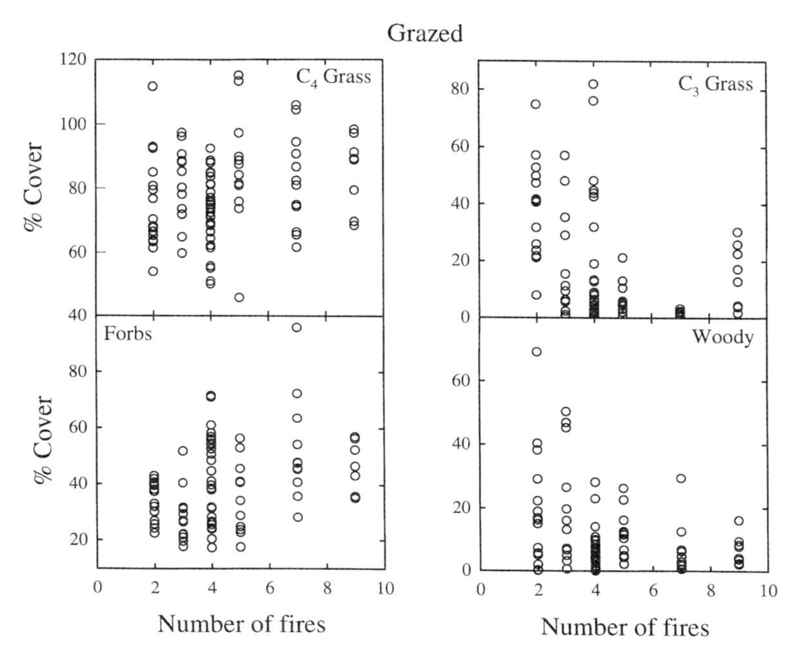

Figure 9.9. The relationship between percent cover of C_4 grasses, forbs, C_3 grasses, and woody species and the number of times a site has been burned during the interval from 1972 to 1993 in grazed prairie. Again, the clear patterns exhibited on ungrazed prairie (Fig. 9.3) are less evident in areas currently grazed by *Bos bison*. From Steinauer and Collins (unpublished data).

aboveground biomass because these patches serve as the foci for intensive grazing events that lead to the development of larger grazing lawns (Vinton et al. 1993; Steinauer 1994).

As keystone species, *B. bison* dramatically alter the clear patterns of plant community structure that develop in their absence. In contrast to ungrazed areas, fire frequency has no influence on C_4 grass cover, but forb cover increases with fire frequency in grazed prairie (Steinauer and Collins unpublished data; Fig. 9.9). In addition, no relationship occurs between plant species richness and fire frequency in the areas of Konza Prairie grazed by *B. bison* (Fig. 9.10). Instead, grazing tends to create considerable heterogeneity in both grass and forb cover. Overall, plant species richness typically is higher in grazed than in comparable ungrazed sites at Konza Prairie (Hartnett et al. 1996; Steinauer and Collins unpublished data), a pattern found in many other grasslands (Peet et al. 1983; Looman 1983; Collins and Barber 1985; Collins 1987; Milchunas et al. 1988; Huntly 1991).

In effect, *B. bison* serve as a significant top-down influence on the abundance and distribution of primary production that would be predicted in a system dominated by herbivores and lacking top carnivores (Oksanen et al. 1981; Fretwell 1987; Carpenter and Kitchell 1993). Thus, although strong bottom-up resource

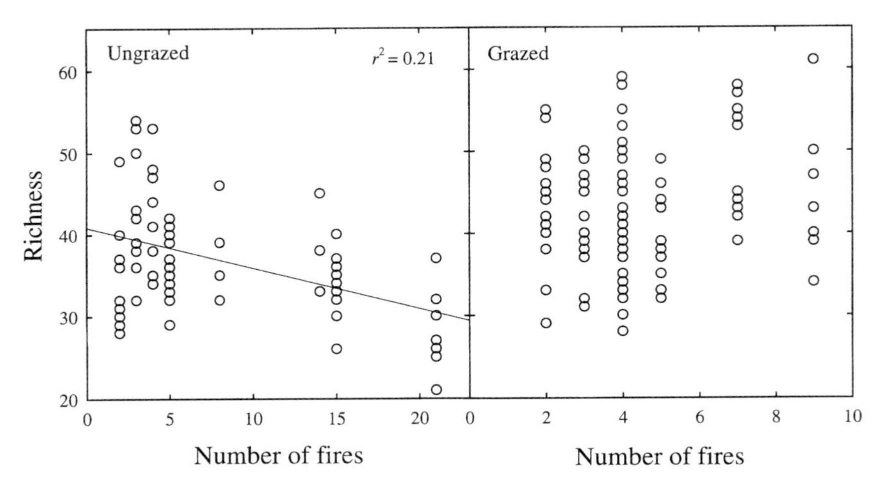

Figure 9.10. Relationship between the number of times a site has been burned during the interval from 1972 to 1992 and the number of species on grazed and ungrazed portions of Konza Prairie. Although a general decline in species richness occurs with increasing fire frequency in unburned prairie, little relationship between fire frequency and plant species richness occurs in the area currently grazed by *Bos bison*. From Steinauer and Collins (unpublished data).

drivers occur in this tallgrass system (light, soil moisture, N dynamics), these drivers are moderated by the myriad activities of large herbivores such as *B. bison*.

The impacts of consumers other than *B. bison* (e.g., belowground arthropods, grasshoppers, small mammals, lepidoptera) on plant community structure in tallgrass prairie are less well understood. An interesting experiment at Konza Prairie by Gibson et al. (1990) demonstrated the challenges of assessing the role of above- and belowground consumers in tallgrass prairie ecosystems. In this factorial experiment, levels of above- and belowground invertebrate herbivores and small mammals were manipulated. After 4 years, however, few significant differences were noted in plant biomass or plant species composition among treatments. Some evidence indicated that numbers of C_4 grasses were reduced by small mammals, whereas some annual forbs were lowest when aboveground herbivory (small mammals plus insects) was reduced. These results provide a dramatic contrast to the changes observed in desert grasslands (Brown and Heske 1990), where dominance shifted from forbs to grasses in treatments where small mammals were absent. Additional long-term experiments along these lines are needed at Konza Prairie to further address the impacts of invertebrates and small mammals on primary producers in tallgrass prairie. However, experiments that include large herbivores, such as *B. bison*, must be conducted in order to fully assess the relative roles of large and small herbivores in tallgrass prairie ecosystems.

Summary

The general pattern that emerges is one in which disturbance drives community structure and dynamics in tallgrass prairie ecosystems. Fire appears to have an organizing and stabilizing effect on plant community structure. Frequent burning tends to reduce species diversity, increase dominance by C_4 grasses, reduce temporal variability, and decrease community heterogeneity. Because the response to burning by the dominant grassland species violates assumptions of the IDH (Connell 1978), no intermediate burning frequency exists that maximizes plant species diversity at Konza Prairie.

Grazing by *B. bison*, on the other hand, is a disorganizing force in this grassland that tends to reverse the impacts of frequent burning. Grazing reduces the abundance of the dominant C_4 grasses and leads to higher plant species diversity and community heterogeneity, greater temporal variation in plant community composition, and increased nutrient retention. In addition, *B. bison* redistribute N in feces and urine and reduce N losses during burning.

Large herbivores, in combination with fire, modulate the intensity of plant competition in this tallgrass prairie. Fire increases both above- and belowground competition in such a way that it favors the growth of a few long-lived C_4 grasses that thrive under the high-light, low-N conditions created by frequent burning. Because *B. bison* preferentially graze these same grasses, the interaction of fire and grazing maximizes species diversity and community heterogeneity (Collins and Barber 1985; Collins 1987, 1992). Overall, plant species diversity at Konza Prairie is a function of a combination of bottom-up and top-down forces that are regulated primarily by disturbances such as fire.

The linkages between primary production and higher trophic levels at Konza Prairie are weak compared with the relationship between primary producers and fire. Although bird response to fire frequency is similar to that of plants, small mammal and grasshopper responses are not. No group of taxa conforms to the predictions of the IDH. Both grasshopper and small mammal populations appear to exhibit 3- to 4-year cycles on Konza (Fig. 9.11), which may override the effects of fire frequency. A longer time series will be needed to assess the effects of fire or climate on these cycles. Nevertheless, the linkages among producers, herbivores, and predators appear weak, and some evidence indicates that at the scale of Konza as a whole, abundances of producers and consumers are decoupled in time (Fig. 9.11). This is surprising, given the logical trophic relationships among these groups.

Considerable debate occurs among ecologists regarding the relative role of bottom-up versus top-down forces in community ecology (Power 1992; Hunter and Price 1992; Strong 1992; Carpenter and Kitchell 1993). Although much of this debate has focused on aquatic food webs (Strong 1992), increasing evidence indicates that top-down and bottom-up forces have important impacts in terrestrial systems as well (Brown and Heske 1990; McLaren and Peterson 1994; Pastor et al. 1993). Understanding how bottom-up and top-down factors control community structure is currently of wide ecological interest (Polis and Strong 1996), yet this has not been an explicit focus of recent work at Konza Prairie.

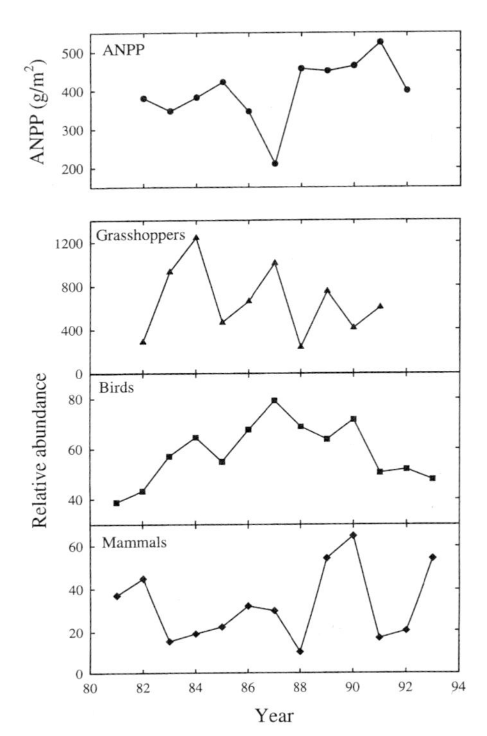

Figure 9.11. Relative abundance of grasshoppers, birds, small mammals, and aboveground net primary production (ANPP) on Konza Prairie as a whole. Although ANPP and animal populations vary considerably from one year to the next, little correlation occurs among fluctuations of different taxa.

This is surprising because early contributions to this debate by Fretwell (1977) and Oksanen et al. (1981) were developed with Konza Prairie in mind (Power 1992). Nevertheless, the interactions between bottom-up and top-down forces on primary production at Konza Prairie make this site an ideal model system for testing predictions regarding trophic structure and biotic regulation of communities (e.g., Hairston et al. 1960; Oksanen et al. 1981).

In general, these broad-scale models of community regulation deal with total abundance of organisms in different trophic levels but do not explicitly address effects of trophic dynamics on species diversity. In a recent theoretical analysis that incorporated trophic structure and grazing as a form of disturbance (Moen and Collins 1996), a bimodal pattern of plant species diversity was predicted along a productivity gradient. This differs from the widely held notion that species diversity is related monotonically to productivity (Grime 1973; Huston 1979; Tilman 1982). Field tests of this model clearly are needed, however, and future efforts will be directed toward comprehending trophic structure and dynamics within the complex fire, climate, and grazing regimes that characterize this tallgrass prairie ecosystem.

Acknowledgments We thank all of those who have helped to collect the long-term species composition data on Konza. This chapter has benefited from numerous constructive comments by and discussions with John Blair.

Part III

Hydrology and Aquatic Ecology

10

Hydrology and Aquatic Chemistry

Lawrence J. Gray
Gwendolyn L. Macpherson
James K. Koelliker
Walter K. Dodds

Konza Prairie includes parts of several stream drainages (Chapter 3, this volume), but the Kings Creek drainage basin lies entirely within the boundaries of the site and has been the focus of aquatic research efforts. The Kings Creek basin encompasses 1,059 ha, with 660 ha in the South Branch, the most intensively studied portion. Since 1979, Kings Creek has been included in the U.S. Geological Survey (USGS) Benchmark Network (station number 06879650) and is the only basin within the network that exclusively drains pristine tallgrass prairie. The key features of the aquatic systems are highly variable surface flow and complex subsurface flow patterns. These features constrain biotic parts of the systems.

Little research on the hydrology and aquatic chemistry of Kings Creek was conducted before the inclusion of Konza Prairie in the LTER Program. Since then, research has focused on temporal and spatial patterns in nutrient concentrations and transport in both surface flows and groundwater (Dodds et al. 1996b; Pomes and Thurman 1991; Tate 1985, 1990) and characterization of groundwater–surface water interactions (Macpherson 1992a; 1996; Macpherson and Schulmeister 1994). The goals of this chapter are to review and synthesize information from studies of surface water and groundwater hydrology, and water chemistry in tallgrass prairie.

Hydrology

Factors Affecting Streamflow

Surface flow in Kings Creek is influenced by climate, geology, and vegetation. Important aspects of these three influences are presented here, along with results of other studies. A more complete discussion of climate, hydrology, and geology at Konza Prairie can be found in Koelliker et al. (1985) and Chapters 2 and 3, this volume.

Annual precipitation at Konza Prairie averages 835 mm but has a once-per-hundred-year chance of being less than 460 mm or over 1,400 mm. Most of the annual precipitation occurs during warmer months; thus snow is not an important part of the water budget (Chapter 2, this volume). Annual evaporation averages 1,360 mm; the annual moisture deficit (evaporation minus precipitation) averages 525 mm but varies from −280 to 1,140 mm (Chapter 1, this volume). Such extreme variation can result in annual water yield varying from zero to over 510 mm, based on records from similar areas in Kansas.

Important geologic influences on surface flows are the high infiltration capacity and relatively high soil water storage capacity of Konza Prairie soils. Because of the shrink-swell character of the clays, these soils crack when dry and swell when wet, producing relatively high infiltration rates when the soils are dry to moist. Duell (1990) showed no runoff from dry soil when 50 mm of artificial rainfall was applied at a rate of 60 mm/h, but the same artificial rainfall event applied on the next day at the same site resulted in 10 to 20 mm of runoff. Bartlett (1988) found that an additional 120 mm of precipitation is needed to return the stream system to normal conditions following an extended dry period, even after soils become saturated.

Limestones and alluvium behave as aquifers at Konza Prairie. Fractures in limestones and grain size of alluvium control aquifer hydraulics. Fracturing in limestone beds and intervening shales permits rapid infiltration of precipitation and substantial storage and attenuation of infiltrating water. If storms occur when the watershed is wet, flooding can occur because of the low infiltration capacity of wet soils and the steep slopes. The thin limestones do not develop extensive karst, but outcrops show evidence of solution-enhancing joints that result in numerous springs and seeps along the main channels. The alluvium consists of either a limestone-chert or clay-silt mixture that has been described as reworked loess (Jewett 1941).

Streamflow also is affected by water demands of the native vegetation. Four years of irrigation experiments designed to provide adequate water to native prairie vegetation during the growing season indicated that an average of about 350 mm/yr of additional water is needed to satisfy evapotranspiration requirements (Koelliker and Lewis unpublished data). This high demand by the vegetation, combined with low rainfall in later summer and early fall, leads to low flows in Kings Creek. When the vegetation is not actively transpiring, the residual layer of dead plant material acts as an effective mulch on the soil. Soil evaporation is very low, although interception losses remain substantial. The combination of a short transpiration season and reduced soil evaporation leads to more percolation of infiltrated water and higher amounts of total water yield than from other watersheds in the area. For example, water yield from Kings Creek is about 15% more than from a nearby watershed with mixed land use, even though that watershed receives about 5% more precipitation than Konza Prairie. Most of this difference is attributed to the mulch effect on unburned

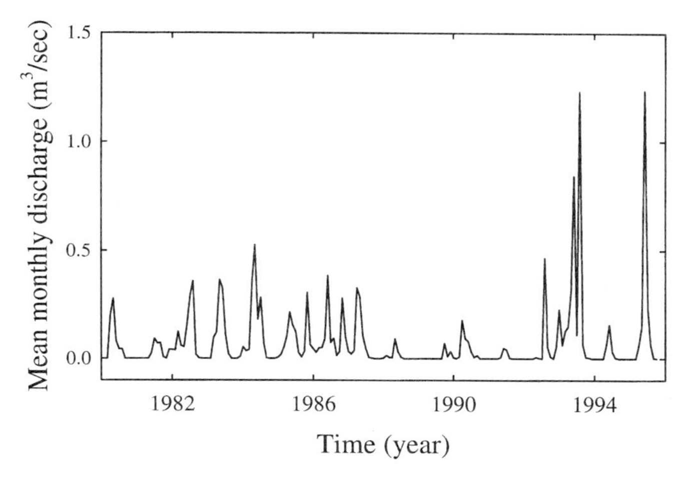

Figure 10.1. Monthly means of daily average discharge at Kings Creek near Manhattan, Kansas (USGS 1995). Discharge measurements began in 1979 and continue to the present. Kings Creek is part of the USGS Benchmark Network (station number 06879650) and is the only basin within the network that exclusively drains tallgrass prairie.

areas. These vegetative conditions provide almost complete and continuous cover for the soil.

Despite steep slopes throughout the Kings Creek watershed, little sheet or rill erosion occurs (Chapter 3, this volume). Field experiments on burned and unburned tallgrass prairie have shown that soil losses from rainfall or overland flow immediately following burning were very low (Duell 1990). Because the bases of plant stems do not burn completely to the ground, burned prairie has nearly the same surface roughness as unburned prairie; thus, overland flow velocities remain low. Most erosional losses occur either in larger channels where velocities are sufficient to transport channel sediments or from gradual mass slumping of soils from steep slopes.

Streamflow Characteristics

A high degree of variability in discharge is evident between seasons and years (Fig. 10.1); a summary of the streamflow characteristics is presented in Table 10.1. Hydrographs of the four smaller watersheds are given in Figure 10.2 (see Chapter 1, this volume, for watershed designations). The total water yield measured at the USGS gauging station is about 25% of the total precipitation on the drainage basin. Lowest average flows occur in late summer and again in winter, when the stream dries completely except for isolated pools near springs and seeps. The longest period of continuous, measurable flow occurred from September 1985 to July 1987 (673 days), and the longest period of no flow occurred from June 1988 to September 1989 (442 days). On average, the discharge pattern

Table 10.1. Selected streamflow characteristics for Kings Creek near Manhattan, Kansas (USGS hydrologic benchmark station 06879650) for Water Years 1979–1995 (USGS 1996)

Characteristic (unit)	Value	Date Occurred
Annual mean (m^3/s)	0.00706	
Highest annual mean(m^3/s)	0.0253	1993
Lowest annual mean(m^3/s)	0.000534	1989
90% highest daily mean (m^3/s)	1.24	13 May 1995
Lowest daily mean (m^3/s)	0.0	Many days
Annual 7-day minimum (m^3/s)	0.0	Many years
Instantaneous peak flow (m^3/s)	27.2	13 May 1995
Instantaneous peak stage (m)	4.20	13 May 1995
Instantaneous low flow (m^3/s)	0.0	Many years
Annual runoff (m^3)	223,000	
10% exceeds (m^3/s)	0.0155	
50% exceeds(m^3/s)	0.000561	
90% exceeds (m^3/s)	0.0	

Note: All values except instantaneous peak stage are reported per km^2, drainage area above station is 10.6 km^2.

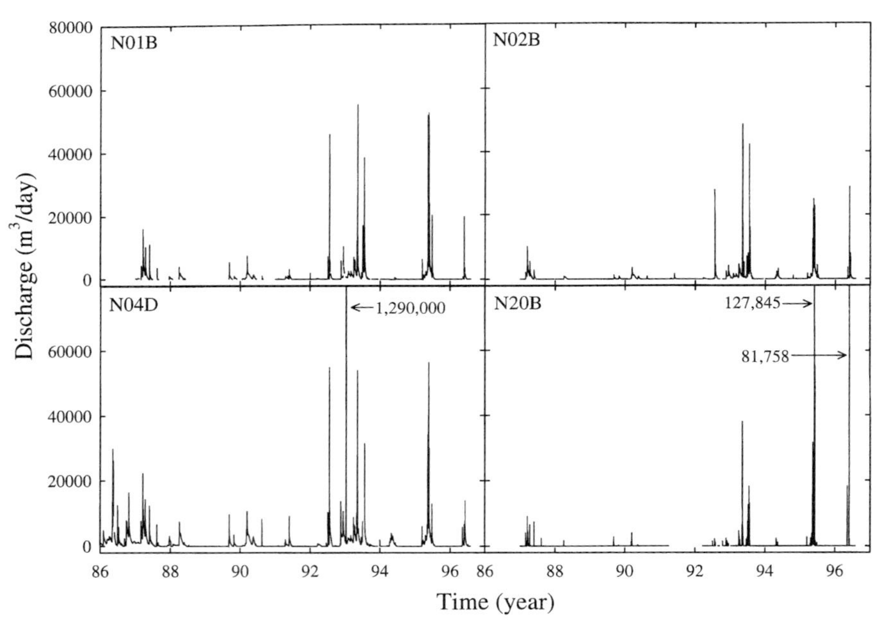

Figure 10.2. Hydrographs for the four smaller drainage basins with weirs in the Kings Creek basin (see Chapter 1, this volume, for watershed designations). Note that large floods do not occur in all watersheds at the same time. This is a result of highly localized thunderstorms; the watersheds are all within several kilometers of each other.

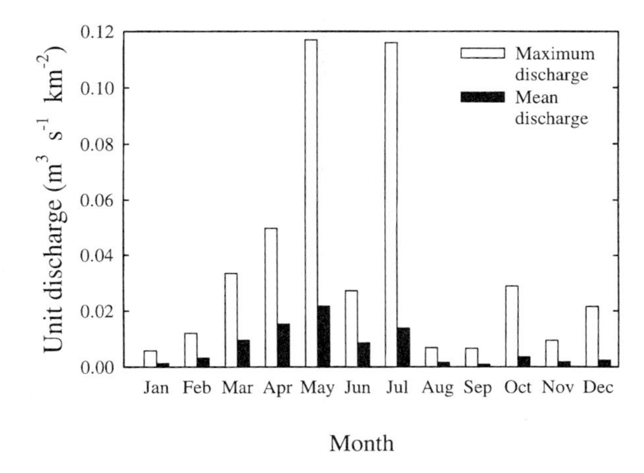

Figure 10.3. Mean and maximum monthly streamflow for Kings Creek near Manhattan, Kansas, for the period of record (USGS 1995). Minimum streamflow for every month has been zero.

in Kings Creek at the USGS gauging station consists of about 200 days of measurable flow, with no flow for the rest of the year. For each month, there has been at least one year of record in which no flow occurred during that entire month (Fig. 10.3).

The greatest amount of discharge generally occurs in April, May, and July. April and May are periods when normal precipitation exceeds evapotranspiration. Although average precipitation in July is less than half the evapotranspiration demand, the three largest floods have occurred during this month. Peak flows of these floods, measured at the USGS station, occurred on 1 July 1982 (128 m^3/s), 22 July 1992 (164 m^3/s), and 17 July 1993 (233 m^3/s). A discharge of 153 m^3/s represents an estimate of a 100-year peak discharge for a drainage area the size of Kings Creek (Clement 1987). Only the 1982 flood was produced by a single storm event. Although rainfall was not measured, simulation results indicate that the storm would have been a 100-year event of 1-hour duration (Humbert 1990). The floods of 1992 and 1993 resulted from more modest storms following extended wet periods that saturated soils and filled depressional storage in the watershed.

Kings Creek has been characterized only moderately with regard to transient storage zones. These zones are important in influencing solute transport and are quantified by the A_s/A (Fig. 10.4) ratio, which increases in value with increasing storage (D'Angelo et al. 1993). The values for the A_s/A ratio determined from solute releases in Kings Creek (Fig. 10.4) are typical of streams studied previously (J. Webster personal communication).

In summary, Kings Creek is a small stream that is intermittent in the upper reaches. However, it does have greater duration of flow than is typical of surrounding areas. It can produce extremely high discharges because of the nature of the soils, the steep topography, and the potential of the climate to produce large, high-intensity storms or extended wet periods that overwhelm the storage capacity of the drainage basin.

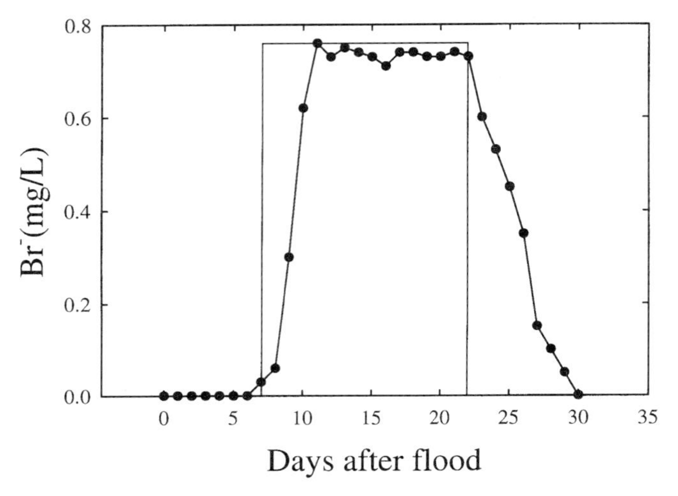

Figure 10.4. Conservative solute (Br^-) release diagram for Kings Creek for 18 June 1996 below the N02B and N20B weirs. Actual data are compared with a theoretical representation showing the effects of downstream transport (advection) only (D'Angleo et al. 1993). Differences between the data curve and the theoretical curve (solid line, no symbols) indicate the influence of dispersion and transient storage (Gray et al. unpublished data). For the fitted curve, the transient storage zone subsurface area $A_s = 0.02 m^2$, dispersion $D = 0.30$, and exchange with subsurface $\alpha = 0.138$. The ratio of A_s/A was 0.2 and discharge was 0.013 m^3/s.

Average Annual Water Budget

Based on a water-budget computer simulation study of the Kings Creek drainage basin by Bartlett (1988), the average annual dispensation of the 835 mm of precipitation varied with landscape position. Infiltration was highest on the slopes (25% of annual precipitation), followed by ridges (17%) and valleys (11%). Actual evapotranspiration (AET) was highest in valleys (65% of annual precipitation), followed by ridges (57%) and slopes (49%). Surface runoff is slightly larger on the slopes because of the shallow soils but overall does not vary much in the landscape. About two-thirds of the water yield is from infiltration, which appears as base flow at the USGS station.

The three landscape positions demonstrate the influence of soil characteristics on the amount of AET and infiltration. The difference in AET between landscape positions is explained primarily by the relative ability of the soils to store water for subsequent use, which, in turn, influences the difference in aboveground biomass production at various landscape positions.

Results from the water budget model presented here are for average conditions. It is important to remember that great variations occur in the precipitation and potential evaporation at Konza Prairie (Chapter 1, this volume), but many field experiments are short-term in nature. Given this variability, long-term data sets are a necessity for the study of streamflow.

Groundwater Hydrology

The hydrology and chemistry of groundwater on Konza Prairie have been studied in two portions of the Kings Creek basin. Thirty monitoring wells were installed in selected limestones and the alluvium of the N04D watershed. The wells are arranged in four transects perpendicular to the stream channel in the lower one-fourth of the watershed. Wells are grouped to occupy up to six sites (nests) along each transect. Each group of wells samples groundwater from one to three limestone units (Eiss, Bader, and Morrill Limestones; Chapter 3, this volume). Soil lysimeters were installed in the thin soils near each of the well nests and sample soil horizons A through C, if present. This network of wells and lysimeters is used to monitor long-term changes in water chemistry and hydrology and to study naturally occurring dissolved organic carbon and nitrate. A second study area contains two nests of six wells and six lysimeters installed in the silt-clay alluvium adjacent to Kings Creek. One cluster is located in an agricultural field and the other in the same alluvium under an oldfield/grassland. These wells are used to compare and contrast N dynamics under different types of vegetation and different conditions of land management.

Measurements of water levels in the local limestone and alluvial aquifers in watershed N04D since 1991 showed that the response of aquifers to recharge events (rainfall) is uneven and presumably related to the frequency of rainfall events. Several periods of precipitation close together were more likely to affect recharge to the aquifers than an isolated rainfall, and significant recharge was less likely during the growing season because of extensive evapotranspiration (Fig. 10.5). Water-level response to recharge was generally very fast and may be less than a day (Macpherson 1992b, 1996). This rapid response may indicate either the fast propagation of a pressure wave in the aquifer system from a recharge event or rapid physical movement of the recharge water to the observation point. A good hydraulic connection between the aquifer and areas of recharge was indicated in either case.

Water in alluvium, whether the alluvium is gravel or silt, is linked closely to adjacent streams. Alluvial aquifers fill before the stream begins flowing and empty fully after streamflow ceases. Although most recharge to limestone aquifers occurs vertically through overlying units, it is also known to occur from the stream itself. Recharge can occur in outcrop areas away from streams, but a prominent groundwater mound in the lowest monitored limestone in watershed N04D was just downstream from a perennial pool in the upper part of the study area. The mound, which had a peak elevation 1 to 2 m below the elevation of the stream valley, was supplied by drainage from the upstream pool into permeable alluvium (Fig. 10.6). The alluvium probably occupied a scoured area in bedrock and acted as a feeder system for transmitting water from the pool to the lowest monitored limestone. Thus, the groundwater in this limestone was a mixture of stream water that had moved nearly vertically through the alluvium to the aquifer and then laterally within the aquifer and some water leaking vertically from overlying units (Pomes 1995; Macpherson 1993, 1996). Based on observations of water-level elevations, the limestone aquifers behaved as either unconfined or confined units.

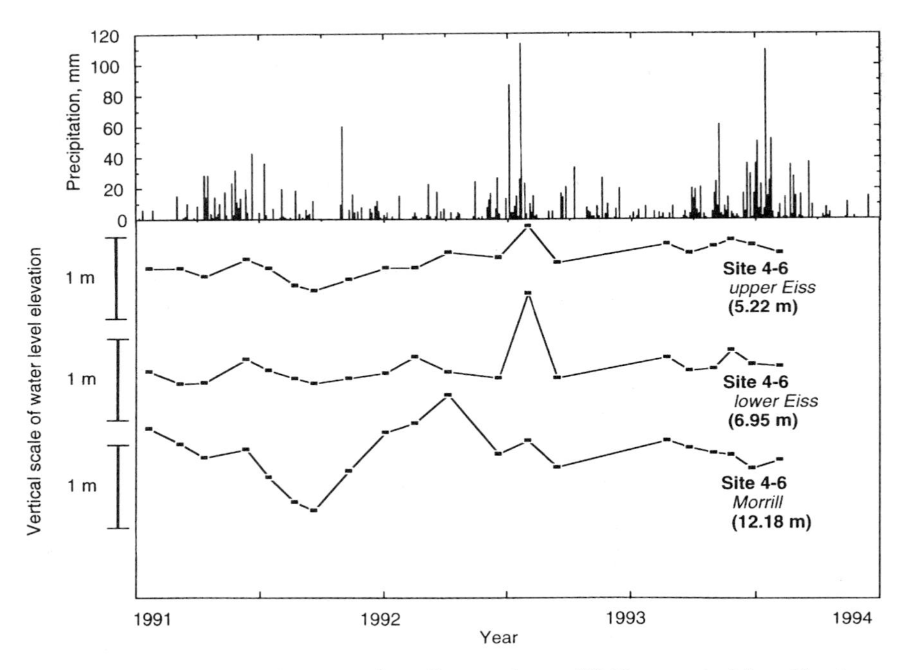

Figure 10.5. Water-level response in wells to recharge, N04D watershed (see Chapter 1, this volume, for watershed designations).

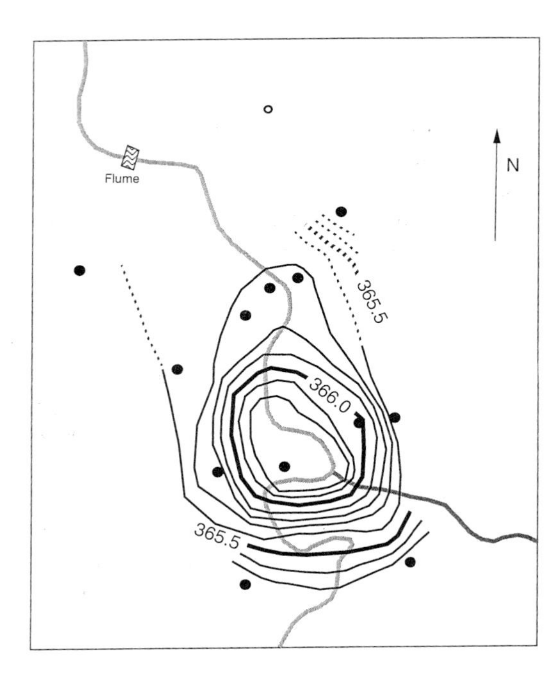

Figure 10.6. Potentiometric surface in the lowermost monitored limestone (Morrill Limestone; see Chapter 3, this volume) in the N04D basin during April 1992 that showed a pronounced groundwater mound beneath the stream. Contours are in meters above mean sea level. Filled circles are well locations, the open circle is a dry well within the Morrill Limestone, the rectangle is the basin flume (weir), and gray lines represent stream channels.

Unconfined alluvial aquifers showed large changes in water level throughout the year, suggesting a strong hydraulic connection between the alluvium and the alluvial aquifer. Discharge areas were springs and streams. Springs and seeps that occur along valley walls at outcrops intercepted movement of groundwater toward the stream. This water either moved back into the subsurface and rejoined a groundwater system, evaporated, or became surface flow.

Water Chemistry

The relations between surface water and groundwater at Konza Prairie are complicated by episodic and irregularly sized precipitation events; highly variable stream discharge; and unsteady, delayed response of aquifers to recharge. Consequently, the water chemistry in the streams and aquifers is related to the history of precipitation events, including magnitude; temporal relation to previous events (thus controlling residence time in the groundwater and surface water reservoirs); and activity of terrestrial vegetation, especially for nutrients.

This section summarizes the current knowledge of water chemistry in surface waters and groundwater at Konza Prairie. The discussion of nitrate and dissolved organic carbon are presented as an integration of results from surface water and groundwater studies because of the ecological focus of the research.

Surface Water Chemistry: Streams

The chemical characteristics of surface water in Kings Creek are monitored at the USGS gauging station and at the V-flumes on the four smaller watersheds in the South Branch. The USGS records summarized here included seasonal grab samples collected between April 1980 and April 1993. Sampling of surface water at the V-flumes began either in 1986 (watershed N04D) or 1987 (watersheds N02B, N01B, and N20B).

Mean values and ranges for concentrations of inorganic constituents and other aspects of surface water from the USGS records are presented in Table 10.2. Dissolved species concentrations reflect the sedimentary (limestone and gypsum) strata of the basin. Major ions include calcium, bicarbonate, magnesium, and sulfate; all show significant negative correlations with discharge. Dissolved oxygen is at saturation when the stream is flowing, and pH is typically basic. Turbidity is generally low, except during storm runoff events, and reflects the historical condition of clear streams in the Flint Hills (Cross and Moss 1987).

Mean annual surface water temperature is 13°C. Stream reaches with inflows of groundwater generally do not freeze in winter and stay below 30°C in summer. Total annual degree-days with continuous flow is about 5,000.

Surface Water Chemistry: Springs

The major-ion chemistry of spring water at Konza is not well known. Using one to two determinations of the chemistry of six springs, Whittemore (1980) postulated

Table 10.2. Concentrations of dissolved inorganic constituents and other water quality parameters in Kings Creek at the U.S. Geological Survey gauging station

Parameter	Mean Value	Minimum	Maximum	Correlation Coefficient r
pH	7.9	6.6	8.3	0.02 n.s.
Conductivity (mS/cm)	485	5	698	−0.62 *
Hardness (mg/L as $CaCO_3$)	287	260	330	−0.56 *
HCO_3^- (mg/L)	260	116	364	−0.64 **
Cl^- (mg/L)	2.2	0.8	14	−0.22 n.s.
SO_4^{-2} (mg/L)	33	6	61	−0.60 **
F^- (mg/L)	0.4	0.2	0.6	0.12 n.s.
PO_4^{-3} (mg/L)	0.02	0.0	0.12	0.92 **
NO_3^- (mg/L)	0.09	0.0	0.91	0.07 n.s.
Ca^{+2} (mg/L)	83	29	100	−0.78 **
Mg^{+2} (mg/L)	15.6	4.2	19.0	−0.80 **
Na^+ (mg/L)	5.0	1.6	13.0	−0.59 **
K^+ (mg/L)	1.5	0.71	5.1	0.71 **
$NH4^+$ (mg/L)	0.03	0.0	0.19	0.20 n.s.
Dissolved silica (mg/L)	13.4	10.0	17.0	0.26 *
Turbidity (NTU)	0.9	0.2	4	—
Dissolved O_2 (mg/L)	9.9	7.0	13.4	—

Source: Data from USGS (1995) records for the period April 1980 to April 1993 (N = 80).

Note: Mean discharge when samples were taken was 262 L/s (range: 2 to 7,896 L/s). The last column shows the correlation between concentration or value of the parameter versus discharge. For the correlation coefficient, *r*, * = $P < 0.05$, ** = $P < 0.01$, and n.s. = not significant.

that total dissolved solids, temperature, and major-ion chemistry were related to discharge rates and, thus, recharge events. Major-ion chemistry of springs is similar to that of surface water and groundwater. The dominant ions are calcium and bicarbonate, with various amounts of magnesium and sulfate, depending upon the aquifer host. Specific conductance and sulfate content from two sampling events showed inverse relations to estimated spring discharge rate, suggesting that recharge dilutes groundwater. Macpherson (unpublished data) found only about 7% relative standard deviation for calcium and alkalinity from April 1991 to November 1993 in a spring near one of the well-nest sites at the N04D watershed. Sulfate concentrations in the spring during this time were relatively low, varying from 12 to 26 mg/L, in contrast to the six springs sampled by Whittemore (1980), in which sulfate concentrations ranged from 40 to 200 mg/L. Thus, the response of water chemistry to recharge is large where gypsum or anhydrite is present in the aquifer host and can contribute large amounts of sulfate to the water. If gypsum or anhydrite is absent in the aquifer host, or if the length of the path between recharge point and the spring is long, the response of water chemistry to recharge is small.

Surface Water Chemistry: Persistent Pools within Streams

Persistent pools are maintained by groundwater discharge and, thus, reflect somewhat the chemistry of the aquifer from which the groundwater discharges.

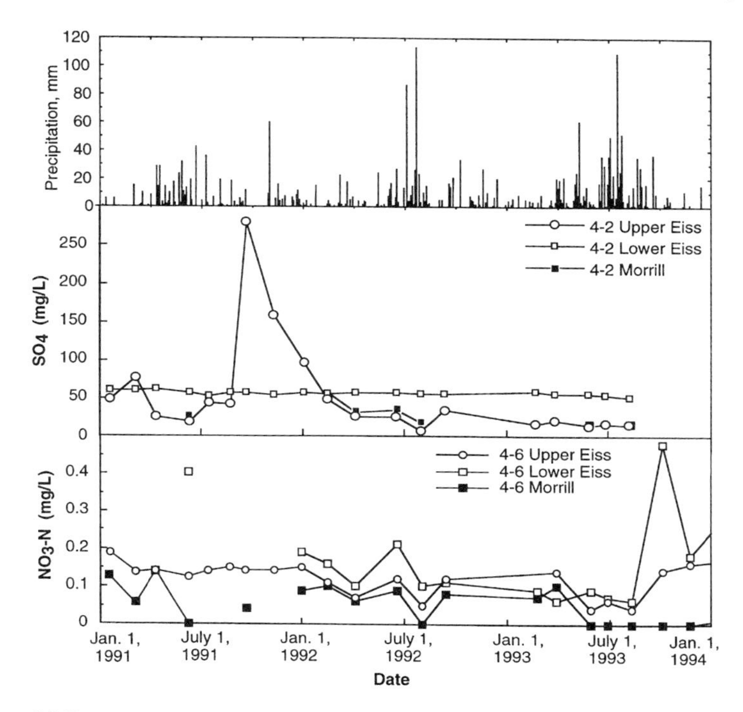

Figure 10.7. Relationship between precipitation, sulfate concentrations, and nitrate concentrations in samples from three wells in the N04D watershed (see Chapter 1, this volume, for watershed designations and Chapter 3 for a description of geologic formations). Points plotted as 0 mg/L NO_3-N were below detection limits. Groundwater sulfate concentrations were most variable in the shallowest limestone (Upper Eiss) and less variable in the deeper limestones. The spike in sulfate concentration in the Upper Eiss limestone is likely due to breaching of a cavity in the limestone that contained gypsum or anhydrite (Macpherson 1996). The variability in nitrate may be attributed to variation in inputs of nitrate, rates of soil organic N oxidation, distribution of areas of nitrate reduction, and mixing of waters with different nitrate concentration.

In addition, the chemistry of the pools is affected by evaporation. The chemistry of water in a perennial pool within the main channel of the N04D watershed exhibited these effects. The pool showed a four-fold increase (by weight) in sulfate during dry periods, as well as increases in calcium and other ions, indicating evapoconcentration. The flushing of the N04D pool after a rainfall event allowed the concentrated water to recharge the lowest monitored aquifer and thus become a useful, natural chemical tracer (Macpherson 1996).

Groundwater Chemistry: General Observations

Water chemistry in all aquifers at Konza Prairie is dominated by calcium and alkalinity (mostly bicarbonate), with magnesium and sulfate content dependent upon host rock and other factors. Total dissolved solids are typically about 300 to 400 mg/L in all units. Wells completed in a relatively low-permeability limestone showed almost no variation in water chemistry over the 3-year monitoring period. Wells in other units produced water with variable chemistry (Fig. 10.7; Macpherson 1996).

Both the limestone and alluvial aquifers contained dissolved oxygen throughout the year (Edler and Dodds 1992; Dodds et al. 1996a). This indicated persistence of a relatively high redox potential and, thus, oxidized chemical forms. The modeled partial pressure of CO_2 (log P-CO_2) in shallow groundwater at N04D, using calcite saturation to constrain pH, was -1.7 to -1.4 (Macpherson 1992b) and is within the normal range for P-CO_2 for shallow groundwater. Seasonal variations in P-CO_2 were evident. CO_2 pressures were highest during the summer and fall and lowest during the late winter, just before the spring thaw.

Sulfate concentrations showed the most variation of all dissolved species in groundwater over the monitoring period. The mean concentration of sulfate in precipitation from 1982 to 1992 was 2 mg/L (range: 0–20 mg/L). Mean sulfate content of groundwater was 20 mg/L (maximum = 120 mg/L). The significantly higher concentrations of sulfate in groundwater compared with precipitation suggest either evapoconcentration of recharge water in the unsaturated zone, dissolution of sulfate-bearing minerals such as gypsum, oxidation of sulfide minerals such as pyrite, or some combination of these factors.

Groundwater sulfate concentrations were most variable in the shallowest limestone and somewhat less variable in the deepest limestone monitored at N04D. In the intermediate limestone, sulfate concentration in groundwater was nearly constant. In one well located in the shallowest limestone, a spike of water with a sulfate concentration more than 10 times higher than all other samples from the same well occurred during an extended period of below-normal precipitation. Because no similar increase occurred in chloride content, the spike probably was due to breaching of a cavity in the limestone that contained gypsum or anhydrite (Macpherson 1996). Other wells have shown similar spikes, although not at the same time. Thus, sulfate can be an indicator of meteoric diagenesis of ancient limestones.

Because sulfate content of wells in the intermediate limestone was relatively constant and the lowermost monitored limestone received variable amounts of recharge from the stream, sulfate content of groundwater in the lowermost limestone can be used to calculate the contribution of water from the overlying, middle limestone. In this setting, sulfate may serve as the best chemical indicator of proportions of laterally derived stream water and vertically recharged groundwater to the lowermost limestone monitored.

Dissolved Organic Carbon in Groundwater

Abundance and characterization of dissolved organic carbon (DOC) in groundwater, soil water, and throughfall reveals information about sources of recharge water and processes of degradation of DOC. Color of DOC is related to concentration of DOC (Black and Christman 1963; Packham 1964) and is different in soils from forest and grassland (Stevenson 1985). Pomes and Thurman (1991) and Pomes (1995) extended this early assessment of color and DOC to show that the color is distinctive for grassland and woodland associations at Konza Prairie. The color distinction is apparently due to the near absence of humic acid in grass–soil water humic material.

Naturally occurring DOC at watershed N04D was found in relatively high concentrations in throughfall and progressively lower concentrations in soil water and groundwater. Typical concentrations were 8 to 32 mg C/L in throughfall, 6 to 30 mg C/L in the soil zone, and less than 2 mg C/L in groundwater. Woodlands showed higher DOC in the soil zone (28–32 mg/L) than grasslands (6–29 mg C/L). DOC in both vegetation types declined significantly through the A to C soil horizons (Pomes 1995).

Chemical and isotopic characterization of the humic material showed that a distinct signature of DOC was derived from the two vegetation types (Pomes 1995). Differences occurred in the number of carbon atoms per carboxyl group and $\delta^{13}C$. During the alteration of DOC from that released by vegetation to that found in groundwater, several changes commonly occur, such as reductions of phenolic, oxygen, nitrogen, and carbohydrate content, and aromaticity. However, differences in the alteration of DOC between the two vegetation types occurred in carboxyl content (increased content in woodlands and decreased content in grasslands), as well as in $\delta^{13}C$ trends. DOC characterization, then, is a discriminating tool for assessment of the vegetation association in recharge areas for belowground water and degrades in predictable ways (Pomes 1995). Thus, it is a valuable tracer of recharge areas in ecosystems with nonuniform vegetation, by providing the potential to identify the vegetation source and by allowing a qualitative measure of how much of the traditional flow path (soil water to ground) has been bypassed through fractures or macropores. Such differences in organic material leaching from leaves may have strong influences on the structure of stream microbial communities and their ability to metabolize DOC (McArthur and Marzolf 1986).

Nutrients in Surface Water

Dissolved nutrients, limited here to N and soluble reactive phosphorus (SRP) species, influence the productivity of the prairie (Chapter 12, this volume). The surface water–groundwater connection is important at Konza Prairie because concentrations of nitrate are typically higher in groundwater than in surface water.

The dynamics of N and P cycling in prairie streams have been explored only recently, and data on many of the processes within these cycles are lacking (Tate 1990). In Kings Creek, information has been obtained with V-flumes on the spatial and temporal concentrations of N and P in streamflows and groundwater (Edler and Dodds 1996; Tate 1985, 1990; USGS 1995) and on N transport from the watersheds (Dodds et al. 1996b). Much of the emphasis has been placed on the dynamics of N, the primary limiting nutrient for plants and microbes (Knapp and Seastedt 1986; Seastedt et al. 1991; Chapter 13, this volume).

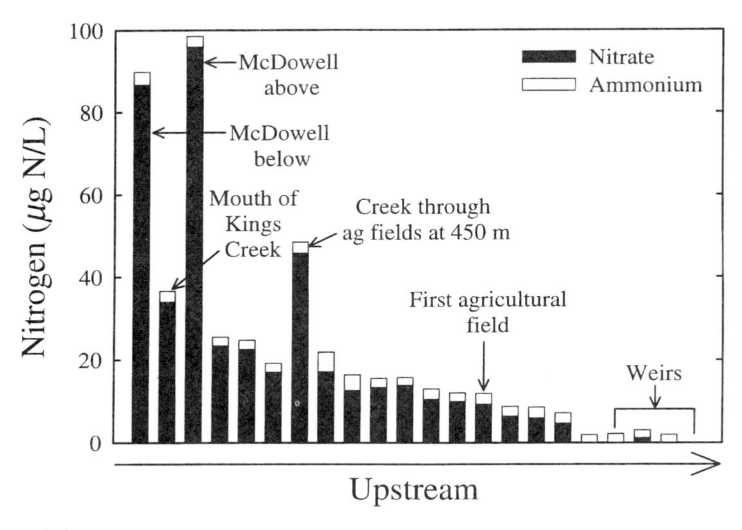

Figure 10.8. Nitrate and ammonium concentrations from a lower (fifth-order) reach of Kings Creek up to third-order prairie reaches (weirs). A small tributary enters at 450 m. Agricultural areas (row crops) border one side of this entire reach and both sides of the small tributary with elevated nitrate content that enters Kings Creek at 450 m. McDowell Creek is a larger creek into which Kings Creek empties (Chapter 3, this volume).

Striking characteristics of Kings Creek surface flows are the low average concentrations of both N and P. Although concentrations of P are comparable, N concentrations are only equal to or one-third of concentrations in streams draining forested watersheds (Meyer et al. 1993). Levels of both nutrients in Kings Creek are only one-fifth (N) to one-tenth (P) of levels found in desert streams (Fisher et al. 1982). Nitrate (8– 28% of total N) and organic N dominate in Kings Creek; ammonium and nitrite concentrations are typically low (Dodds et al. 1996b; Tate 1985).

Spatial variation in streamflow concentrations of N and P is evident in data obtained by Tate (1985, 1990). In general, concentrations of nitrate and SRP decreased downstream from permanent sources of water in the headwaters, although local reaches had elevated concentrations because of inputs from seeps and tributaries. Perennial reaches in the gallery forest typically had much lower concentrations of nitrate (mean = 12 μg/L) than intermittent reaches upstream (mean = 41 μg/L; Tate 1990). Small seeps had a mean nitrate concentration of 46 μg/L, although not all seeps had relatively high nitrate levels. Between watersheds within the Kings Creek basin, nitrate levels show a significant negative correlation with watershed area (Dodds et al. 1996b). Nitrate concentration in the stream reaches below agricultural areas on Konza Prairie are elevated greatly over those found in tallgrass prairie (Fig. 10.8).

Temporal variation in concentrations of N occur as a result of several interrelated factors. These include discharge regimes, seasonal activity and burning of terrestrial vegetation, and diel activity of aquatic biota.

Data collected to date reveal that surface water concentrations of nitrate and total N are correlated negatively with discharge. Within individual watersheds, however, this relationship may not hold (Dodds et al. 1996b). Initial flows in previously dry tributaries and storm flows in perennial channels showed 10-fold increases in nitrate concentrations (McArthur et al. 1985; Tate 1990), although the effect lasted only a few hours or days. Edler and Dodds (1996) reported a strong positive correlation between spring discharge rates and nitrate concentrations. Both of these observations suggest two possible processes. A "first-flush" effect may occur after rainfall in which nitrate stored in the soil zone is flushed into surface water and groundwater. Nitrate concentrations as high as 10.7 mg/L have been reported in soil water in the N04D watershed (Macpherson and Schulmeister 1994), although concentrations are typically two to three orders of magnitude lower (Macpherson and Schulmeister 1994; Seastedt and Hayes 1988). Highest soil-water concentrations have been found in the late fall through early spring period and may be associated with certain types of vegetation (Macpherson and Schulmeister 1994). An alternative explanation is that the higher nitrate concentration in precipitation provides a spike in nitrate concentrations in surface water or groundwater. This spike then follows a short flow path between recharge and discharge points. When rainwater infiltrates rapidly through soil and aquifers, less nitrate is removed by microbial immobilization.

Nitrogen levels in Kings Creek and its tributaries are correlated with the activity of terrestrial vegetation and time since burning. Tate (1990) found that mean nitrate, organic-N, and total-N concentrations in surface flows were greater during the dormant season (September–March) than during the growing season (April–August).

Nitrate exhibited the greatest change, with a two- to threefold increase in mean concentrations in intermittent reaches, perennial reaches, and seeps in the dormant season compared with the growing season. Dodds et al. (1996b) found that N concentrations in stream water decreased with time since burning. They speculated that N increased after fire because of less removal of N in precipitation by vegetation and more rapid overland flows resulting from decreased aboveground biomass. Nitrogen entering streams thus decreased over time because of immobilization by increased surface litter. Biotic uptake of nitrate by aquatic biota also may be important in the trend of decreasing N levels downstream. Tate (1985, 1990) found that daytime levels of nitrate in stream water were lower than nighttime levels; furthermore, experimental additions of nitrate showed that uptake rates by algae were significantly higher in the day than at night.

The low nutrient levels found in Kings Creek reflect the relatively "tight" cycling of these nutrients by the tallgrass prairie. Dodds et al. (1996b) reported that total-N concentrations in stream water from Konza Prairie watersheds averaged only 150 μg/L, corresponding to an export of only 0.16 kg N ha^{-1} y^{-1} and 1.5% of the N input from precipitation. These values are considerably lower than those reported for other terrestrial systems, except a mature deciduous forest.

Nutrients in Groundwater

Nitrate concentrations in groundwater and soil water of the limestone aquifers at the N04D study site exhibit spatial and temporal variability (Macpherson 1993; Macpherson and Schulmeister 1994). Nitrate concentrations in most soil waters ranged from less than 50 μg N/L to 200 μg N/L, with most samples containing less than 50 μg N/L; two samples contained more than 3,000 μg N/L. This range was also typical of groundwater at N04D, in which measured concentrations ranged from less than 50 to about 3,000 μg N/L. Most groundwater samples under prairie vegetation contained less than 300 μg N/L. Nitrate was not distributed uniformly within soil horizons or geologic units, and highly variable concentrations were apparent in time-series analyses of individual wells and lysimeters (Fig. 10.7). The variability in distribution of nitrate in Konza soil water and groundwater may be attributed to differential inputs of nitrate, supplied mainly from rainwater; variable rates of soil organic N oxidation; highly restricted areas of nitrate reduction; mixing of waters with different nitrate-N concentrations; or a combination of these factors.

Stable isotopes of N ($\delta^{15}N$) can be used to examine the origin and fate of nitrate. Nitrate derived from fertilizer typically is more depleted in ^{15}N (and thus the $\delta^{15}N$ value is more negative) than rainwater and soil organic N, and these, in turn, typically are more depleted in ^{15}N than nitrate derived from animal waste (Kreitler 1975). Nitrate that has been reduced by bacteria to N gases or ammonium will have a $\delta^{15}N$ value that is enriched in ^{15}N relative to, and thus more positive in $\delta^{15}N$ than, the nitrate before reduction (e.g., Korom 1992). Values for soil water at the N04D site (Macpherson and Schulmeister 1994) compared well with published values of $\delta^{15}N$ for nitrate in rain and nitrate derived from soil organic matter (Heaton 1986), suggesting that soil water nitrate might be derived from both of these sources. The similarity between the $\delta^{15}N$ value in groundwater nitrate at one well and that given by Heaton (1986) for rainwater may indicate direct delivery of rainwater to parts of the aquifers. Highly enriched (positive) $\delta^{15}N$ in nitrate from another well suggested that nitrate reduction had occurred in the aquifer or in the overlying soil zone before transport to the aquifer. Thus, differing recharge paths deliver rain-derived pristine nitrate or bacterially altered nitrate to the aquifers at N04D.

Nitrate concentrations in alluvium at the prairie and agricultural sites (Macpherson 1994), like concentrations at the N04D site, exhibited spatial and temporal variations. At both sites, $\delta^{15}N$ was slightly enriched or within the high range of nitrate derived from soil organic N. No significant difference occurred between the $\delta^{15}N$ values of nitrate at the prairie and agricultural sites, even though nitrate concentrations were about 10 times higher at the agricultural site and a fertilizer-type $\delta^{15}N$ value was expected (Macpherson 1994). The relation between $\delta^{15}N$ and nitrate concentrations was complicated in that no two wells behaved similarly. For example, one well showed a clear antithetic relation between the $\delta^{15}N$ value of nitrate and NO_3-N concentration during part of the year, supporting denitrification as on ongoing process (Fig. 10.9). However, this relation did not hold for the rest of the year, and other wells showed still different

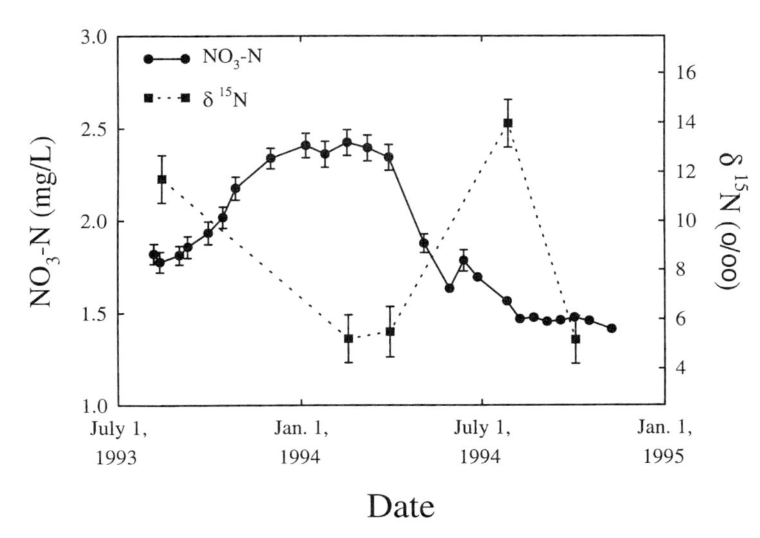

Figure 10.9. $\delta^{15}N$ and nitrate through time in one well at the agricultural monitoring site. The lack of a consistent pattern between nitrate and $\delta^{15}N$ complicates the interpretation that denitrification resulted in higher $\delta^{15}N$ coincident with lower nitrate.

patterns of $\delta^{15}N$ values of nitrate and nitrate concentrations. Denitrification rates were low at these subsurface sites (Sotamayor 1996).

In summary, nitrate in water at Konza Prairie is sourced from precipitation and, to a lesser extent, from mineralization of organic N. The degradation of nitrate as it is transported along either above- or belowground flow paths is dependent upon time of year, size of the precipitation event driving the flow, and time since last burning of the prairie. Nitrate concentrations are generally low in both surface water and groundwater, although some high concentrations have been measured in samples collected during the late fall to early spring.

Summary

Groundwater hydrology and geochemistry at Konza Prairie are dynamic and complicated by the relatively complex geology and unpredictable climate. The need for clearer understanding of exact groundwater recharge and discharge zones and stream-aquifer interactions and of their influences on both water levels and groundwater chemistry drives ongoing research. Because the geology of this region is typical of a large part of the central part of the United States and Canada, Konza Prairie is a valuable site for this kind of investigation.

Future research on aquatic chemistry will include continued characterization of the chemical processes occurring in the saturated subterranean ecosystem and a more complete characterization of the movement of nutrients (especially N) between subsurface, surface, and biotic compartments. This research will describe what is possibly the most variable subsystem in the highly variable prairie ecosystem.

Acknowledgments We thank all of the students, technicians, investigators, and funding agencies who have supported this research. This is contribution no. 97-183-B from the Kansas Agricultural Experiment Station, Kansas State University, Manhattan.

11

Structure and Dynamics of Aquatic Communities

Lawrence J. Gray
Walter K. Dodds

North American prairie streams are characterized by variable flow regimes (Jewell 1927; Matthews 1988). Low-order streams alternate between stable flows during spring and early summer and intermittent flow to completely dry during late summer and winter. At any time, conditions may be interrupted by scouring floods (Chapter 10, this volume). Thus, the biotic communities occurring in prairie streams constantly are adjusting to a patchy environment created by the temporal and spatial variations in streamflow. The biota of the prairie is influenced heavily by high variation in abiotic factors. Perhaps stream communities experience the most variable conditions of all the prairie ecosystems.

Initial research on the biota and ecosystem characteristics of Kings Creek primarily concerned characterization of inputs and storage of allochthonous organic material (Gurtz et al. 1982, 1988); leaf litter decomposition (Killingbeck et al. 1982; Smith 1986); and surveys of aquatic biota (Gurtz et al. 1982). Within the LTER Program, research has focused on additional studies of in-stream leaf decomposition (Gurtz and Tate 1988; Hooker and Marzolf 1987; Tate and Gurtz 1986); macroinvertebrate community dynamics (Gray 1989, 1993; Gray and Johnson 1988; Kavenaugh 1988); and characterization of subsurface invertebrate and microbial communities and processes (Edler and Dodds 1992, 1996; Eichem et al. 1993). A common emphasis of these studies has been to relate biotic processes to the surface flow extremes and the interactions between surface and subsurface flows that characterize the complex hydrology of the Kings Creek basin (Chapter 10, this volume).

Figure 11.1. View of a tallgrass prairie stream with associated riparian vegetation consisting of grasses, shrubs, and small trees. The open aspect of this reach is typical of Konza Prairie streams; the gallery forest provides a closed-canopy environment downstream. (A. Knapp)

Stream Channel Characteristics

The riparian vegetation along Kings Creek occurs in three distinct zones: (1) prairie grasses and forbs in headwater channels (stream orders 1–2); (2) grasses, forbs, shrubs, and small trees along the middle reaches (stream order 3; Fig. 11.1); and (3) gallery forest along fourth- and fifth-order channels dominated by bur oak (*Quercus macrocarpa* Michx.), chinquapin oak (*Q. muhlenbergii* Engelm.), hackberry (*Celtis occidentalis* L.), and elms (*Ulmus* spp.). The extent of riparian shading of the channel varies from 0 to 10% in the headwaters to 50 to 100% in the gallery forest (Gurtz et al. 1982). A recent historical analysis of the extent of the gallery forest on Konza Prairie has shown that it has expanded in area by 54% since 1939 (Knight et al. 1994).

Benthic sediments in Kings Creek consist of gravel and cobble riffles and runs separated by silted pools. The gravel and cobble consists of broken particles of limestone and shale that break into flat, platelike pieces. Sediment depth varies from approximately 1 m in unconsolidated bars to only a thin veneer of fine particles on bedrock. Pools are formed in depressions within gravel bars or at road crossings. Gallery forest reaches contain a few logs and other organic debris dams where the stream has undercut tree roots along high banks. Nearly all debris dams and logs in the lower reaches of Kings Creek present during the 1980s were swept out of the channel during the floods of July 1992 and July 1993. However, the expansion of the gallery forest may influence the future extent of debris dams and organic matter dynamics in general (Gray 1997).

Fishes

Four species of fishes are common in the main reaches and headwaters of Kings Creek, including the creek chub (*Semotilus atromaculatus* [Mitchill]), southern redbelly dace (*Phoxinus erythrogaster* [Rafinesque]), common stoneroller (*Campostoma anomalum* [Rafinesque]), and orangethroat darter (*Etheostoma spectabile* [Agassiz]). Qualitative collections and field observations indicate that dace are the most common fish present throughout the drainage basin. Several other species common to the Kansas River have been collected at the confluence with Kings Creek, but none have been found in upstream reaches. The species assemblage in Kings Creek is common for clear streams sustained by springs in eastern Kansas (Cross and Moss 1987), and these fishes exhibit remarkable resilience to the extremes in flows in Kings Creek. During periods of extreme drought when the main channels are dry, these species persist in small numbers in headwater springs. Reproduction in springs and subsequent dispersal after resumption of flow result in rapid recolonization of downstream reaches.

Macroinvertebrates

Nearly 200 species of aquatic macroinvertebrates have been recorded from Kings Creek, although the number is conservative because many groups have not been identified past the family or genus level (Gurtz et al. 1982). Common species of aquatic insects in Kings Creek include the mayflies (Ephemeroptera) *Fallceon quilleri* (Dodds), *Stenacron interpunctatum* (Say), *Stenonema femoratum* (Say), and *Caenis delicata* Traver; the stoneflies (Plecoptera) *Perlesta placida* (Hagen), *Allocapnia vivipara* (Claassen), and *Zealeuctra* sp.; the caddis flies (Trichoptera) *Hydroptila* sp., *Cheumatopsyche pettiti* Banks, and *Ironoquia punctatissima* (Walker); the beetle (Coleoptera) *Stenelmis crenata* (Say); and the true flies (Diptera) *Tipula abdominalis* (Say), *Simulium* spp., and Chironomidae spp.

Over one-half of the known taxa of aquatic insects are midges (Chironomidae). Kavenaugh (1988) identified more than 58 species of midges from Kings Creek. He found 50 species in reaches with permanent flow and 35 to 44 species in reaches with intermittent flow. Common midges include *Cricoptus* spp., *Orthocladius* spp., *Corynoneura* spp., *Paratanytarsus* spp., *Tanytarsus* spp., *Thienemanniella* sp., *Dicrotendipes* spp., and *Zavrelimyia sinosa*. *Corynoneura* spp. and *Paratanytarsus* spp. dominated the midge fauna in perennial reaches, whereas *Orthocladius abiskoensis* was dominant in ephemeral reaches.

Common noninsect invertebrates include small oligochaetes (*Nais* spp.), turbellarians (*Dugesia* sp.), snails (*Physa hawnii* Lea), and various crustaceans.

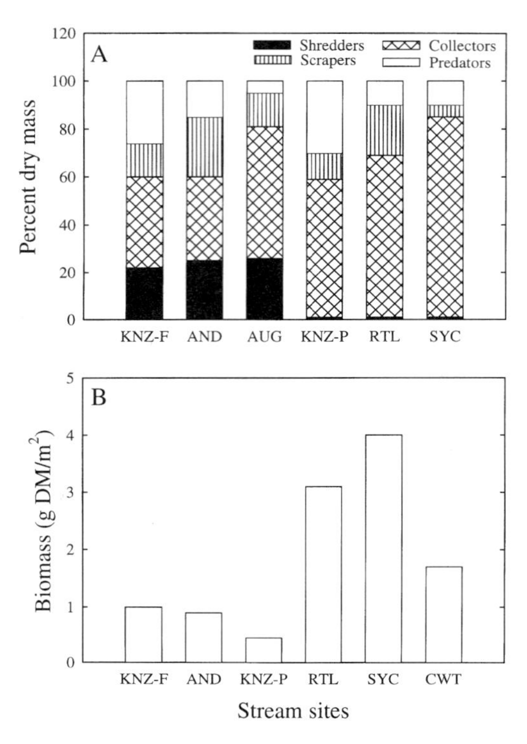

Figure 11.2. Comparison of trophic structure (A) and mean biomass (dry mass; B) of macroinvertebrate communities in Kings Creek with other North American streams. Site abbreviations and data sources: KNZ-F = Konza Prairie, gallery forest reaches; AND = H. J. Andrews Experimental Forest, OR (Douglas fir–coniferous forest; Anderson and Sedell 1979); AUG = Augusta Creek, MI (mixed deciduous forest; Cummins and Klug 1979); KNZ-P = Konza Prairie, prairie reaches (Gray and Johnson 1988); RTL = Rattlesnake Creek, KS (midgrass prairie; Gray 1984); SYC = Sycamore Creek, AZ (lowland Sonoran Desert; Fisher et al. 1982); and CWT = Coweeta Hydrologic Laboratory, NC (mature deciduous forest; Wallace 1988).

Crayfish (*Orconectes* spp.) are common throughout the watershed, particularly *O. neglectus* (Faxon). Like the common fishes, crayfish survive in headwater springs during dry periods, then move downstream when flow resumes.

Trophic Structure and Life Histories

Trophic structure categories for stream macroinvertebrates are based on mode of feeding and particle size ingested (Cummins and Merritt 1984). Shredders feed on large particles (over 1 mm), typically leaf and woody materials. Scrapers or grazers have adaptations that allow them to feed on diatoms and other fine particulates attached to surfaces. Collectors feed on fine particulates either by filtering material from the passing water or by "vacuuming" it from the stream bottom. Predators feed on living animal tissue.

The trophic structure of stream macroinvertebrate communities in gallery forest and grassland reaches of Kings Creek is dominated by collectors (38% and 58% of total biomass in the gallery forest and grassland reaches, respectively; Fig. 11.2A). Common species of collectors are sediment-feeding chironomids (*Tanytarsini* spp., *Cricotopus* spp., *Corynoneura* spp., and *Dicrotendipes* spp.) and small mayflies (*Fallceon quilleri* and *Caenis delicata*). Filter-feeding collectors, mainly hydropsychid caddis flies, constituted only 4% of total biomass in gallery forest reaches and 1% in grassland reaches. The dominance of collectors in the trophic structure of both riparian types is due to their generalist food habits and rapid life cycles. At summer water temperatures of 20 to 25°C, *F. quilleri*, *C. delicata*, and chironomids are able to complete their life cycles in 2 to 3 weeks in Kings Creek. Rapid life cycles reduce the probability of direct mortality from floods or drying, unlike species with longer developmental periods, such as *Tipula* (Gray 1989).

The most common shredder taxa are the crane fly *Tipula abdominalis* and the stone flies *Allocapnia vivipara* and *Zealeuctra* sp. The caddis fly *Ironoquia punctatissima* is rare in main channels but locally common in headwater springs. Shredders constituted 22% of total biomass in gallery forest reaches, compared with only 1% of total biomass in grassland reaches (Fig. 11.2A). This difference is correlated with higher allochthonous inputs and storage of detritus in gallery forest channels compared with prairie channels (Gurtz et al. 1982; see section on Leaf Decomposition).

Dominant scrapers were the snail *Physa hawnii* in grassland reaches and heptageniid mayflies (*Stenonema interpunctatum* and *Stenacron femoratum*) in gallery forest reaches. Scraper contributions to trophic structure were similar in both habitats and to other streams (Fig. 11.2A). Snail densities were highly variable, reflecting frequent discharge extremes and possibly predation by crayfish.

Tanypodinae chironomids and ceratopogonids (Diptera) were the most abundant predators in grassland streams. The stone fly *Perlesta placida* contributed most of the predator biomass in the gallery forest. An interesting aspect of the trophic structure of Kings Creek compared with the other streams was the greater proportion of predators. Predators averaged 27% of macroinvertebrate biomass in Kings Creek but only 5 to 15% in other streams (Fig. 11.2A). Considering the other predators present in Kings Creek that were not included in benthic biomass, such as crayfish, orangethroat darters, and creek chubs, predators may have an important role in structuring the benthic community.

Streamflow patterns influence the trophic structure of stream macroinvertebrate communities by eliminating species with long life cycles and changing the types and quantities of food resources present. The trophic structure of stream communities in gallery forest is very similar to that of other forested streams during extended periods of continuous flow without major flooding, particularly with respect to shredder abundance (Fig. 11.2A). During more typical periods when floods and drought are common, the trophic structure of gallery forest reaches more closely resembles that of grassland reaches, and thus other prairie and desert streams, because of restrictions on life cycles and reduced retention of allochthonous inputs (Gurtz and Tate 1988; Hooker and Marzolf 1987; Smith 1986).

Biomass, Density, and Production

Densities and standing biomass of benthic macroinvertebrates in Kings Creek show wide variations spatially and temporally. During periods of extended flows without flooding, mean density and biomass in gallery forest reaches averaged 17,100 individuals/m and 1.0 g DM (dry mass)/m, respectively. In grassland reaches, corresponding values were 5,500 individuals/m and 0.45 g DM/m (Gray and Johnson 1988). A comparison of biomass values from Kings Creek and other North American streams is presented in Figure 11.2B.

Discharge extremes greatly reduce macroinvertebrate populations. Floods scour substrates and reduce macroinvertebrate populations by more than 95%, and densities gradually decline as channels dry (Gray 1989). With continuous flow without flooding, highest densities occur in late spring to early summer (primarily midges and mayflies), with a secondary peak occurring in winter (mainly winter stoneflies).

Total emergence productions of aquatic insects in Kings Creek during periods of continuous flow in spring and summer were 51 mg m^{-2} d^{-1} in gallery forest reaches and 12 mg m^{-2} d^{-1} in grassland reaches (Gray 1993). Emerging aquatic insects were mostly chironomids (84% of emergence) and mayflies (7%). In other temperate, low-order streams with continuous flow, spring and summer emergence averages 15 to 35 mg m^{-2} d^{-1} (Illies 1975; Harper 1978). Kings Creek is not as productive as Sycamore Creek, a lowland Sonoran Desert stream, where mean emergence was 111 mg m^{-2} d^{-1} from May to September (Jackson and Fisher 1986). However, insect production leaving Kings Creek appears to be significant compared with production in surrounding terrestrial habitats. Daily aboveground production of terrestrial arthropods in the tallgrass prairie during the growing season is 5.8 mg DM/m^2 (calculated from data in Scott et al. 1979). Emerging aquatic insects are an important source of food for terrestrial predators, such as birds (Gray 1993).

Aquatic Flora and Primary Production

Algae present in Kings Creek include filamentous greens (*Cladophora, Spirogyra, Chaetophora, Stigeoclonium*, and *Zygnema* spp.); diatoms (*Cymbella, Cocconeis, Pinnularia, Navicula*, and *Gomphonema*); and cyanobacteria (e.g., *Nostoc*) (Gurtz et al. 1982; Tate 1990). Algal abundance in Kings Creek varies spatially and seasonally in response to stormflows, nutrient levels, and activity of terrestrial vegetation (Tate 1990). In general, algae (particularly filamentous greens) are abundant in fall, winter, and spring, often forming extensive mats in channels. During summer, mats of filamentous green algae are replaced by thin mats of periphytic cyanobacteria and diatoms. These filamentous mats are more luxuriant in regions where hyporheic water flows from under gravel back into the stream.

Aquatic macrophytes found in Kings Creek include *Veronica catenata* Penn. and *Nasturtium officinale* R. Br. Both species are distributed throughout the watershed but are most abundant near springs and other groundwater sources.

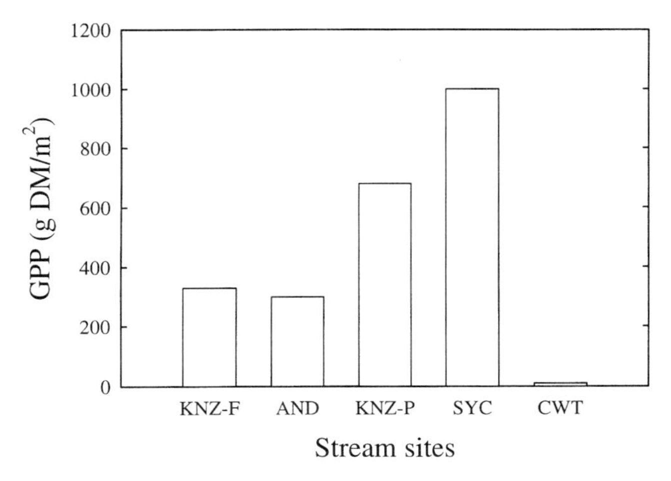

Figure 11.3. Comparison of annual gross primary production (GPP, dry mass) in Kings Creek with other North American streams. Site abbreviations and data sources: KNZ-F = Konza Prairie, gallery forest reaches; AND = H. J. Andrews Experimental Forest, OR (Douglas fir–coniferous forest; Naiman and Sedell 1980); KNZ-P = Konza Prairie, prairie reaches (Gurtz et al. 1982); SYC = Sycamore Creek, AZ (lowland Sonoran Desert; Fisher et al. 1982); and CWT = Coweeta Hydrologic Laboratory, NC (mature deciduous forest; Webster et al. 1983).

Emergent plants along the margins of springs and seeps include sedges (*Carex* and *Eleocharis* spp.), cattails (*Typha latifolia* L.), and prairie cordgrass (*Spartina pectinata* Link) (Gurtz et al. 1982).

Primary production in Kings Creek varies with habitat and discharge regime. Prairie reaches typically have higher gross primary production (GPP) than gallery forest reaches because of reduced shading and possibly higher levels of nutrients (Dodds et al. 1996c). Overall, rates of primary production in gallery forest reaches are similar to those in other forested streams (Fig. 11.3). Rates in prairie reaches are less than rates in open, desert streams with higher nutrient levels (Fig. 11.3). Daily rates of gross primary production in Kings Creek (1.0–1.7 g O_2 m^{-2} d^{-1}) are similar to those reported for other prairie streams (Matthews 1988). Both N and P can limit algal productivity in Konza streams (Tate 1990), but the system appears balanced between N and P limitation.

Recovery of Biota from Scouring Floods

Streamflows greater than 500 L/s have been observed to cause scouring of sediments in Kings Creek (Dodds et al. 1996c). The influence of a flood in May 1990 on macroinvertebrate biomass at a gallery forest site is shown in Figure 11.4. The flood reduced all populations by more than 90%, but biomass recovered to pre-

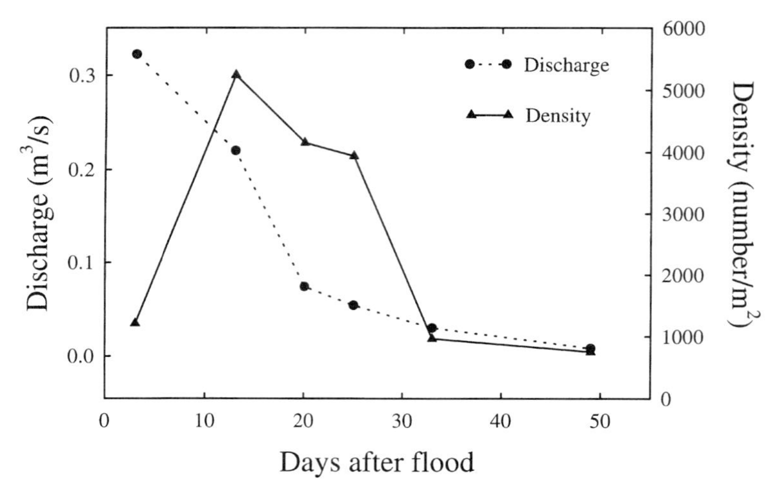

Figure 11.4. Relationship between steam discharge and macroinvertebrate densities after a flash flood in a gallery forest reach of Kings Creek. X-axis represents the period of 18 May to 3 July 1990. Data from Gray (1993 and unpublished).

flood levels within 2 weeks largely because of rapid recolonization and growth of midge populations. Biomass then decreased sharply as discharge declined because of a lack of recruitment. Recovery after the 100-year flood in July 1992 was slower; original benthic macroinvertebrate densities had not been reached after 60 days following the flood (Fritz and Dodds 1996).

The recovery of primary production in watershed N04D (prairie/shrub riparian) after the flood of 22 July 1992 was examined by Dodds et al. (1996c). Before the flood, midday rates of gross primary production were 60 mg O_2 m^{-2} h^{-1} in emergent vegetation (*Nasturtium*; Fig. 11.5A) and 41 mg O_2 m^{-2} h^{-1} for sediment periphyton (Fig. 11.5B). After floodwaters receded, rates of production were 13 mg O_2 m^{-2} h^{-1} in emergent vegetation and less than 1 mg O_2 m^{-2} h^{-1} in sediments. Production steadily increased thereafter and reached preflood levels within a month. Biomass (as chl *a*) followed a different trend (Fig. 11.5C). Preflood concentrations of chl *a* in sediments were 17 to 47 mg/m and decreased to zero after the flood. However, chl *a* increased to only 0.1 to 5.6 mg/m in 48 days after flooding and may reflect nutrient limitations (Tate 1990). Thus, production rates are able to recover more rapidly than biomass.

Leaf Decomposition

Direct litterfall and lateral movement inputs of tree leaves in Kings Creek vary greatly with stream reach. In gallery forest reaches, Gurtz et al. (1982, 1988) found that the total input of leaves was 519 g AF (ash free) DM/m^2, an amount comparable to that for other forested streams. Principal tree species, in order of

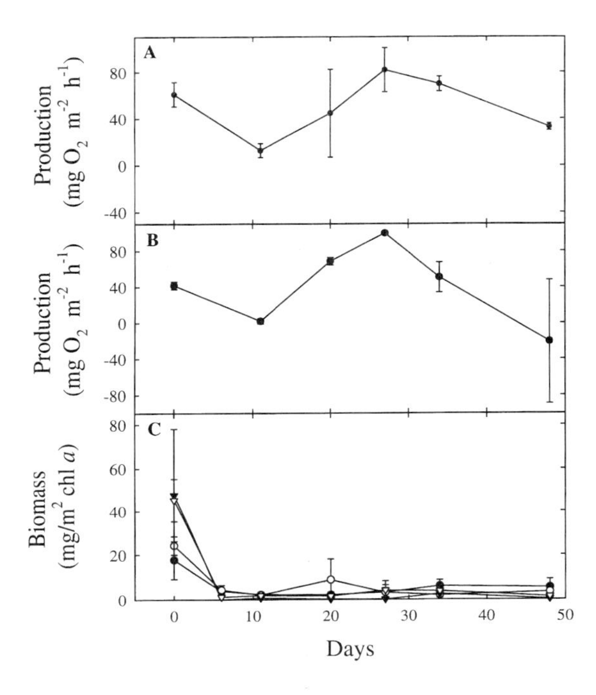

Figure 11.5. Primary production for sediments with macrophytes (A) and sediments alone (B) in watershed N04D following the 22 July 1992 flood. C shows periphyton biomass (as chl *a*) in sediments at four permanent, spring-fed pools in watershed N04D before and after this flood. Error bars = 1 standard deviation. The flood occurred on day 6 of these graphs. From Dodds et al. (1996c), with permission of Kluwer Academic Publishers.

importance, were *Q. macrocarpa*, *C. occidentalis*, *Q. muhlenbergii*, green ash (*Fraxinus pennsylvanica* Marsh.), and *Ulmus* spp. Prairie reaches received only 68 g AFDM/m^2, primarily from *Ulmus* spp. and *Q. muhlenbergii*.

Several studies have examined the rate of decomposition of leaves from riparian trees in Kings Creek; results from these studies are summarized in Table 11.1. Leaf decomposition was examined using either leaf packs attached to bricks (Killingbeck et al. 1982; Smith 1986; Gurtz and Tate 1988) or leaf packs in mesh bags with 2-mm openings (Hooker and Marzolf 1987; Tate and Gurtz 1986). Values for k, the rate coefficient, were computed from ash-free dry mass after leaching in all studies.

These studies indicate that rates of leaf decomposition in Kings Creek are dependent on leaf species, flow regime, and abundance of shredder insects. *Ulmus* spp. and *C. occidentalis* exhibited the fastest rates of decomposition, generally disappearing within a year, whereas *Q. macrocarpa*, *Q. muhlenbergii*, and sycamore (*Platanus occidentalis*) were "slow" species (Peterson and Cummins 1974), requiring 2 to 3 years to decompose. Rates of decomposition of *C. occidentalis* leaves were similar in gallery forest and prairie/shrub reaches when perennial flow was present. During intermittent flows, decomposition rates of *Ulmus* spp. leaves were significantly slower than in perennial reaches because of the effects of drying and lessened degree of physical breakdown by current. Shredder insects were rare in the leaf packs of Smith (1986) and Tate and Gurtz (1986) but common in leaf packs during the studies by Hooker and Marzolf

Table 11.1. Decomposition rates of tree leaves in Kings Creek.

Reference	Habitat	Ulmus spp.	Celtis occidentalis	Quercus macrocarpa	Quercus muhlenbergii	Platanus occidentalis
Smith (1986)	Gallery forest pool	0.0061 (378)	0.0066 (349)	0.0034 (677)	0.0035 (658)	0.0018 (1,279)
	Gallery forest riffle		0.0182 (127)		0.0048 (480)	
Killingbeck et al. (1982)	Gallery forest riffle		0.0072 (320)	0.0023 (1,001)	0.0016 (1,439)	
Hooker and Marzolf (1987)	Gallery forest riffle	0.0199 (116)	0.0252 (91)			
	Prairie/shrub riffle	0.0211 (109)	0.0250 (92)			
Tate and Gurtz (1986)	Perennial prairie/shrub	0.0101 (228)				
	Intermittent prairie/shrub	0.0076 (303)				
Gurtz and Tate (1988)	Gallery forest riffle		0.0177 (130)	0.0020 (1,151)		

Note: Rate coefficients, k (per day), calculated from the negative slopes of the regression of the natural logarithm of percent original ash-free dry weight remaining versus number of days leaves were in the stream. Values in parentheses represent the number of days for 90% of the original leaf material to be lost.

(1987) and Gurtz and Tate (1988). Hooker and Marzolf (1987) found that weekly decomposition rates increased after colonization by *Tipula*.

The importance of leaf inputs to energy flow in Kings Creek depends on flow conditions. Drought and flooding effectively remove leaf material from processing by aquatic organisms by drying and exporting it before microbial or invertebrate colonization can occur (Tate and Gurtz 1986; Gurtz and Tate 1988). "Slow" leaf species constitute one-half of the inputs into gallery forest reaches of Kings Creek (Gurtz et al. 1988). Given the flows experienced in Kings Creek during the period from 1979 to 1992, no time periods without channels going dry or being scoured by floods have been long enough for these leaves to completely decompose. *Ulmus* spp. and *C. occidentalis* leaves could have completely decomposed in gallery forest reaches in 8 of the 13 years of discharge records at the USGS station. Periods of drought also affect shredder abundance. Shredders are common in Kings Creek only after at least a year of continuous flow (Gray unpublished data).

Microbial communities depending upon C input from decomposing leaves also reflect the landscape position (McArthur et al. 1985). Bacteria isolated from stream sediments in grassland reaches grew only on grass leachates. Bacteria from downstream gallery forest reaches were able to grow on leaf litter and grass leachates. Presumably, these observations apply to microbes on the actual leaves as well.

The decomposition of *Quercus* and *Ulmus* leaves in groundwater leaving springs on watershed N04D of Kings Creek was studied by Eichem et al. (1993). *Ulmus* leaves decomposed faster than *Quercus* leaves, similar to findings of in-stream decomposition.

Hyporheic and Groundwater Biota

The biota and food webs within groundwater habitats of the Kings Creek basin depend on the structure of the habitat and, as for most groundwater habitats (Ghiorse and Wilson 1988), inputs of organic C from the surface. In limestone aquifers, large cracks and channels within the rock allow for movement of invertebrates and organic detritus. Organic C in the limestone aquifers ranges from 1 to 6 mg/L in the dissolved form and from 0.6 to 62 μg/L in the particulate form (Edler and Dodds 1996; Eichem et al. 1993) and is adequate to support a significant invertebrate community, including subterranean isopods, amphipods, and planarians. The isopod *Caecidotea tridentata* (Hungerford) and the amphipod *Bactrurus hubrichti* Shoemaker are nonpigmented, subterranean crustaceans found in these habitats. *C. tridentata* is locally abundant; more than 21,000 individuals were collected in the discharge from one artesian spring during one 17-month period (Edler and Dodds 1996). These isopods exhibit a stable age distribution, with reproduction occurring throughout the year. Microbial production in these aquifers may be limited by N. This is suggested by significant positive correlations between numbers of bacteria and nitrate concentrations (Edler and Dodds 1996). The food webs of these aquifers have yet to be delineated fully.

The small pore spaces found in alluvial aquifers exclude movement of all but the smallest organisms and particles through the water table. Water movement through these pores may control microbial metabolism and N cycling (Dodds et al. 1996d). In alluvial aquifers underneath native prairie, bacteria may be limited by N and C. However, in aquifers under till agriculture, the amount of soluble organic C in the sediments decreases and the amount of dissolved inorganic N (specifically nitrate) increases, suggesting that microbial production in the groundwater is more likely to be limited by C under the agricultural fields and N in the prairie. Further study is necessary to confirm this trend. The largest potential consumers of bacteria in these sediments are protozoa. They occur throughout the hyporheic habitat and may have a potential impact on microbial biomass and activity (Strauss 1995).

Summary

A significant difficulty in studying the ecology of Kings Creek is the lack of a model for prairie streams in general. As Matthews (1988) commented, much of the research on prairie streams is still at the stage of initial ecological exploration, particularly when compared with studies of streams draining forested watersheds and streams in true desert. The standard River Continuum Concept (Vannote et al. 1980) does not necessarily apply to prairie streams. The studies that have been done on Kings Creek suggest that it shares a variety of characteristics with streams in other regions, depending on the stream reach examined and flow regimes.

In the gallery forest reaches of Kings Creek during periods of stable flows, the stream is similar to other streams bordered by deciduous forest (e.g., Cummins 1974) in that allochthonous inputs, especially leaf litter, predominate. Instream primary production is low because of shading by riparian trees and low nutrient levels that reflect a high degree of nutrient retention by the terrestrial ecosystem (Chapters 10, 13, this volume). Even with stable flows, however, gallery forest reaches differ from forested streams elsewhere in the paucity of retention structures, such as debris dams, and the dominance of fine particulates in transport (Gurtz et al. 1982). With the variable flows typical of Kings Creek, many processes, from leaf decomposition to insect life cycles, cannot go to completion.

The prairie reaches of Kings Creek, in general, resemble desert streams (Fisher 1986). The energy source for prairie reaches is likely to be dominated by autochthonous production because of the lack of shading and lower inputs of allochthonous organic matter. Macroinvertebrate communities are dominated by collector-gatherers and have few shredders. The biota is dominated by small, rapidly growing species that can recolonize quickly after disturbance. However, unlike desert streams, productivity is limited by low nutrient levels, and exchanges between surface flows and groundwater are more spatially and temporally variable. In addition, the terrestrial system has a much greater influence on inputs of nutrients and energy into the stream, as well as influencing flow regimes.

Future research on the biotic communities of the Kings Creek ecosystem will include continued characterization of the biological processes occurring in the saturated subterranean ecosystem, direct and indirect influences of fire and grazing on community composition and abundance, and trophic interactions occurring in surface streams.

Acknowledgments This chapter is dedicated to the memory of Chris Edler. We thank all of the students, technicians, investigators, and funding agencies who supported this research. This is contribution no. 97-184-B from the Kansas Agricultural Experiment Station, Kansas State University, Manhattan.

Part IV

Ecosystem and Landscape-Level Analysis

12

Patterns and Controls of Aboveground Net Primary Production in Tallgrass Prairie

Alan K. Knapp
John M. Briggs
John M. Blair
Clarence L. Turner

An analysis of the patterns and control of energy input into tallgrass prairie through primary producers can provide a basis for more comprehensive assessments of the causes and consequences of dynamics in populations, communities, and ecosystem-level components of this grassland. In this chapter, we focus on aboveground net primary production (ANPP). Belowground productivity is less well studied, but what is known is covered in a later chapter (Chapter 14, this volume). Our objectives are to describe the spatial and temporal patterns of ANPP documented at Konza Prairie; to assess the effect that fire has on ANPP from the ecophysiological to the ecosystem level; to evaluate climatic controls of ANPP; and to present a conceptual model (based on a non-equilibrium paradigm; Chapter 1, this volume) that incorporates multiple limiting resources (light, water, and N) of productivity and can account for the dynamics of ANPP in tallgrass prairie. The interactive effects that large ungulate herbivory may have on ANPP will be discussed briefly at the conclusion of this chapter.

Generalizations about Patterns and Controls of ANPP Prior to LTER

Prior to the initiation of the Konza Prairie LTER Program, a number of generalizations were made regarding patterns and controls of ANPP in grasslands, some of which are supported by our analyses of Konza Prairie LTER data and others of which are not consistent with our results. Some of these generalizations are reviewed in the following pages.

Variability in ANPP and Precipitation

Risser et al. (1981) summarized their own studies and the research of others that focused on ANPP in tallgrass prairie. They estimated annual ANPP in tallgrass prairie to be about 400 g/m^2 across a wide range of sites in North America. Surprisingly, little geographic variability was noted by Risser et al., or in the earlier studies reviewed. However, year-to-year variability was described as very substantial, with maximum versus minimum levels of ANPP varying as much as fourfold (Harper 1957, cited in Risser et al., 1981). Even greater interannual variability was noted by Towne and Owensby (1984). Risser et al. (1981) also concluded that ANPP in tallgrass prairie was correlated strongly with simple meteorological variables, such as precipitation, potential evapotranspiration, solar radiation, and temperature. Indeed, coefficients of determination ranged from 0.88 to 0.96 when these variables were correlated with ANPP. Sala et al. (1988) also noted that across grassland sites from shortgrass to tallgrass prairies, precipitation was related strongly to ANPP ($r^2 = 0.90$). The relationship between ANPP and precipitation that they provided predicted that annual ANPP should average 510 g/m^2 at Konza Prairie.

Fire, Nitrogen, and Herbivores

Since the early studies of Weaver (1954), there has been much interest in the response of grassland ANPP to fire. After reviewing studies that spanned 50 years, Risser et al. (1981) concluded that, when water availability was adequate, fire increased ANPP in tallgrass prairie for 2 or even 3 subsequent years. However, they noted that annual burning did not increase ANPP, and in dry years ANPP was reduced by fire. Indeed, they reviewed several accounts of large reductions in ANPP (over 50%) after fire but also noted reports of postfire increases in ANPP up to fourfold relative to sites protected from fire. Clearly, generalizations were difficult to make, variability in the postfire response of ANPP was substantial, and the causes of this variability were unknown.

Risser et al. (1981) also noted that ANPP in tallgrass prairie generally is considered to be limited by both water and N (Owensby et al. 1969), with N limitations being more important in more mesic sites or years. Finally, grazing by ungulates was reported to increase ANPP in tallgrass prairie, as long as moisture conditions were favorable and grazing intensities were not severe (Risser et al. 1981).

In summary, studies completed prior to the Konza Prairie LTER Program led Risser et al. (1981) to conclude that ANPP in tallgrass prairie is best described as quite variable temporally but consistent across large geographic gradients. ANPP was correlated strongly with meteorological variables, especially precipitation, and was water- and N-limited. The strong interaction among C, water, and N flows through grasslands also was emphasized by McNaughton et al. (1982). Fire or moderate grazing may increase ANPP, but annual burning or fire in dry years may decrease ANPP.

Spatial Patterns in ANPP at Konza Prairie

Because of the importance of accurate estimates of ANPP and the labor-intensive nature of data collection, considerable effort has been expended evaluating sampling methods and intensities in tallgrass prairie as well as other grasslands (Knapp et al. 1985; Briggs and Knapp 1991; Biondini et al. 1991). These studies indicated that, at a given topographic position, harvesting 20 0.1-m^2 quadrats at the time of peak biomass (mid- to late August) and combining the current year's dead biomass (separated from previous years' dead biomass in sites not subjected to fire) with green biomass provides an acceptable estimate of annual ANPP with a standard error of less than 10% of the mean. Moreover, such estimates have sufficiently low variance to statistically detect differences between treatments when mean values differ by 20% or more (Briggs and Knapp 1991). Although herbivory by insects is unaccounted for in these estimates, field manipulations and insecticide treatment of belowground and foliage insects did not alter peak biomass levels in this system (Gibson et al. 1990; Chapter 8, this volume). Moreover, even though some decomposition of tissues senescing early in the growing season may occur by August, decomposition is slow when litter is lodged within the plant canopy (Seastedt 1988) and thus affects estimates of ANPP minimally.

Fire and Topographic Position

Given the experimental design at Konza Prairie (Chapter 1, this volume) with watersheds burned at various frequencies, with or without large ungulates, and with ANPP sampling sites located at a variety of topographic positions, a single average value of annual ANPP for Konza Prairie occludes significant spatial and temporal variability (Plates I and II). Nonetheless, for ungrazed sites (herbivory will be discussed later) irrespective of fire history or topographic position, ANPP at Konza Prairie has averaged 417.1 ± 18.7 g/m^2 during the period 1975–1996, with maximum ANPP measured in 1993 (589.9 ± 24.6 g/m^2) and minimum ANPP measured in 1980 (231.2 ± 18.6 g/m^2). Forbs (largely C_3 perennial herbs) constituted about 20% of total ANPP. This long-term average at Konza Prairie is similar to the value reported by Risser et al. (1981) and about 20% lower than that predicted by the Sala et al. (1988) relationship between ANPP and precipitation for Great Plains grasslands.

More informative are long-term average values for ANPP from uplands (with shallow Florence soils) and lowlands (deep Tully soils; Chapter 4, this volume) in annually burned watersheds, watersheds burned only 1 in 4 years, and long-term "unburned" watersheds (defined as sites in which no more than two or three fires have occurred in 20 years, "low fire frequency" in Fig. 12.1A). Clearly, topographic position and fire frequency have important and interactive effects on the productivity of the system, with the highest levels of annual ANPP in annually burned lowlands (1975–1996 average = 527.5 ± 26.9 g/m^2). This is primarily due to the increased productivity of the dominant C_4 grasses in annually burned sites (Towne and Owensby 1984). In contrast, forb production was greater in unburned watersheds (Fig. 12.1B).

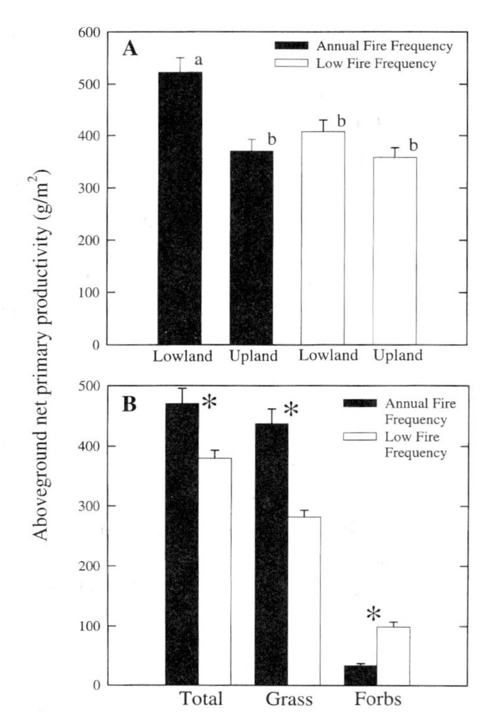

Figure 12.1. Effect of fire (annual spring fire versus low fire frequency = two to three fires in 20 years) and topographic position on annual aboveground net primary production in ungrazed watersheds at Konza Prairie. Values are means, with vertical bars denoting the standard error or the mean. Different letters or * indicate significant differences at $P < 0.05$.

Upland and lowland sites span most of the range of topographic and resource variability on Konza, but much of the landscape lies at intermediate elevations. ANPP at these intermediate locations is generally more similar to that of lowland than upland sites. For example, we have sampled ANPP at intermediate hillside locations for 4 years, and the means of these values were not significantly different from lowland values for the same time period. More intense sampling at 11 locations along transects that spanned upland-lowland-upland sites in both annually burned and unburned watersheds also showed that ANPP at intermediate topographic positions was more similar to that of lowlands than uplands (Knapp et al. 1993a). Moreover, in all data sets comparing ANPP at various topographic positions, greater watershed-level spatial variability in ANPP has been found in annually burned watersheds than in those subjected to a low fire frequency (Fig. 12.2).

In contrast to conclusions based on topographic gradients, at a given topographic position within a watershed, small-scale (patch level = 50-m transects) spatial variability in ANPP was lowest under an annual fire regime and greater if fire was infrequent or excluded (Fig. 12.3C). This pattern is expressed more strongly if just variability in grass biomass is considered (Fig. 12.3A). Small-scale spatial variability in forb biomass was three to six times greater than in the grasses, reflecting the patchy distribution of these subdominant species (Fig. 12.3B; Chapter 6, this volume). These data are consistent with visual impressions of annually burned watersheds in which a uniform "sea of grass" is encountered.

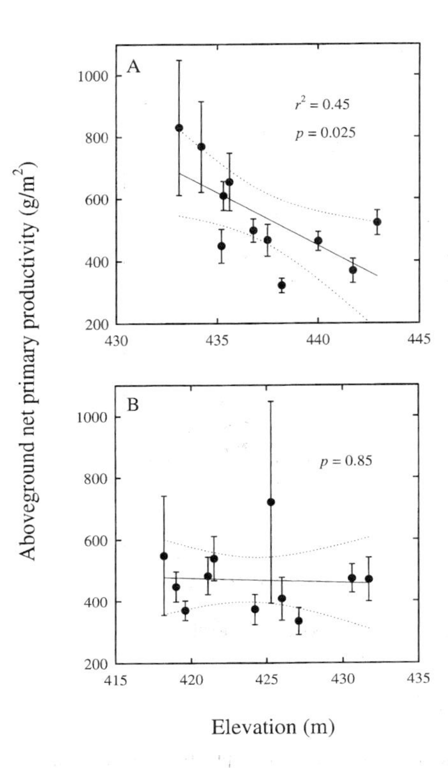

Figure 12.2. The relationship between annual ANPP at Konza Prairie and topographic position. In watersheds subjected to annual fire (A), substantial variability occurs in ANPP from lowlands to uplands; thus, ANPP is related significantly to elevation. Such a relationship does not exist in watersheds with a low fire frequency (B). Dashed line indicates 95% confidence interval.

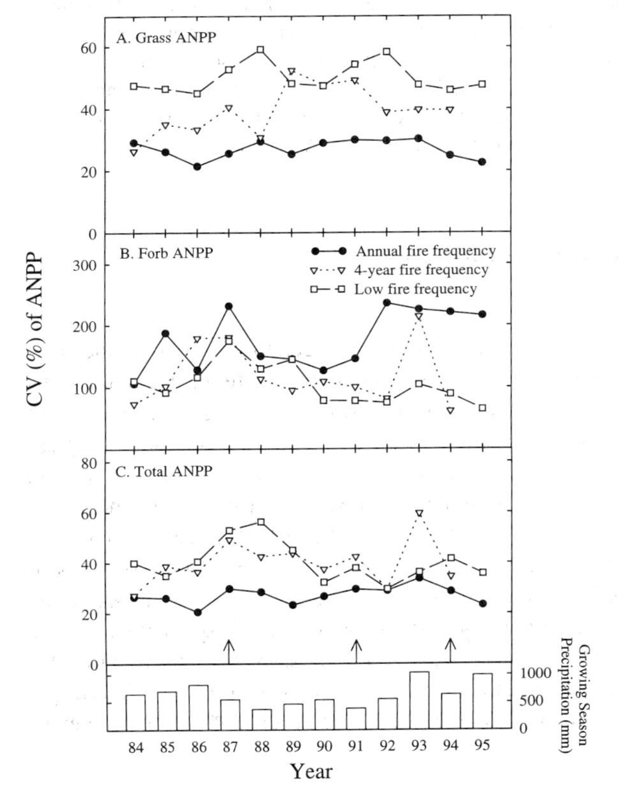

Figure 12.3. Comparison of coefficients of variation (CV) for ANPP (total, grass, and forb components) based on 20 to 40 plots harvested on Konza Prairie each year over a 12-year period. Sites were in watersheds burned annually, at 4-year intervals (years with fire indicated by arrows), or infrequently (two to three fires in 20 years). Thus, the CV for any year indicates how spatially variable ANPP is under each of the fire regimes. Also shown is growing season (April–September) precipitation.

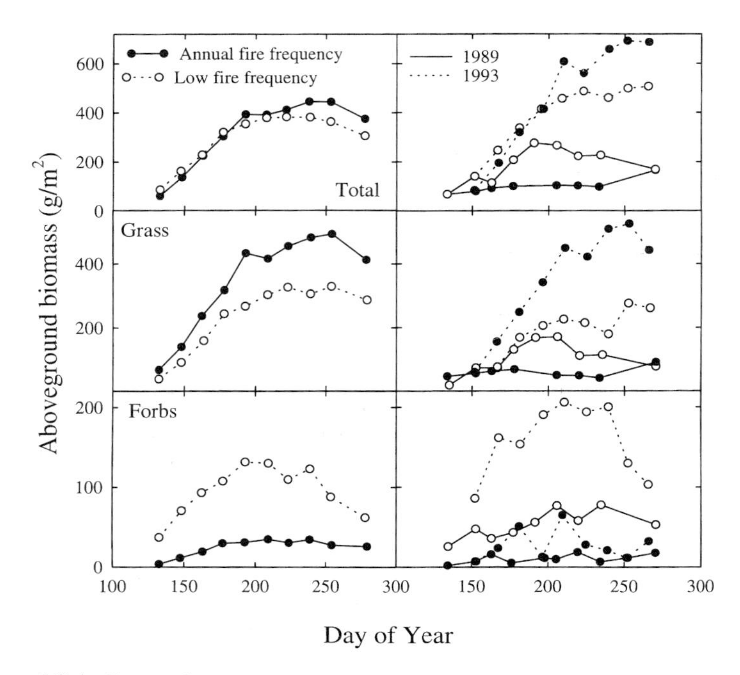

Figure 12.4. Seasonal course of aboveground biomass in two adjacent sites (annually burned and a site with a low fire frequency) on Konza Prairie. Panels on the left show mean values for total, grass, and forb biomass based on 12 years of sampling at approximately biweekly intervals. Panels on the right show the seasonal course of biomass accumulation in the years with maximum ANPP (1993, the wettest year) and minimum ANPP (1989, second year of a 2-year moderate drought). Note that in the wettest year, fire enhanced ANPP, but in the dry year ANPP was greater in unburned prairie.

Temporal Patterns in ANPP at Konza Prairie

Intra-annual

At a site on Konza Prairie with soils intermediate between the deep Tully and shallow Florence series, ANPP has been measured at approximately 2-week intervals throughout the growing season. This seasonal record encompasses 12 years, with values for adjacent annually burned and unburned (low fire frequency) tallgrass prairie. Sites were separated by a 10-m mowed fireguard, and the unburned site had been protected from fire for 5 years prior to this study. Comparison of the mean seasonal course of biomass accumulation at these sites shows that greater than 80% of annual ANPP in any year occurred during the initial 2 months of the growing season (Fig. 12.4). The rate of biomass accumulation (3.9 $g/m^2/d$) during this period did not differ significantly between annually burned and low fire frequency sites. Peak biomass (standing crop) occurred by 15 August at both sites. In contrast with sites with deep soils, annual fire had no effect on ANPP at this site over the 12-year period. However, in a year with abnormally high precipitation (1993, 110 cm versus

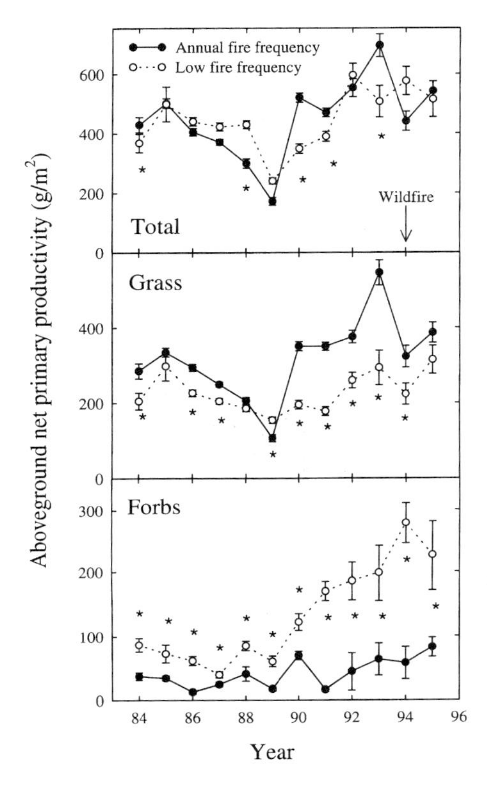

Figure 12.5. Twelve-year record of ANPP in adjacent sites exposed to an annual fire regime or a low fire frequency on Konza Prairie. Vertical bars indicate ± 1 SE of the mean, and means in years with * are significantly different at $P < 0.05$.

the 82-cm 30-year mean), ANPP was significantly higher in the burned site (Fig. 12.4). Conversely, in the year with low precipitation and ANPP (1989, 35 cm of precipitation), ANPP was significantly higher in the unburned site (Fig. 12.4). With each year considered independently, ANPP was significantly increased by annual fire in 4 of 12 years and was significantly reduced in 1 of 12 years (Fig. 12.5). In contrast, ANPP of the grass component was enhanced significantly by annual burning in 8 of 12 years and reduced significantly only once (1989; Fig. 12.5). Forb ANPP was always significantly higher in the unburned site (Fig. 12.5).

Interannual

A longer-term record of annual ANPP (22 years) exists for upland and lowland sites in annually burned and unburned watersheds on Konza Prairie (Fig. 12.6). In 16 of 20 years (all sites were burned in 2 of those years), ANPP was significantly higher in annually burned lowlands compared with unburned lowlands. Conversely, only in a single year was ANPP higher in the unburned than in the annually burned lowlands (Fig. 12.6). Annual ANPP in uplands seldom was affected by fire. Perhaps what is most significant about this record is the amount of temporal variability in ANPP in this tallgrass prairie. The ranges of ANPP in burned sites were 279 to 785 g/m^2 in lowlands (mean = 527.5 ± 26.9) and 178 to

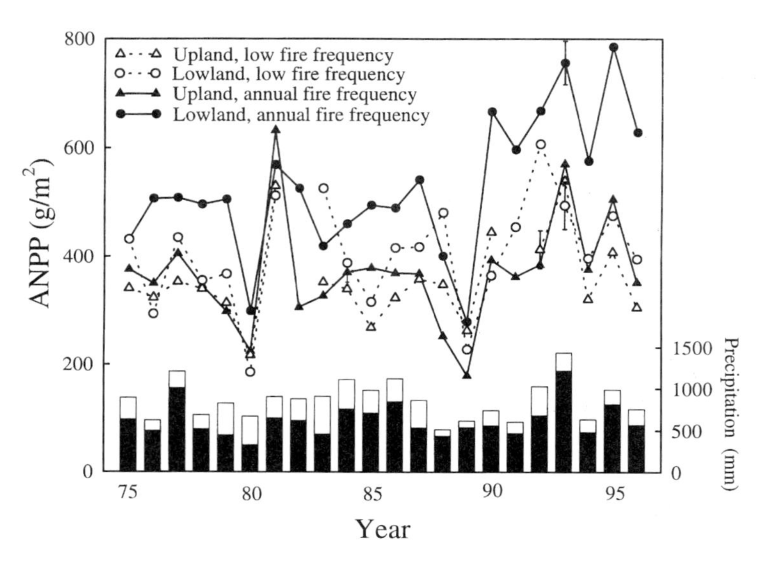

Figure 12.6. Twenty-two year record of annual ANPP in upland and lowland sites in watersheds subjected to an annual or low frequency fire on Konza Prairie. Also shown is annual precipitation and growing season precipitation (April–September) for each year. Vertical bars indicate ± the maximum SE of the mean for each fire treatment and topographic position. Years without data for unburned sites were the result of wildfires. Updated from Briggs and Knapp (1995).

570 g/m^2 in uplands (mean = 369.1 ± 21.5). In sites with a low fire frequency, ANPP ranged from 185 to 606 g/m^2 in lowlands (mean = 406.7 ± 21.6) and from 217 to 540 g/m^2 in uplands (mean = 355.2 ± 17.9). In a watershed burned every 4 years, a wide range in ANPP also occurred, with annual biomass production varying from 187 to 618 g/m^2 in lowlands versus 143 to 501 g/m^2 in uplands. Coefficients of variation of annual ANPP calculated for annually burned, unburned, and 4-year fire frequency sites confirmed that the highest degree of temporal variability occurred in annual and 4-year fire sites (25% and 29%, respectively); those sites protected from fire had more constant levels of productivity (CV of 10–12%).

Thus at Konza Prairie, the effect of fire is dependent on topographic position, with spring fires usually increasing annual ANPP in lowlands (by about 28%) but not in uplands. In contrast to the conclusions of Risser et al. (1981), annual fire increased ANPP in all but the driest years, and sites subjected to fire at 4-year intervals showed no evidence that an individual fire increased ANPP for more than the immediate growing season after the fire. Moreover, variability in ANPP was greater both spatially and temporally in burned watersheds than in those subjected to a low fire frequency, but at a given topographic position, patch-scale (10–50 m) spatial variability was greater in unburned sites. Discussion of the factors responsible for these dynamics and a mechanistic explanation for postfire responses in ANPP follow.

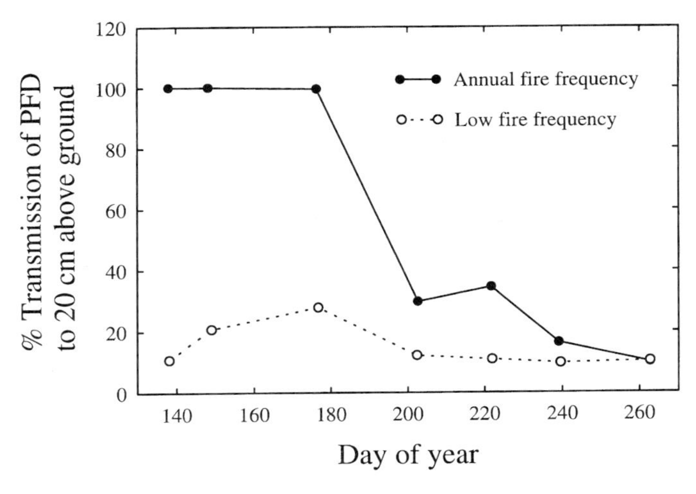

Figure 12.7. Temporal patterns of light (photon flux density, PFD) available to plants in tallgrass prairie as influenced by fire. In burned sites, the detritus layer is removed. In May, differences in light penetration between burned and unburned sites are greatest. For example, at the soil surface, virtually no light penetrates unburned canopies, whereas over 80% reaches the soil surface in burned sites. Throughout the growing season, as the canopy of burned sites closes, differences in canopy light penetration between burned and unburned areas diminish. Data from Knapp (1984b).

Causes of Postfire Responses in ANPP

Role of Detritus

The primary mechanism by which fire increases ANPP in tallgrass prairie is through the removal of the accumulation of detritus produced in previous years (Hulbert 1969; Rice and Parenti 1978; Towne and Owensby 1984; Knapp and Seastedt 1986). Standing dead biomass has been reported to accumulate to levels of up to 1,000 g/m^2 in tallgrass prairie (Weaver and Rowland 1952), and a steady state is achieved about 3 years after a fire. The specific effects of this blanket of dead biomass on ANPP are numerous and manifest at the organismic through the ecosystem levels (Knapp and Seastedt 1986).

In the more productive lowland sites on Konza Prairie, detritus may accumulate to over 30 cm deep, and this nonphotosynthetic biomass intercepts or reflects a substantial fraction of the energy potentially incident on the soil surface and on emerging shoots. This reduction in light available to shoots in sites without fire occurs for up to 2 months (Fig. 12.7). Knapp (1984b) estimated that 59% of the energy potentially incident on emerging shoots in unburned sites is lost because of detrital shading in the first 30 days of the growing season, and a 14% reduction in energy occurs over the entire growing season. Because soil moisture is usually high in the spring, loss of energy at this time is especially critical for primary production. Indeed, production in the first 30 days after fire was 55% greater in burned lowland sites compared with ANPP for the same period in unburned sites (Knapp 1984b). In concert with reductions in light available to plants, the early spring temperature environment is much different between burned and unburned sites. Soils are 2 to 10°C warmer in burned sites, depending on depth (Hulbert 1969, 1988; Rice and Parenti 1978; Old 1969), and leaf temperatures of *Andropogon gerardii* may be cooler initially in unburned sites. However, as leaves grow through the detrital layer, their temperatures may reach levels 5 to 7°C higher than those in burned sites (Knapp 1984b). This latter response is attributed to low air movement and a consequent lack of convective heat exchange between the detrital layer and the atmosphere above, and may even lead to lower water potentials in *A. gerardii* in unburned sites (Knapp 1984b).

Fire and Water Availability

The detritus layer in unburned sites generally can be viewed as an inhibitor of ANPP in tallgrass prairie. However, in years with low rainfall, the detritus layer reduces evapotranspiration, allowing soils to remain moist longer into the season, prolonging growth, and enhancing ANPP in unburned sites. For example, long-term records of soil moisture at Konza Prairie show that soils are almost always drier in burned sites than in sites subjected to a low fire frequency, and these differences are accented during dry years and during the driest portions of any particular year (Fig. 12.8). Similar results have been reported in other tallgrass prairie sites (McMurphy and Anderson 1965; Anderson 1965; Old 1969). Furthermore, in more xeric mixed grasslands to the west, fire almost always results in a reduction in ANPP; this has been attributed to greater water stress in grasses in burned sites because of the removal of the protective "mulch" (Redmann 1978; Engle and Bultsma 1984). This interaction between fire and moisture availability is also consistent with the lack of an increase in ANPP in burned upland sites (with shallow soils) on Konza Prairie (Figs. 12.1, 12.6). At the other extreme of the moisture gradient on Konza Prairie, fire results in much greater increases in annual ANPP in prairie wetlands dominated by *Spartina pectinata* than in lowland prairie (from about 600 g/m^2 in wetlands without fire to more than 1,500 g/m^2 in burned wetlands; Johnson and Knapp 1993). Clearly, the degree of seasonal, interannual, and spatial variability in water availability characteristic of this ecosystem interacts strongly with fire, strengthening the contention that water is a critical resource with both direct and indirect effects on ANPP in tallgrass prairie (Chapter 1, this volume). Analysis of precipitation as a climatic factor influencing ANPP is presented later.

Ecophysiological Characteristics of *Andropogon gerardii*

Before considering abiotic controls of ANPP from an ecosystem perspective, it is instructive to review the leaf-level responses in the dominant grass on Konza

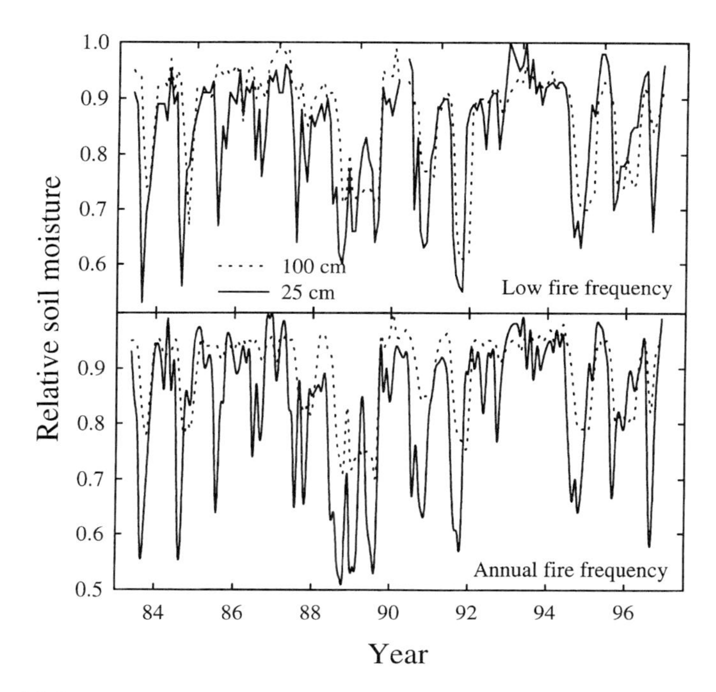

Figure 12.8. Thirteen-year record of soil moisture (expressed relative to the highest volumetric values measured at soil saturation in the spring) at 25- and 100-cm depths in lowland sites. Data are from watersheds subjected to an annual and a low fire frequency on Konza Prairie. In each year, minimum values always occurred in the burned watersheds. Note that during a moderate drought in 1988–1989, soil moisture recharge did not occur even in the winter/spring. Updated from Briggs and Knapp (1995).

Prairie, *A. gerardii* (Fig. 12.9), to key abiotic variables such as light, temperature, water, and N. This C_4 grass constitutes most of the biomass at Konza Prairie, and responses in this species are most likely to be consistent with responses in ANPP. Moreover, limited data on other warm-season grass codominants such as *Sorghastrum nutans* and *Panicum virgatum* indicate that they are similar to *A. gerardii* in many leaf-level ecophysiological characteristics (Knapp 1985b, 1993; Svejcar 1990).

Response of Andropogon gerardii *to Fire*

Because *A. gerardii* constitutes most of the plant biomass in many tallgrass prairies, whole-plant and leaf-level responses to fire and the altered energy environment described earlier are well documented (Peet et al. 1975; Knapp 1984b, 1985b; Knapp and Gilliam 1985; Svejcar 1990). These responses include increases in stomatal size, density, and conductance to water vapor diffusion; greater leaf thickness and specific leaf mass; higher chlorophyll a/b ratios; and,

Figure 12.9. The dominant grass on Konza Prairie, *Andropogon gerardii*, is a C_4, long-lived perennial. This species typically responds to spring fire with an increase in flowering culm height and density. Flowering culms can be over 2 m in height in lowlands. (A. Knapp)

perhaps most important, increased tiller density and higher early season photosynthetic rates in leaves emerging from burned versus unburned prairie (Table 12.1; Knapp 1985b; Knapp and Gilliam 1985). Many of these physiological and morphological differences between plants from burned and unburned sites are consistent with previously documented differences between sun and shade plants (Boardman 1977). These observations reflect the importance of the effect of postfire increase in irradiance near the soil surface on production of warm-season grasses. Leaf N concentration and nitrogen use efficiency (NUE) also may be greater in *A. gerardii* in burned sites (Knapp 1985b; Ojima et al. 1994), but this probably is not due to greater N limitations in unburned sites (Seastedt et al. 1991; Chapter 13, this volume). Instead, greater root uptake activity early in the season may lead to short-term increases in tissue N content in burned *A. gerardii* in the spring.

Table 12.1. Comparative responses of the dominant grass at Konza Prairie, *Andropogon gerardii,* to fire

	Burned	Unburned
Leaf thickness (mm)	249 ± 11	127 ± 3
Specific leaf mass(mg/cm^2)	6.09 ± 0.34	4.82 ± 0.11
Leaf width (mm)	9.46 ± 0.36	3.59 ± 0.21
Leaf N content (%)	2.50 ± 0.04	1.77 ± 0.01
Chlorophyll content (mg/g)	3.49 ± 0.34	5.25 ± 0.59
Stomatal density (#/mm^2, abaxial)	108.5 ± 3.8	89.6 ± 4.6
Stomatal pore length (mm, abaxial)	44.7 ± 0.8	34.8 ± 0.5
Stomatal conductance (mmol m^{-2} s^{-1})	309 ± 26	174 ± 9
Net photosynthetic rate (μmol m^{-2} s^{-1})	41.6 ± 3.8	28.1 ± 4.0
Shoot mass (g/shoot)	1.23 ± 0.22	0.38 ± 0.04
Tiller density (#/m^2)	264.3 ± 23.8	125.7 ± 12.7
Flowering stalk density (#/m^2)	66.4 ± 7.3	8.5 ± 1.6
Leaf area index (m^2 /m^2, estimated)	4.17	2.80

Sources: From Knapp (1984b, 1985b); Knapp and Gilliam (1985); and Knapp and Hulbert (1986).

Note: Values listed reflect maximum differences (± SE) between sites burned versus those protected from fire. These differences often occur in the spring in lowlands. Data for leaf area indices are from sites dominated by *A. gerardii* but may include other species. All values are significantly different between burned and unburned sites ($P<0.05$).

Photosynthesis and Light

Maximum photosynthetic rates in *A. gerardii* can be as high as 40 μmol m^{-2} s^{-1} and usually occur in burned sites in the spring (Knapp 1985b; Svejcar 1990). These rates of CO_2 uptake are similar to those in most other C_4 grasses. Maximum rates of photosynthesis in unburned sites are typically lower (about 22–28 μmol m^{-2} s^{-1}). Differences between burned and unburned sites are present primarily early in the spring, when the light and temperature environment differ maximally. Indeed, photosynthetic rates in *A. gerardii* are strongly dependent on the light environment, with net photosynthesis in shaded, lower canopy leaves being less than 30% of that in the upper canopy leaves (Schimel et al. 1991). Plant N is allocated such that N in the plant canopy is correlated strongly with photosynthetic capacity (Schimel et al. 1991; Turner et al. 1995).

The response of photosynthesis to incident light in *A. gerardii* is also typical of other warm-season grasses and species that grow in high light environments (Boardman 1977). In burned sites, photosynthesis does not appear to light saturate even at full sun levels, whereas in unburned sites, saturation may occur at about half full sunlight (Knapp 1985b; Schimel et al. 1991). However, even in high-light burned sites, photosynthesis is reduced in leaves low in the canopy. In unburned sites, this reduction in photosynthesis near the bottom of the canopy is more pronounced because of the steeper light gradient resulting from the accumulation of detritus (Schimel et al. 1991).

Photosynthesis and Temperature

Andropogon gerardii can maintain photosynthetic rates at 90% of maximum over a relatively wide range of leaf temperatures (28–41°C; Knapp 1985b). Moreover, photosynthesis in *A. gerardii* is much more tolerant of high temperatures than low temperatures. For example, at 48°C, photosynthesis was at 70% of maximum rates, but at 21°C photosynthesis was reduced to less than 30% of maximal rates (Knapp 1985a, b). This sensitivity to cool temperatures helps explain patterns of early season ANPP in topographically complex grasslands such as Konza Prairie. In lowland sites, where wind speeds are reduced and soils and plants are warmer in the spring than at upland sites, ANPP during the initial 30 days of the growing season may be increased almost 30% compared with upland sites (Knapp 1985a). This increase occurs before any appreciable differences in plant water stress are detectable between topographic locations.

Photosynthesis and Leaf Water Relations

At optimal levels of temperature and light, water stress often limits photosynthesis in most tallgrass prairie species (Hake et al. 1984; Martin et al. 1991). Photosynthesis in *A. gerardii* begins to decrease significantly between leaf xylem pressure potentials of –1.8 and –2.1 MPa (Fig. 12.10). However, positive CO_2 uptake may still be measurable below –3.5 MPa when leaves have folded because of loss of turgor (Knapp 1985b). Moreover, even after a drought period in which xylem pressure potentials were decreased to –6.6 MPa, photosynthesis recovered to 28 to 48% of predrought levels after water stress was relieved (Knapp 1985b).

Significant osmotic adjustment occurs in *A. gerardii* (Knapp 1984c, 1985b) in response to drought. For example, the osmotic potential at the point of turgor loss decreased by 1.33 MPa during a drought period in 1983. How this adjustment in tissue water relations might affect the relationship between leaf xylem pressure potential and photosynthesis is unknown, as is the impact of this adjustment on photosynthetic recovery after drought. However, similar levels of osmotic adjustment have been reported only for desert species or grasses from much more xeric regions (Knapp 1984c). In more xeric habitats, osmotic adjustment is considered a critical acclimatory response in plants that allows them to maintain photosynthetic and metabolic activity during periods of water stress (Smith and Nowak 1990). Rainfall at Konza Prairie is bimodal during the growing season (Chapter 2, this volume); when seasonal water stress does occur, it is usually in July through August, when temperatures are high (Borchert 1950). As noted previously, the tallgrass prairie environment is one in which substantial variability in the water status of plants occurs both within a season and between years. For example, in 1982, irrigation had no effect on plant water status of *A. gerardii* or ANPP of this grassland (Fig. 12.10). However, irrigation in the next year resulted in a significant increase in ANPP, and tallgrass prairie grasses without supplemental water experienced levels of water stress similar to those endured by desert plants (Knapp 1984c). Thus, the combination of high spring rainfall in this grassland, leading to relatively high leaf area, and the probability of

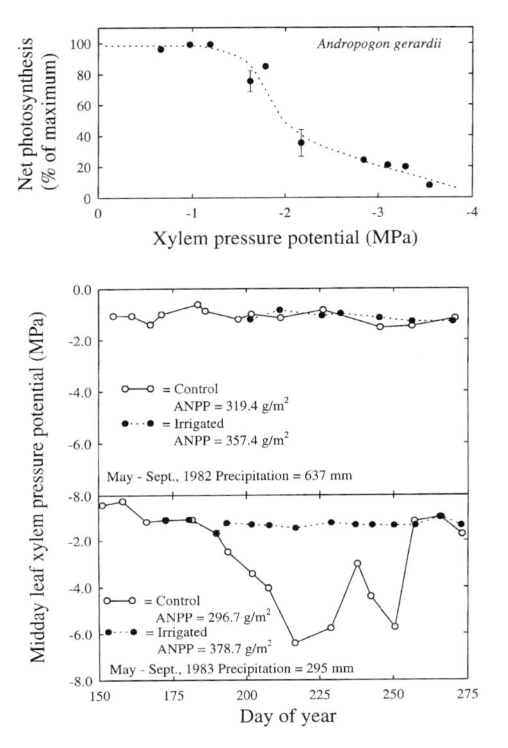

Figure 12.10. Top: Responses in net photosynthesis in the dominant tallgrass prairie C_4 grass, *Andropogon gerardii*, to water status (xylem pressure potential). Responses to water status are for mature leaves from tillers transplanted into 50-cm-deep cylinders and grown outdoors. Water was withheld from these plants to vary water status. Bottom: An example of the degree of interannual variability in water availability that can occur in tallgrass prairie. Data represent the seasonal course of midday leaf xylem pressure potential in *A. gerardii* in irrigated and control unburned plots during 1982, a year with abundant growing season precipitation, and 1983, a relatively dry year. Plots were protected from fire in both years. In 1982, irrigation with an additional 61 cm of water did not affect leaf water status, but the addition of a similar amount of water in 1983 had a dramatic impact on water status and ANPP. From Knapp et al. (1993a) and Knapp (1984c), with permission of the Ecological Society of America and Springer-Verlag.

severe water deficits occurring at midseason (caused in part by low rainfall and the high leaf area exhausting soil moisture reserves) may have selected for species capable of greater osmotic adjustment than would be predicted based upon the annual amount of precipitation that falls in this region.

Photosynthesis and Nitrogen

Photosynthesis and leaf N concentration are correlated positively in *A. gerardii* (Turner et al. 1995), as is typical of most plants. As noted earlier, fire may increase leaf N concentration for a short time (Knapp 1985b), but in general leaf N concentrations are similarly low (under 2.0%) in plants in burned and unburned sites, although NUE may be higher in burned sites (Knapp 1985b; Ojima et al. 1994). Although N requirements are low for *A. gerardii*, N availability in the soil is also low because of low rates of net mineralization and the substantial immobilization potential of *A. gerardii* litter (Knapp and Seastedt 1986; Wedin and Pastor 1993). The frequent short-term (several day to weeks) droughts characteristic of the growing season (Fahnestock and Knapp 1994) lead to significant drought-induced translocation of N from shoots to belowground organs (Hayes 1985; Heckathorn and DeLucia 1994) and a reduction in N volatilization (Heckathorn and DeLucia 1995). As a result, postdrought leaf N and photosynthetic rates are reduced even after leaf water stress is alleviated. Although such a strategy may constrain C gain for much of the season, it probably reflects selective pressures imposed by the low availability of soil N and the potential for N loss through herbivory and fire (Heckathorn and DeLucia 1994).

In summary, as the dominant species on Konza Prairie, *A. gerardii* has leaf-level ecophysiological characteristics that enable it to respond positively to a postfire environment characterized by high light levels, low N, warm temperatures, and variable moisture levels. In the following we assess how well these leaf-level responses in photosynthesis scale to ecosystem-level responses in ANPP.

Climatic Controls of ANPP in Tallgrass Prairie

Precipitation

Although climatic variables such as solar radiation and temperature have been correlated with ANPP in tallgrass prairie (Risser et al. 1981), variability in water relations measures (e.g., precipitation, evapotranspiration) is thought to be the primary determinant of both seasonal and interannual dynamics in ANPP in grasslands (Sala et al. 1988; Risser et al. 1981; Abrams et al. 1986). We have assessed the climatic controls of interannual variability in ANPP at Konza Prairie by correlating ANPP with all the standard meteorological variables (solar radiation, temperature, precipitation); potential evapotranspiration (estimated by pan water evaporation); and soil moisture measurements. Annual as well as growing season (May–September) and monthly subsets of these data also have been assessed. Analyses have been performed on all ANPP data collected from Konza Prairie since 1975 and on data subdivided by fire frequency, topographic position, and specific watersheds (Briggs and Knapp 1995).

From 1975 to 1995, annual ANPP at all annually burned and unburned sites combined was correlated significantly with annual precipitation (Fig. 12.11), but much less of the interannual variability in ANPP was accounted for in this data set ($r^2 = .19$) than was reported in other studies ($r^2 = .90$). The differences between the slopes in Figure 12.11 can be attributed to the inability of relationships based on regional gradients in moisture availability to scale to site relationships based on interannual variability in precipitation (Laurenroth and Sala 1992). At Konza Prairie, some of this unexplained variance is due to differences in burning regime and the variety of topographic positions of sampling sites. When the data

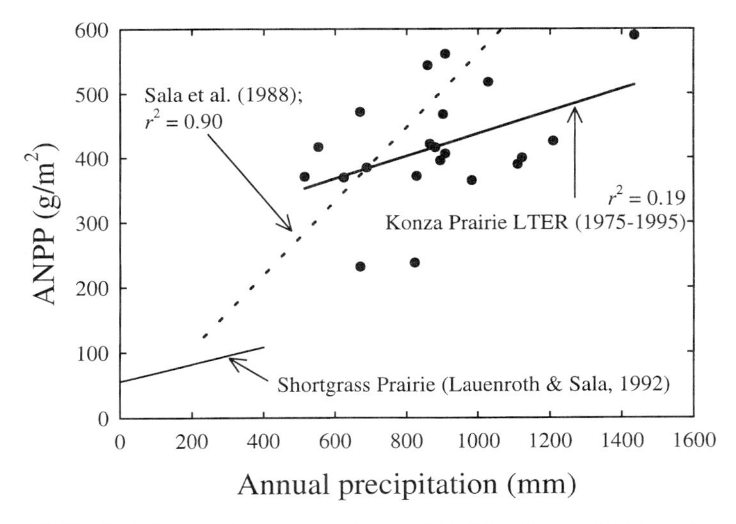

Figure 12.11. Relationship between total annual ANPP at Konza Prairie and annual precipitation. Data are from all sites sampled from 1975 to 1995 and include both burned and unburned sites at a variety of topographic positions. Also shown are the relationships developed by Sala et al. (1988) for the central Great Plains based on regional gradients in precipitation and by Lauenroth and Sala (1992) for shortgrass steppe.

were subdivided into annually burned, low fire frequency, upland and lowland sites, ANPP in all burned locations was correlated significantly with a variety of measures of water availability (Table 12.2). In contrast, ANPP in lowland sites with a low fire frequency was not correlated significantly with any precipitation or evapotranspiration variables. The amount of variance explained also was usually lower in upland, unburned sites than in annually burned sites (Table 12.2). No other meteorological variables (e.g., solar radiation, temperature) were correlated significantly with ANPP.

This analysis provides two insights into the level of control that water availability has on ANPP in tallgrass prairie. First, water is an important limiting re-

Table 12.2. Amount of variance (r^2) explained by meteorological variables as determined by linear regression of total ANPP from upland and lowland sites at Konza Prairie subjected to an annual fire frequency or low fire frequency, 1975–1995

Treatment	Total Precipitation (mm)	Growing Season Precipitation (mm)	Total Pan Water Evaporation (mm)	Summer Pan Water Evaporation (mm)
Annual fire lowlands	n.s.	0.21	0.33	0.28
Annual fire uplands	0.32	0.23	0.47	n.s.
Low fire frequency lowlands	n.s.	n.s.	n.s.	n.s.
Low fire frequency uplands	0.20	0.24	0.27	n.s.

Note: Growing season precipitation and pan water evaporation were calculated for 1 April to 30 September. Summer pan water evaporation was calculated for 1 July to 30 September. n.s. = not significant.

source primarily in burned sites and uplands. In tallgrass prairie subjected to a low fire frequency, other factor(s), such as the substantial reduction in light availability to shoots from detrital shading (Fig. 12.7), may constrain ANPP such that water will limit production only in extremely dry years (Knapp 1984c) or in dry locations (such as uplands). Second, the amount of the interannual variability in ANPP explained by precipitation or pan water evaporation is surprisingly low, even in burned uplands (Table 12.2).

Additional evidence that the detrital layer in unburned tallgrass prairie reduces the control that water availability has on ANPP comes from studies of the relationship between spatial variability in ANPP and soil moisture gradients along transects spanning upland-lowland-upland topographic positions (Knapp et al. 1993a). As noted earlier, in an annually burned watershed, ANPP varied substantially from uplands to lowlands, whereas in the unburned watershed, ANPP was relatively uniform from uplands to lowlands (Fig. 12.2). In contrast, soil water content was significantly more variable from upland to lowland sites in the unburned watershed. Although spatial variability in ANPP was related to patterns in soil moisture in both watersheds, the slope of this relationship was almost twice as great in the annually burned watershed compared with the unburned watershed. Why is the relationship weaker in the unburned watershed? Most likely it is because the relatively uniform detrital layer in unburned watersheds (Knapp et al. 1993a) reduces the role that gradients in soil moisture availability play in determining landscape patterns in ANPP. Indeed, the relationship between soil water and ANPP may saturate at lower levels of water availability in unburned sites.

Regarding the second insight above, that a relatively small portion of the interannual variability in ANPP at Konza Prairie can be explained by precipitation or pan water evaporation variables, it is interesting to note that when a 12-year subset of the 22-year data set was analyzed, coefficients of determination varied considerably from the longer-term record (Table 12.3). This observation illustrates the value of a long-term record. Because of the magnitude of ANPP responses to relatively infrequent climatic-fire interactions, increasing the sample size (from 12 to 21 years) in this system actually decreased the variance explained by precipitation in some cases. The 12-year analysis was performed separately because biomass harvested was not sorted by growth form (grass versus forbs) until 1984. Although total ANPP in burned uplands and grass ANPP in burned lowlands were correlated relatively strongly with growing season precipitation and summer pan water evaporation, respectively (Fig. 12.12), the general conclusion that water is a much more important control of ANPP in burned than in unburned sites is still supported by this 12-year analysis (Table 12.3). Typically, the grass component of ANPP was related more strongly to precipitation or pan water evaporation than was total ANPP, and forb ANPP was never related significantly to any variable (Table 12.3).

Soil Moisture

Soil moisture may be a better predictor of ANPP in grasslands because the soil provides a storage component that can reflect previous years' excesses or

Table 12.3. Amount of variance (r^2) explained by meteorological variables for total ANPP, grass ANPP, and forb ANPP from upland and lowland sites at Konza Prairie exposed to either an annual or low fire frequency, 1984–1995

Treatment	Total Precipitation (mm)	Growing Season Precipitation (mm)	Total Pan Water Evaporation (mm)	Summer Pan Water Evaporation (mm)
Annual fire lowlands				
Total ANPP	n.s.	0.31	0.34 (−)	0.50 (−)
Grass ANPP	0.14	0.35	0.41 (−)	0.56 (−)
Forb ANPP	n.s.	n.s.	n.s.	n.s.
Annual fire uplands				
Total ANPP	0.33	0.61	0.47 (−)	0.54 (−)
Grass ANPP	0.35	0.59	0.53 (−)	0.59 (−)
Forb ANPP	n.s.	n.s.	n.s.	n.s.
Low fire frequency lowlands				
Total ANPP	n.s.	n.s.	n.s.	n.s.
Grass ANPP	0.17	0.19	n.s.	n.s.
Forb ANPP	n.s.	n.s.	n.s.	n.s.
Low fire frequency uplands				
Total ANPP	n.s.	0.16	n.s.	n.s.
Grass ANPP	n.s.	n.s.	n.s.	n.s.
Forb ANPP	n.s.	n.s.	n.s.	n.s.

Note: Growing season precipitation and pan water evaporation were calculated for 1 April to 30 September. Summer pan water evaporation was calculated for 1 July to 30 September. n.s. = not significant, (–) = negative relationship.

deficits in water availability and use. Since 1984, soil moisture has been measured to 200-cm depths at 25-cm increments in annually burned and unburned lowlands at Konza Prairie. When these data were correlated with the corresponding lowland ANPP data, again no significant relationships were found for ANPP in unburned sites (Briggs and Knapp 1995). In burned lowlands, however, growing-season soil moisture at 50 cm was the best predictor of grass ANPP, with an r^2 of .77 (Fig. 12.12), but most other depths less than 100 cm also were correlated significantly with total and grass ANPP. Forb ANPP also was related significantly to soil moisture but only at depths lower in the soil profile (100–200 cm). Although roots of the dominant prairie grasses are known to extend to 2 m in sites with deep soils, many forbs position a greater proportion of their roots deeper in the soil profile than do the grasses (Weaver 1954). As expected, the amount of variability in forb ANPP explained was very low, but surprisingly, the relationship was negative (Fig. 12.12). This negative relationship may have been due to asymmetrical competitive interactions between the forbs and grasses in this grassland (Silvertown et al. 1994). Indeed, forb ANPP is negatively correlated with grass ANPP, and forb abundance is the lowest in the most productive annually burned sites on Konza Prairie (Chapter 6, this volume). Combinations of soil moisture and other meteorological variables in multiple regression analyses failed to improve any of the single-factor relationships.

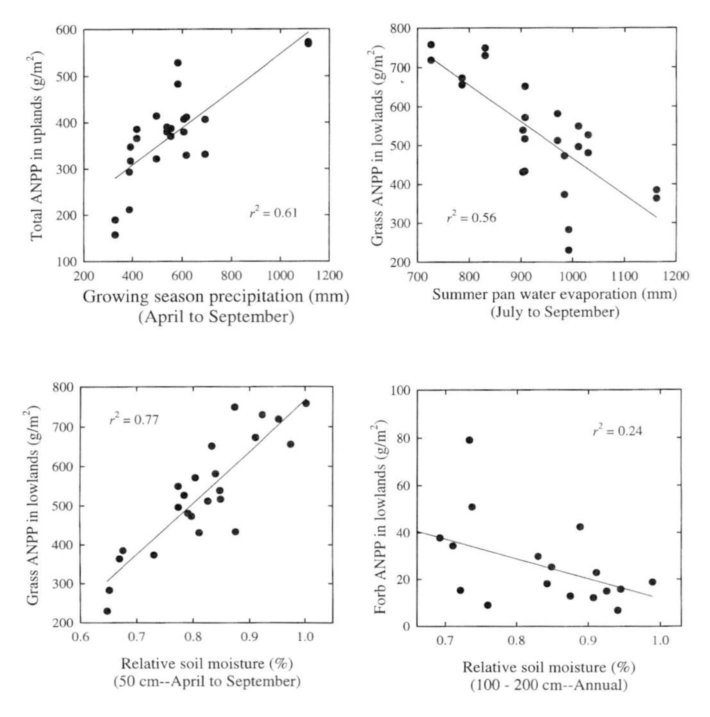

Figure 12.12. Relationships between ANPP (total, grass and forb components) and measures of water availability in burned sites on Konza Prairie. All are significant at the $P < .05$ level. Fewer significant relationships occurred in sites subjected to a low fire frequency (see Table 12.3). Updated from Briggs and Knapp (1995).

In summary, soil moisture may be the best predictor of ANPP in annually burned, lowland tallgrass prairie, but a substantial amount of variability still is left unexplained in other sites. Certainly, light limitations are important in unburned sites and perhaps late in the season in very productive burned sites as well. A third factor potentially limiting ANPP in tallgrass prairie that has merited much attention is nutrient availability.

Nutrient Controls of ANPP in Tallgrass Prairie

Tallgrass prairie soils are considered to be some of the most fertile in the world. The large-scale conversion of this grassland to agricultural production that initially required no input of fertilizer is evidence of this high fertility. As noted in Chapter 1, only the rocky soils and relatively steep topography of the Flint Hills spared this portion of the prairie from the plow. These fertile soils support a vegetative community that is dominated by C_4 grasses, species that have relatively low nutrient requirements (Wedin and Tilman 1990; Turner et al. 1995). Thus, it is somewhat surprising that several studies have demonstrated nutrient limitations to ANPP in tallgrass prairie (Owensby et al. 1970; Seastedt et al. 1991). This apparent paradox is resolved with the recognition that these soils are high in organic forms of essential nutrients, but available inorganic forms often are present only in very low concentrations in the soil solution (Seastedt and Hayes 1988; Hayes and Seastedt 1989; Chapter 13, this volume).

Nitrogen

Nitrogen is considered the nutrient that most limits ANPP in tallgrass prairie (Risser et al. 1981). Owensby et al. (1970) fertilized tallgrass prairie with N and P and recorded large production responses to N (46%) in years with adequate rainfall but no response to additions of P. This lack of apparent limitation by P may be a product of the highly dependent and efficient mycorrhizal association with the dominant grasses (Hetrick et al. 1988, 1989). Nitrogen fertilization studies conducted on Konza Prairie are consistent with these earlier studies. Nitrogen added as NH_4NO_3 at the rate of 10 g N/m^2 resulted in an average increase of 68% (range 46–91%) in ANPP in annually burned plots but only a 9% increase (range 2–16%) in unburned sites. Soil water collected at 20 cm in lysimeters from unfertilized burned and unburned sites indicated that greater organic N was present in unburned plots in the spring, but smaller differences were found in NO_3-N (lower in soil from burned sites; Seastedt and Ramundo 1990). Overall, soils from unburned sites have greater available NO_3-N and NH_4-N because of both greater mineralization (Chapter 13, this volume) and reduced plant uptake and demand in unburned sites (Knapp and Seastedt 1986; Seastedt and Ramundo 1990; Ojima et al. 1994). Thus, N limitations to ANPP are thought to be greater in annually burned sites compared with those not subjected to fire.

Nitrogen and Topography

Across topographic gradients in burned and unburned watersheds, N limitations are most pronounced in annually burned lowlands, locations where light and water usually do not limit ANPP early in the season. This conclusion is supported by analyses of topographic patterns of both extractable soil N (Schimel et al. 1991) and in situ N mineralization rates (Turner et al. 1997; Chapter 13, this volume). Because of the rapid early season growth in burned lowlands, N demand also is likely to be greatest at such sites. The effects of topography in unburned watersheds are less clear. Schimel et al. (1991) suggested that N limits ANPP the least in unburned lowlands, based on measurements of extractable soil N. However, Turner et al. (1977) found greater rates of net N mineralization at upland sites of both burned and unburned watersheds. Thus, N limitation may vary with topography most consistently in annually burned watersheds. This conclusion is similar to those made for water limitations.

Nitrogen Enrichment

As a potentially limiting resource, N is distinct from light or water in that, apparently, sites can become enriched in this resource. For example, vegetative uptake of N in unburned sites is not sufficient to exhaust the N mineralized; thus, soil N pools may increase (Ojima et al. 1994). This N is not necessarily leached from the system, however, and can be stored and accrue over several years. This storage phenomenon is best exemplified by fertilization studies conducted in burned sites with different fire histories. In annually burned sites, N fertilizer had a strong impact on ANPP, but in sites that had not been burned for several years, additional N did not enhance ANPP (Fig. 12.13), and sites with intermediate fire histories had intermediate responses to additional N (Seastedt et al. 1991). As a result of this interaction between time since fire and N availability, sites that have remained unburned for several years and then are burned display a "pulse" in ANPP that is greater than the responses to fire in annually burned sites, even though climatic conditions are the same in all sites (Fig. 12.13; Briggs et al. 1994).

A Nonequilibrium Model Explaining Variability in ANPP in Tallgrass Prairie

Recognition of the underlying differences in availability, storage, and demand of the three key resources (light, water, and N) limiting ANPP in tallgrass prairie, and the consequences of these differences is critical for understanding the high degree of variability in productivity of this system. Below we describe a conceptual model that incorporates the unique attributes of these resources and provides a mechanistic basis for the characteristic responses of ANPP to variations in resource availability and demand.

In considering the ungrazed watersheds for which long-term data are available, we have implied that annually burned and long-term unburned conditions are alternative states of the tallgrass prairie ecosystem. However, the unburned condition (long-term exclusion of fire) in tallgrass prairie is not a stable alternative state for comparison with annually or infrequently burned prairie. This is because long-term unburned prairie might be viewed best as a successional ecosystem. Long-term exclusion of fire leads to dominance by C_3 grasses and forbs and ultimately to dominance by woody plants (Bragg and Hulbert 1976; Collins and Adams 1983; Chapter 6, this volume). Thus, when patterns of ANPP are assessed in long-term unburned prairie, it is not surprising that few meteorological variables are correlated strongly with ANPP. Similarly, long-term changes in ANPP in successional systems are driven more by biotic change than by the vagaries of climate (Olff and Bakker 1991). As a result, overall ANPP is lowered, but stability in ANPP from year to year (defined as low variance) is increased in long-term unburned sites.

In sites that experience periodic fire, the tallgrass prairie is a relatively stable ecosystem, at least in terms of continued dominance by C_4 grasses (Chapter 6,

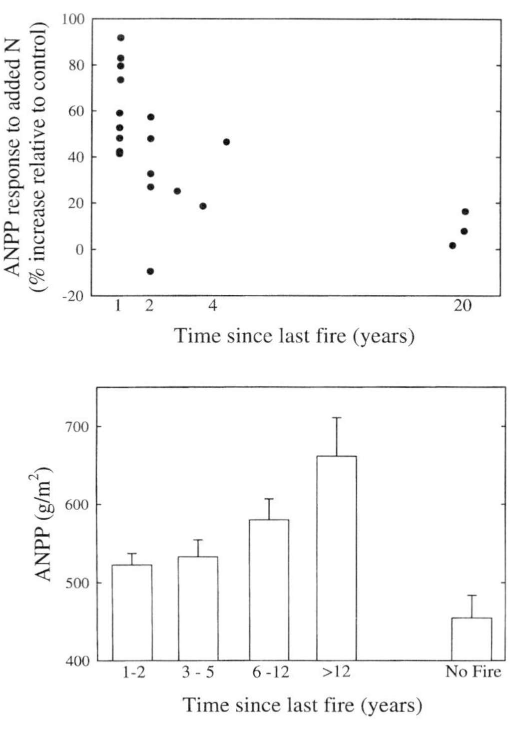

Figure 12.13. Top: Relative response in ANPP to N additions (10 g/m^2) as a function of the time since the site had been burned. Data are from a variety of sites from experiments conducted over several years on Konza Prairie. From Seastedt et al. (1991), with permission of Springer-Verlag. Bottom: Relationship between ANPP and the time since fire had occurred for sites in a number of watersheds on Konza Prairie. All ANPP data were collected in the same year (1991), and all but the "no fire" site were burned in that year (Briggs et al. 1994). Vertical bars depict the SE of the mean for each time interval.

this volume). Yet these sites have the greatest interannual and topographic variability in ANPP. These sites also have the greatest potential for multiple resource limitations to be manifest, with both water and N limiting ANPP in years with fire and light limiting ANPP in years without fire. Indeed, when ANPP data from watersheds with a variety of fire frequencies are combined, variability is very high and single factors such as precipitation are of little predictive value (Fig. 12.11). This variability is a product of the nonequilibrium nature of the system (caused by climatic variability, fire, and grazing; Chapter 1, this volume) and the potential for resources to alternate between limiting and nonlimiting status.

Light, Water, and Nitrogen

Our conceptual model of the tallgrass prairie incorporates the three resources that potentially limit ANPP: light, water, and N (Chapter 1, this volume). Key attributes of these resources differ fundamentally under nonequilibrium conditions. Radiant energy is viewed best as a discontinuous resource in tallgrass prairie (because of the presence versus absence of detritus). In burned sites, light limitations are relaxed when fire removes the detrital layer, and light does not strongly limit ANPP until perhaps late in the season, when leaf area index reaches a maximum. After only one growing season without fire, however, enough litter has accrued so that light limitation to ANPP occurs in the next year.

In contrast to light, water can be considered a limiting resource with more continuous levels of variation and can be stored by the system. Soil water storage also can exist in a deficit state, such that soil recharge must occur before plant use, or water can accumulate and carry over from an earlier wet year. This storage capability may help explain why interannual variability in ANPP is greater than variability in rainfall in tallgrass prairie and in many other grasslands (Le Houerou et al. 1988). Ultimately, however, water cannot accumulate or be stored beyond soil saturation in any one year. Additional inputs beyond soil saturation are lost to groundwater or runoff.

Finally, N also is continuous in its ability to limit ANPP and can be stored by the system (Seastedt and Ramundo 1990; Ojima et al. 1994). A characteristic alluded to earlier that sets N apart from soil water is that N can accumulate and be stored over several years. Thus, soil N enrichment can occur. This enrichment is beyond the level that is possible for the system to accrue in a single year. Thus, these three critical resources differ both in the nature of limitation they impose (discontinuous versus continuous) on ANPP and in the amount and time course of storage by the system (none to several years).

The key to understanding the high degree of variability in ANPP in tallgrass prairie lies in the recognition that multiple limiting (and interacting) resources potentially can affect ANPP; that important differences in the properties of these resources occur under nonequilibrium conditions; and that fire, grazing, and climatic variability, which are all requisite for the maintenance of this ecosystem (McNaughton et al. 1982; Axelrod 1985), strongly influence the availability of, and demand for, each of these resources.

Nonequilibrium Characteristics

It has long been known from studies of successional systems that those ecosystems not in equilibrium will exhibit unique states and behaviors relative to systems in equilibrium (Odum 1969). Two examples will illustrate this nonequilibrium perspective in tallgrass prairie. Previously, we noted the characteristic "pulse" in ANPP that occurs when a site that has been protected from fire for several years is burned (Fig. 12.13). ANPP can be as much as 30% higher in these sites relative to annually burned sites, but such high ANPP cannot be sustained beyond one year. Subsequently, ANPP decreases to annually burned levels if the site is burned the next year and thereafter. Clearly, climatic factors do not explain this pattern in ANPP, so what is the mechanism for this "transient maximum" (Seastedt and Knapp 1993) in ANPP? It is the result of two limiting

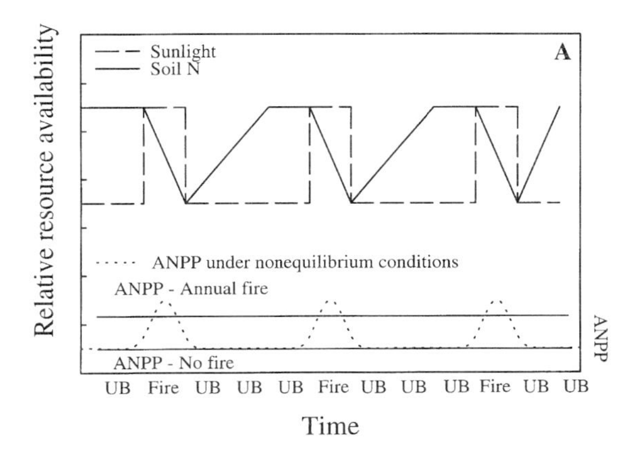

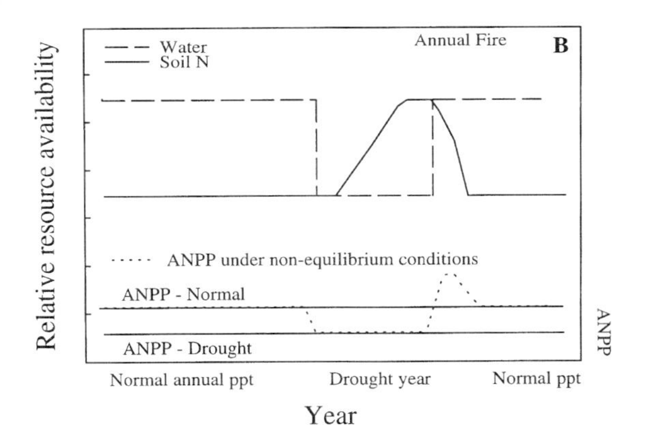

Figure 12.14. Idealized view of how multiple limiting resources (light, N, or water) may interact to produce transient maximum responses in net primary production in tallgrass prairie. (A) Under an annual fire regime, soil N limits production in all but drought years, whereas light availability limits production with fire excluded. Under nonequilibrium conditions of intermittent fire, light and N availability vary at different temporal scales. N accrues slowly in the soil when fire is excluded, whereas light limitations are alleviated rapidly by fire but become important in years when fire is absent. Thus, in years with fire that follow a period of fire exclusion, both light and N are available at nonlimiting levels, and productions responds maximally. (B) A similar scenario occurs with water, N availability, and drought years. When water limits production in drought years, soil N increases. When normal precipitation returns, both water and N availability are high and production is maximal. In both cases, the maximum ecosystem response occurs not at equilibrium conditions of annual fire, no fire, or "normal" precipitation, but under nonequilibrium conditions when the availability of resources is altered by variability in fire or climate.

resources, N and light, being available at nonlimiting levels simultaneously (Fig. 12.14). Recall that under equilibrium conditions either resource can limit ANPP (N in annually burned sites, light in unburned sites). But under nonequilibrium conditions, in which fire occurs in some years and not others, soil N accumulates in years without fire because light is the primary limiting resource (Ojima et al. 1994). Subsequently, when light limitation is removed by a spring fire, N and light are both available in nonlimiting amounts and a transient maxima in ANPP results (Seastedt and Knapp 1993).

A similar, nonequilibrium phenomenon may occur after drought years in tallgrass prairie (Fig. 12.14). Inspection of the 21-year record of ANPP indicates that 2 of the 3 years with highest ANPP in annually burned lowlands were not in years with high precipitation but in those that followed droughts (see 1982 and 1990 in Fig. 12.6). Drought also may lead to soil N enrichment, even in annually burned sites, because water is the primary factor limiting ANPP during these periods. However, after adequate rainfall returns, ANPP increases in the following year beyond levels expected based on precipitation or soil moisture levels (Fig. 12.14).

The postdrought response of tallgrass prairie is visually most evident by the tremendous increase in *A. gerardii* and *S. nutans* flowering culms (Knapp and Hulbert 1986). Two significant "pulses" in flower culm production have been noted on Konza Prairie since 1980 (Chapter 6, this volume), and both followed drought periods (1980 and 1988–1989).

In each of these examples, consideration of single variables, such as precipitation or evapotranspiration, that relate to ANPP so well regionally and in other grasslands cannot account for the dynamics in ANPP in tallgrass prairie. Instead, the entire suite of potentially limiting resources and the history of the site need to be considered before the patterns and controls of ANPP can be understood. This array of factors, including soil moisture, growing season precipitation, and drought and fire history, implicitly includes light and N limitations. Although our research provides for a conceptual understanding of the dynamics of ANPP in sites burned at least infrequently, determining the strength and the consequences of interactions among these factors will require a longer-term database. Similarly, Hobbs and Mooney (1995) concluded that the complex climatic and disturbance regime characteristic of a California annual grassland made prediction and modeling difficult and that long-term studies were essential in grasslands.

ANPP and Herbivory

To this point, the focus on patterns and controls of ANPP has been in tallgrass prairie not subjected to herbivory by large ungulates. As noted in Chapter 1 (this volume), grazing activities by native large ungulates, such as *Bos bison*, were certainly an important ecological factor in the presettlement tallgrass prairie (Axelrod 1985). Today, domesticated cattle have replaced these native ungulates in numbers and perhaps impact on managed sites. Research at Konza Prairie has incorporated *B. bison* only since their reintroduction in 1987 and cattle since 1994. Nonetheless, we can draw on preliminary data from Konza Prairie and results from a large literature on the effects of ungulate grazing on grasslands to make some predictions about how patterns and controls of ANPP will be altered by the activities of ungulate herbivores.

Herbivores versus Fire

In many ways, grazing by large ungulates may mimic the effects of fire in tallgrass prairie. Grazers will remove or markedly reduce the plant canopy and the detritus layer, allowing light to penetrate to the soil surface more readily. This will warm the soil and position plants in a higher light environment than in ungrazed sites (Fahnestock and Knapp 1993). Unlike fire, which provides for more uniform canopy removal, grazing activities, at least by *B. bison*, are spatially variable or patchy (Vinton et al. 1993). Moreover, *B. bison* grazing can be selective, with grasses being consumed preferentially (Hartnett et al. 1997). As a result, growth and reproduction in some forbs may increase in grazed patches (Fahnestock and Knapp 1993, 1994).

Variability Induced by Herbivore Activities

Bos bison grazing activities potentially influence energy flow into the tallgrass prairie at a number of scales. As mentioned previously, *B. bison* grazing is patchy, selective, and influenced by fire, with burned areas serving as preferred sites for grazing activities (Vinton et al. 1993). The establishment of grazing patches with selectively grazed species represents one level of variability in resource availability and demand, and the deposition of resource-rich urine and dung patches is another (Steinauer and Collins 1995). Overlaid on this spatially and temporally variable pattern of resources, and consequently ANPP, is the issue of compensation and even overcompensation by the grazed plants, a biotically mediated control of ANPP (McNaughton 1983a). Wallace (1990) showed that *A. gerardii* may have higher net photosynthetic rates after grazing, and studies by Vinton and Hartnett (1992) have shown that this species, the dominant grass on Konza Prairie, could compensate aboveground for herbivory in sites that had not been grazed in the previous year. However, grazing for 2 years in succession resulted in a lack of compensation for biomass removed in the second year. The use of belowground reserves to compensate for aboveground tissue lost in year one, and a subsequent reduction of stored resources in year 2, was the mechanism proposed to explain this pattern.

At the ecosystem level, fenced exclosures on Konza Prairie have protected sites in grazed watersheds for several years. When adjacent grazed sites were provided short-term protection, ANPP in uplands was reduced in the first year after protection relative to long-term ungrazed sites (Fig. 12.15) but recovered in year 2. Thus, the potential productivity of some sites is reduced by grazing for at least 1 year, but this potential recovers rapidly if the site is not grazed again. At an even larger scale, Turner et al. (1993) measured ANPP in a number of sites with different grazing histories and found that compensatory regrowth does occur in sites with little history of grazing but not in sites heavily grazed in pre-

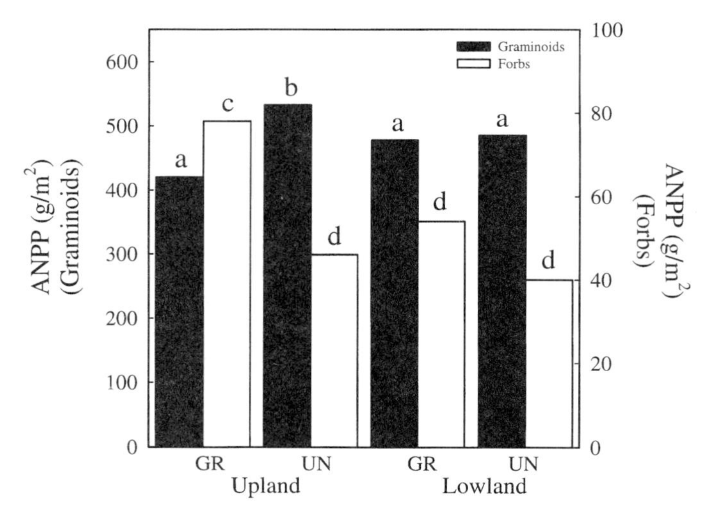

Figure 12.15. Responses in ANPP to grazing history at upland and lowland sites on Konza Prairie. "GR" refers to sites previously grazed by *B. bison* but protected from grazing during the year of measurement; "UN" refers to adjacent sites protected from grazing for more than 10 years. At upland sites, graminoid ANPP (and total ANPP, data not shown) was reduced significantly (noted by different letters) during the first year after grazers were excluded, but forb production was increased.

vious years. Again, the role of belowground storage reserves in the recovery of lost tissue appears to be critical. This "historical" component to understanding ANPP responses to herbivory is consistent with our earlier conclusions from ungrazed sites. Finally, shifts in species composition that result from long-term grazing activities (Hobbs and Mooney 1995; Chapters 6, 9, this volume) will further complicate predictions about *B. bison* grazing on ANPP in tallgrass prairie.

This brief overview of the impacts of large herbivores on ANPP serves primarily to strengthen our arguments for the necessity of viewing the tallgrass prairie as a nonequilibrium system. Incorporating grazing history and intensity, as well as the role of smaller herbivores (e.g., nematodes, aboveground invertebrates) in regulating ANPP in tallgrass prairie (Seastedt et al. 1988b; Todd et al. 1992; Chapters 8, 14, this volume), will increase the utility of our conceptual model of controls of productivity.

Summary

This review and synthesis of information on the dynamics and controls of aboveground ANPP in tallgrass prairie can be summarized as follows. Annual ANPP has averaged about 416 g/m^2 on Konza Prairie over the last 21 years. Spring fire increased ANPP at lowland sites in all but the driest years, and, contrary to earlier generalizations, annual burning also led to increases in ANPP.

However, the greatest response of ANPP to fire occurred in burned sites that previously were protected from fire for several years. Topographic position strongly influenced ANPP in burned watersheds, but in many unburned watersheds, upland and lowland ANPP was not significantly different, or differences were small compared with burned watersheds. In unburned watersheds, significant increases have occurred in the contribution that forbs make to ANPP, whereas in annually burned sites, grasses contribute over 80% of the ANPP.

Factors controlling ANPP varied depending on fire history and topography. In general, ANPP in long-term unburned watersheds (particularly in lowlands) was not correlated strongly with any meteorological variable. These systems are perhaps best viewed as successional, with ANPP controlled by biotic turnover as they shift from dominance by C_4 grasses to dominance by C_3 forbs and woody plants. In annually burned watersheds, water relations variables (precipitation, pan water evaporation, soil moisture) were correlated better with ANPP but not to the extent in grasslands farther west. In intermittently burned watersheds, which probably best reflect the unmanaged state of the ecosystem, variability in ANPP was much higher than could be accounted for by meteorological variables. Our view of the controls of ANPP in tallgrass prairie has progressed to the realization that multiple resources limit ANPP, with light being important in years when fire does not occur, and N and water being more important in sites that are burned frequently. Moreover, fire, climatic variability, and activities by herbivores all are expressed differentially across the landscape. These influence the pattern of temporal switching among limiting resources and result in nonequilibrium behaviors in which transient periods occur when none of these resources are limiting. This is when ANPP is maximized, and such periods are neither detectable nor predictable without a long-term record and a nonequilibrium perspective.

Acknowledgments This is contribution no. 97-187-B from the Kansas Agricultural Experiment Station, Kansas State University, Manhattan.

13

Terrestrial Nutrient Cycling in Tallgrass Prairie

John M. Blair
Timothy R. Seastedt
Charles W. Rice
Rosemary A. Ramundo

The cycling and availability of nutrients, especially N, affect both the structure and the dynamics of tallgrass prairie ecosystems and, in turn, are affected by changes in ecosystem processes such as production and decomposition. Nutrient availability can limit productivity (Owensby et al. 1970; Seastedt et al. 1991; Chapter 12, this volume) and alter species composition of tallgrass prairie plant communities (Tilman 1987; Gibson et al. 1993; Wedin and Tilman 1993, 1996; Chapters 6, 9, this volume), affect plant physiological responses to the environment (Turner and Knapp 1996); determine nutritional quality for herbivores (Allen et al. 1976; Chapter 14, this volume); and influence rates of litter decomposition (Pastor et al. 1987; Seastedt et al. 1992). In addition, nutrient cycling processes are altered directly by management practices (fire, grazing, soil disturbance) and site history, as well as external factors, such as climate and topography (Turner et al. 1997). These changes in intra-ecosystem nutrient transformations affect the retention of nutrients, which is reflected in ecosystem-level nutrient budgets and has long-term consequences for tallgrass prairie productivity and community composition. Thus, nutrient cycling studies can provide the basis for evaluating the influence of natural and anthropogenic factors on ecosystem function; prairie plant and animal communities; and abiotic ecosystem characteristics (e.g., soil and solution chemistry). Our intentions in this chapter are to summarize the focus and status of nutrient cycling research in tallgrass prairie prior to the initiation of the Konza Prairie LTER Program; to briefly review studies done on Konza Prairie during the last 15 years; and to present a synthesis of our current understanding of nutrient cycling processes, patterns, and controls in tallgrass prairie ecosystems.

An Overview of Nutrient Cycling in Tallgrass Prairie

Much of what was known about nutrient cycling in tallgrass prairie ecosystems prior to the initiation of the LTER Program in 1982 was summarized by Risser et al. (1981) and Risser and Parton (1982). These syntheses, which were based largely on research done at the Osage Prairie Site in northeastern Oklahoma, highlighted several key characteristics of tallgrass prairie ecosystems relevant to nutrient cycling. For example, the importance of soil pools as major reservoirs of C and N in these ecosystems was emphasized. Even at peak plant biomass, greater than 98% of total ecosystem N can be found in soil pools (Risser and Parton 1982). These tremendous stores of soil organic matter and nutrients, defining characteristics of tallgrass prairies (Seastedt 1995; Chapter 1, this volume), made them very well suited for agricultural use, which contributed to the demise of much of the original extent of this ecosystem (Samson and Knopf 1994, 1996). Results summarized in Risser et al. (1981) and other studies in tallgrass prairie (e.g., Kucera 1981) also demonstrated the importance of N as the nutrient most limiting to plant productivity and most likely to influence plant species composition and consumer responses. In undisturbed prairie, the majority of soil N exists in organic forms unavailable for plant uptake, and new plant production must be supported by a relatively small pool of available N with rapid turnover times. Prairie plants, with their extensive root systems, are well adapted to growing under conditions of limiting nutrients, and some of the N required for new plant production is recycled within the plant itself. Resorption of N from senescing plant shoots and translocation to belowground storage organs (rhizomes and roots) allow perennial prairie plants to store and reuse previously captured N in subsequent growing seasons (Heckathon and DeLucia 1996). This is an important adaptation in an ecosystem where N in aboveground tissues is subject to loss by fire or removal by grazers. However, storage and reuse of N appear to provide only about half of the N needed by plants during a growing season (Hayes 1985; Hayes and Seastedt 1987; Seastedt and Ramundo 1990), and plant uptake of N released from soil pools is essential for maintaining productivity. Mineralization is the biological process whereby organic N is converted to inorganic forms available for plant uptake. Although Jenny (1930) summarized some of the large-scale patterns and climatic controls of soil organic C and N accumulation in grasslands, few data on N mineralization rates in tallgrass prairie soils were available prior to the Konza Prairie LTER Program, and many of the specific factors controlling N accumulation and turnover in tallgrass prairie soils were not well understood. Likewise, little information existed on decomposition and the release of N from decomposing plant litter (but see Koelling and Kucera 1965). As a result, early attempts at simulation modeling of the tallgrass prairie N cycle (i.e., the ELM grassland model used in the International Biological Program) resulted in significant overestimation of N uptake into aboveground biomass and subsequent movement into detrital biomass and soil pools (Risser et al. 1981). Presumably, this was related to an underestimate of the importance of internal N fluxes, such as nutrient immobilization in soils and translocation of nutrients in plants, and a lack of sufficient data on decomposition/mineralization processes in these systems.

Significant gaps also existed in understanding the effects of fire on nutrient cycling processes, particularly with respect to N. In spite of its importance, N limitation is not a universal feature of tallgrass prairie, and its relative impact varies with fire, grazing, topography, soil texture and depth, and precipitation. Although fire was known to generally increase annual aboveground productivity in tallgrass prairie under certain conditions (Daubenmire 1968; Vogl 1974), the linkages between fire, nutrient availability, and productivity were not well developed. Hulbert (1969) concluded that the effect of fire was due primarily to removal of standing litter and subsequent changes in the soil/light environment, as opposed to short-term changes in nutrient availability. However, Sharrow and Wright (1977) noted that fire volatilized some elements, such as N and sulfur bound in dead biomass, so frequent fires could alter the balance of nutrient inputs and outputs, potentially leading to long-term changes in nutrient storage and availability. Early attempts to include fire in the ELM model (Parton and Risser 1979; Risser et al. 1981; Risser and Parton 1982) suggested that frequent fires (return intervals of 3 years or less) could stimulate net primary production (NPP) in the short term but also would decrease annual plant uptake of N and result in long-term losses of N (ranging from 0.6 g m^{-2} yr^{-1} for 3-year fire return intervals to 2.1 g m^{-2} yr^{-1} under annual burning). In the short term, these fire-caused N losses were thought to be offset by increased rates of soil N mineralization (Kucera and Ehrenreich 1962; Daubenmire 1968; Old 1969). In the long term, however, the accrual of these annual N losses was predicted to result in reduced storage of soil organic N and decreased productivity on frequently burned sites, although results of the few available long-term fire studies remained equivocal. Grazing by ungulates also was known to impact N cycling in a variety of ways, and there was speculation about the interactions of grazing and fire. Model predictions suggested that annual N losses would be somewhat higher in a grazed prairie that is burned frequently because of increased volatilization losses (Risser et al. 1981). However, at the time these model predictions were published, few data sets of long enough duration were available to determine their validity. It was from this background that the Konza Prairie LTER Program developed. What follows is a synopsis of research results from this program, along with relevant studies from other sites, which provide the framework for our current understanding of patterns and controls of nutrient cycling processes and nutrient retention in tallgrass prairie ecosystems.

Nutrient Inputs and Outputs at Konza Prairie

From its inception, one objective of the nutrient cycling research at Konza Prairie has been to quantify "patterns of inorganic input and movement through soils, groundwater, and surface waters to evaluate the interaction of geochemical and biological processes" as defined in the original LTER nutrient cycling "core area" (Callahan 1984). Data related to this objective include information on an-

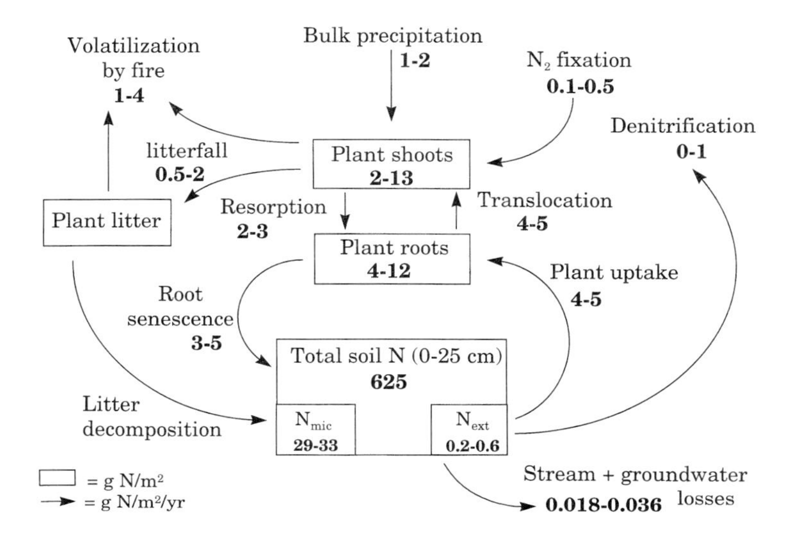

Figure 13.1. A conceptual model of N cycling in ungrazed tallgrass prairie, synthesized from several studies completed at Konza Prairie. Standing stocks (boxes) are reported as g/m^2 and fluxes (arrows) as $g/m^2/yr$. N_{mic} is microbial biomass N; N_{ext} is KCl-extractable NH_4^+- and NO_3^--N. The wide range of values for plant N mass and volatilization losses are due primarily to differences in fire frequency (annual versus infrequent fires).

nual nutrient inputs and outputs, essential for the development of ecosystem-level nutrient budgets (Fig. 13.1) and for evaluation of the impact of management or land-use practices on long-term nutrient storage. A long-term record of nutrient inputs also will be critical for assessing any directional changes in nutrient loading (e.g., Lynch et al. 1995).

Nutrient Inputs

Atmospheric deposition is the major flux of "new" nutrients to tallgrass prairie ecosystems. As might be expected in a semihumid grassland, both dryfall and wetfall inputs are important. Precipitation chemistry and wetfall inputs have been documented at Konza Prairie since 1983, in conjunction with the National Atmospheric Deposition Program (NADP). Mean concentrations of the dominant ions in precipitation at Konza Prairie (NH_4^+, Ca^{2+}, NO_3^-, and SO_4^{2-}) are presented in Table 13.1. Annual inputs of these and other key nutrients in wetfall at Konza Prairie were presented by Gilliam (1987) and are included in a recent NADP summary report (Lynch et al. 1995). Inputs of ions in precipitation at Konza Prairie partly reflect regional factors. For example, relatively high inputs of Ca^{2+} likely are derived from Ca-rich soils characteristic of the region (Gilliam 1987). Likewise, the high NH_4^+ concentrations in wetfall at Konza Prairie are probably due to a combination of regional agricultural activity and relatively

Table 13.1. 10-year data set of volume-weighted mean annual concentrations (mg/L) of dissolved ions in wetfall samples collected at Konza Prairie

Ion	$\bar{x} \pm$ SE	Range
NO_3^-	1.278 ± 0.040	1.080–1.510
NH_4^+	0.347 ± 0.022	0.210–0.460
SO_4^{2-}	1.334 ± 0.047	1.080–1.580
Ca^{2+}	0.338 ± 0.020	0.230–0.490
Mg^{2+}	0.032 ± 0.002	0.022–0.048
K^+	0.032 ± 0.002	0.023–0.049
Na^+	0.082 ± 0.004	0.065–0.119
Cl^-	0.125 ± 0.005	0.110–0.170
H^+	0.011 ± 0.001 (pH=4.98)	0.007–0.016

Note: Wetfall chemistry was provided by the NADP based on samples collected at the main Konza Prairie weather station.

high-pH soils (Chapter 4, this volume), both of which promote ammonia volatilization in the Midwest (Junge 1958; Munger and Eisenreich 1983). Interestingly, a recent analysis of trends in wetfall chemistry at Konza Prairie indicated a significant trend of increasing NH_4^+ concentrations between 1985 and 1993 (Lynch et al. 1995). The extent to which this may be related to changing land-use practices and increased agricultural impact in the region is unclear. Annual inputs of N in both wetfall and dryfall have been summarized in several publications (Seastedt 1985a; Knapp and Seastedt 1986; Ramundo and Seastedt 1990; Ramundo et al. 1992). Inorganic N inputs in wetfall from 1983 to 1995 averaged 0.46 g N m^{-2} and ranged from 0.22 to 0.64 g N m^{-2} yr^{-1}, with approximately equal amounts of NO_3-N and NH_4-N. Year-to-year variability in wetfall N inputs (approximately threefold) is highly dependent on interannual variability in precipitation amounts, and a significant positive correlation occurs between input (volume) of annual precipitation and N in wetfall (Fig. 13.2). On-site estimates of inorganic N in bulk precipitation (wetfall + dryfall) have been obtained using open funnel collectors and are approximately double those of wetfall only (Gilliam 1987). A comparison of annual inputs of N in bulk precipitation and in wetfall only (Fig. 13.3) indicates that dryfall inputs are significant sources of inorganic N deposition at Konza Prairie. Inputs of organic N in bulk precipitation add another 0.6 g N m^{-2} yr^{-1}, raising our estimates of mean annual total N inputs to about 1.6 g N m^{-2} yr^{-1}. We should note that bulk precipitation collectors are not very efficient at sampling dryfall (Lovett 1994), and actual dryfall inputs are likely to be much higher than the values presented here (Gilliam 1987). Another difficulty in interpreting these data is the lack of certainty about how much dryfall N originates locally and how much represents "new" inputs from outside the system.

Seasonal patterns of atmospheric nutrient inputs are linked to seasonal pat-

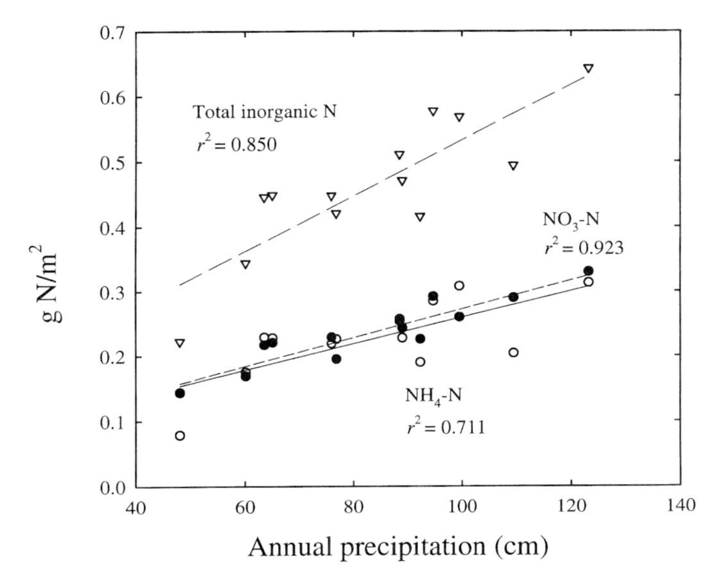

Figure 13.2. Relationship between annual precipitation amounts and annual wetfall inputs of ammonium (closed circles), nitrate (open circles), and total inorganic N (triangles) at Konza Prairie.

terns of precipitation, which is greatest during the growing season (Fig. 13.4). In fact, about 75% of annual precipitation inputs on Konza Prairie occur during the growing season (Chapter 2, this volume). The co-occurrence of maximal precipitation inputs with maximal evapotranspiration and plant uptake likely contributes to the high degree of nutrient retention in tallgrass prairie (Dodds et al. 1996b).

Nitrogen fixation by free-living cyanobacteria and through symbiotic associations of microbes and higher plants is a potentially significant N input in some grasslands (Woodmansee 1978) that has sometimes been cited as a source of replacement for N volatilized in fires (Leopold 1949; Vogl 1974; Ojima et al. 1990, 1994). However, direct measurements of N fixation rates under field conditions in tallgrass prairie ecosystems are lacking. Estimates of potential N fixation by cyanobacteria (*Nostoc muscorum*) isolated from Konza Prairie soils have been obtained under controlled conditions (Eisele et al. 1990). These estimates suggest that fixation by crustal soil cyanobacteria could be significant (up to 1 g N m^{-2} yr^{-1}), provided that light availability and soil moisture remained adequate for a long enough period in the early spring (Eisele et al. 1990). However, these favorable conditions may not persist for long in most years. That study also demonstrated that rates of N fixation by soil cyanobacteria were related positively to P availability, suggesting that P deposited in the ash following a fire may stimulate cyanobacterial N fixation. This does provide a potential mechanism for replacement of N lost during fire. Including the effects of burning in CENTURY model simulations of tallgrass prairie resulted in estimated rates of

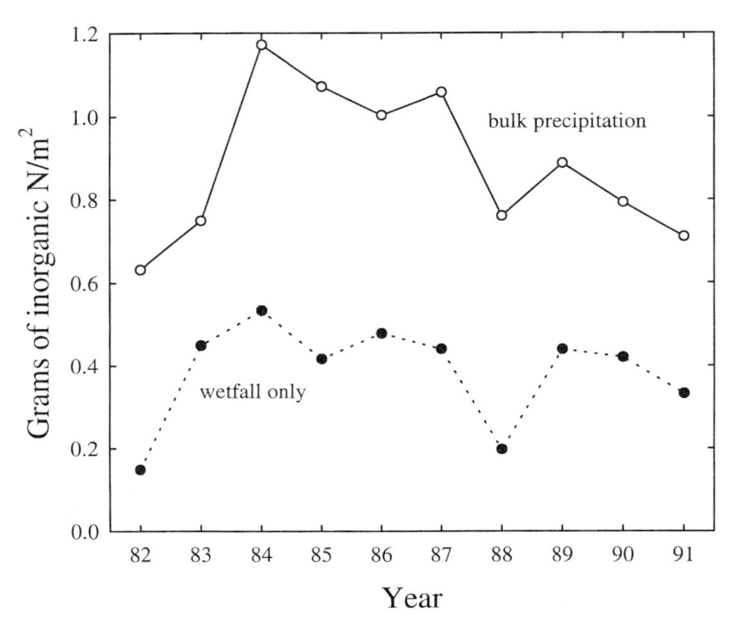

Figure 13.3. Ten-year record of inorganic N inputs at Konza Prairie. Bulk precipitation inputs of inorganic N are approximately double those of wetfall inputs, indicating that dryfall can be a significant source of N in tallgrass prairie of the central Great Plains.

soil N fixation ranging from 0.42 g N m^{-2} yr^{-1} in unburned prairie to 0.52 g N m^{-2} yr^{-1} in annually burned prairie (Ojima et al. 1990). Symbiotic and heterotrophic N fixation in tallgrass prairie also may be important but are in need of further study. Population densities of several legume species (e.g., *Amorpha canescens, Dalea* spp., and others) have increased in response to frequent fires at Konza Prairie (Towne and Knapp 1996), but N fixation has not been documented in these species. Likewise, associative N fixation in the rhizosphere of various C_4 grasses has been reported (Tjepkema 1975; Tjepkema and Burris 1976; Morris et al. 1985; Bredja et al. 1994; Boddey and Dobereiner 1994), but the potential for associative N fixation to quantitatively affect tallgrass prairie N budgets has not been investigated adequately.

Nutrient Outputs

The major potential pathways for nutrient loss from tallgrass prairie in the Flint Hills are hydrologic fluxes (stream and groundwater), gaseous fluxes (primarily denitrification and ammonia volatilization), and fire-induced losses. Given the near equality of potential evapotranspiration and annual precipitation amounts, hydrologic export of nutrients is not expected to be a major pathway of nutrient loss in most grassland ecosystems (Woodmansee 1978). Although stream and groundwater chemistry have been characterized at Konza Prairie (Chapter 10,

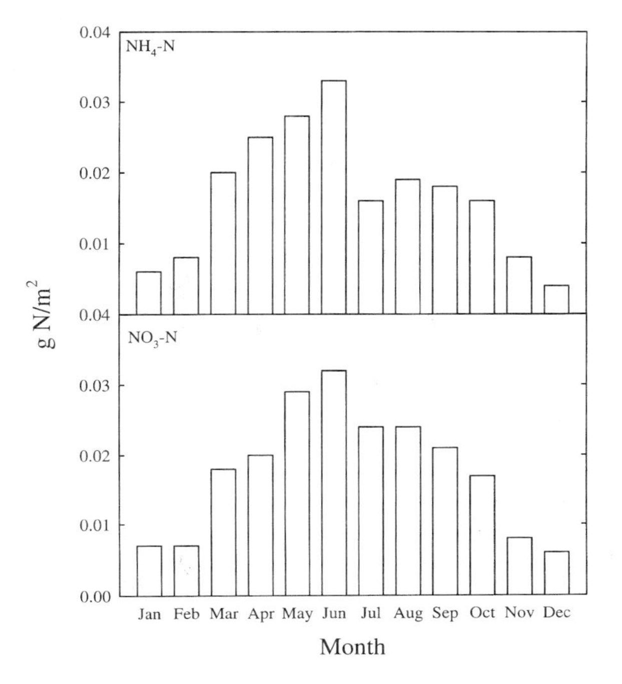

Figure 13.4. Seasonal patterns of inorganic N input in precipitation at Konza Prairie. Mean monthly inputs of ammonium and nitrate are based on 11 years of data (1984–1994) from the National Atmospheric Deposition Program.

this volume), accurately measuring total hydrologic export from individual watersheds is difficult because of the complex geology and subsurface flow patterns characteristic of the region (Chapter 3, this volume). However, we have estimated losses of N by streamflow as a function of flow volumes and water chemistry and combined these with modeled groundwater fluxes and N data to estimate total hydrologic N export from four Konza Prairie watersheds over a 5-year period (Dodds et al. 1996b). These estimates indicate very low annual hydrologic losses of dissolved N (0.02–0.04 g N m^{-2} yr^{-1}), which are equivalent to only 0.01 to 6% of annual precipitation inputs. The large imbalance in annual hydrologic inputs and outputs of N attests to the tremendous immobilization potential of tallgrass prairie. Plant transpiration and N uptake also influence seasonal patterns of N export by streams. Tate (1990) reported significantly lower concentrations of NO_3^- in perennial streams and seeps during the growing season compared with the fall/winter period when plants are dormant. To date, we have not detected any strong effects of fire or grazing by *Bos bison* on hydrologic export of N (Dodds et al. 1996b). This is somewhat surprising, given the effects of fire and grazing on soil N availability as discussed later in this chapter. However, landscape position, the timing and magnitude of precipitation events, and subsequent hydrologic flows apparently are the most important variables affecting

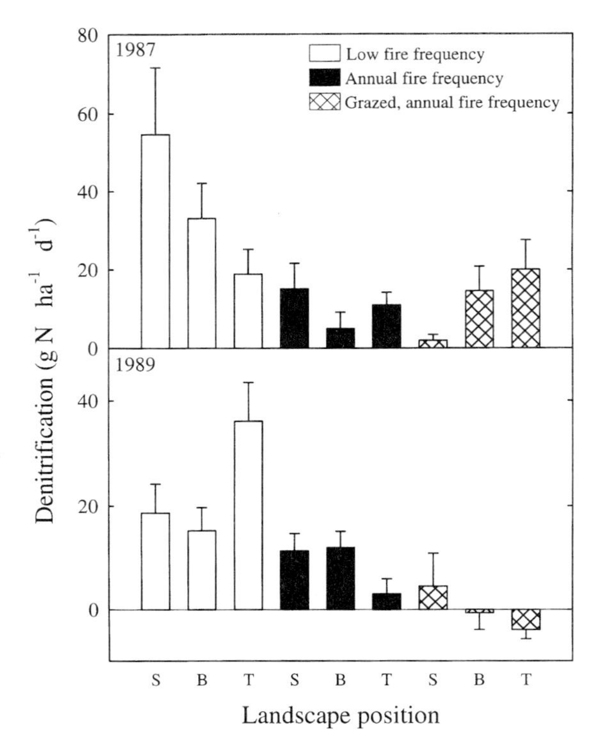

Figure 13.5. Denitrification rates determined from laboratory incubations of soil cores taken from different land-use treatments and different landscape positions at Konza Prairie. S = summit, B = back-slope, T = toe-slope. From Groffman et al. (1993), with permission of the Ecological Society of America.

nutrient output by surface waters in the Flint Hills (Tate 1990; Dodds et al. 1996b; Chapter 10, this volume).

Denitrification, which is the microbial reduction of nitrate (NO_3^-) in soil solution to gaseous forms of N (NO_x and N_2), can be a substantial loss of N from many ecosystems, although it usually has been considered to be insignificant in grasslands (Woodmansee 1978; Seastedt and Hayes 1988). However, recent studies suggest that denitrification losses may be important in some areas of tallgrass prairie. Laboratory-based measurements of denitrification rates were obtained in cores taken from different landscape positions and land-use types across Konza Prairie from 1987 to 1989 (Fig. 13.5; Groffman et al. 1993). Measured denitrification rates were significantly higher in unburned prairie than in burned, burned and grazed, or cultivated sites, and were generally greatest in the spring, coincident with high soil water content. This is consistent with other studies that have reported both greater soil moisture and nitrate concentrations in unburned than in burned prairie (Knapp and Seastedt 1986; Turner et al. 1997). Extrapolation of these rates to the field suggests localized losses of up to nearly 1 g N m^{-2} yr^{-1} in fertile sites with deep soils (Groffman et al. 1993). Indeed, the positive relationship between soil characteristics associated with high plant productivity and potential denitrification (i.e., deep, fine-textured soils; high soil water content; high N mineralization and nitrification rates) has been proposed as a basis for remotely sensed estimates of N trace gas flux across tallgrass prairie landscapes (Groffman and Turner 1995), and landscape-level estimates of denitrification at Konza Prairie using this approach averaged 0.62 g N m^{-2} yr^{-1}. These potential losses are of the same magnitude as N inputs in precipitation or N losses in fire, although verification of actual flux rates under field conditions remains to be done. Additionally, for actual denitrification fluxes to reach such high values, high soil moisture, large quantities of nitrate, and a sufficient supply of labile C must occur together. Such a combination is probably uncommon in undisturbed tallgrass prairie in years of typical climatic conditions.

Ammonia (NH_3) volatilization is another mechanism for gaseous N loss from grassland soils, and such losses can be substantial, particularly in grazed grasslands (Detling 1988; Whitehead 1995; Frank and Zhang 1997). Ungulate grazers can consume a large fraction of aboveground biomass and return much of the ingested plant N to the soil surface in the form of dung and urine (Detling 1988; Frank et al. 1994). Woodmansee (1978) estimated that up to 50% of the N excreted by ungulate grazers could be lost to ammonia volatilization, which was equivalent to 30 to 100% of annual N inputs in bulk precipitation in the grasslands he examined. However, actual rates of ammonia volatilization appear to be less than this in most systems (Schimel et al. 1986; Detling 1988). Ammonia gradients above tallgrass prairie canopies at Konza Prairie were studied as part of the NASA FIFE (First International Satellite Land Surface Climatology Program Field Experiment) (Zachariassen unpublished data, cited in Blad and Schimel 1992). Those results suggest that the tallgrass prairie may act alternately as a source and a sink, and ammonia volatilization from the soil may be matched by canopy uptake of atmospheric ammonia. Volatilization of ammonia also can occur from plant surfaces (Wetselaar and Farquhar 1980), and leaf-level rates of ammonia volatilization have been recorded for three C_4 grass species common to Konza Prairie (Heckathorn and DeLucia 1995). Although the measured ammonia volatilization losses from leaf surfaces during a 2- to 3-week drought period can exceed 5% of total aboveground biomass N, they appear to be small in terms of whole-system nutrient fluxes.

In contrast to many other ecosystems where biologically mediated transformations (i.e., denitrification) or hydrologic fluxes are major sources of N loss, fire represents the biggest loss of N from ungrazed tallgrass prairie (Ojima et al. 1990; Seastedt and Ramundo 1990). The effect of fire on ecosystem-level nutrient outputs varies for different elements and with different fire frequencies and intensities. Most P and cations released from burned prairie vegetation are redeposited in the ash, whereas lighter elements, including C, N, and S, are volatilized and returned to the atmosphere (Raison 1979). However, losses or redistribution of P and cations in windblown ash can be substantial (Ojima et al. 1990). The amount of N volatilized in a specific fire depends on amounts of N accumulated in aboveground biomass and detritus and fire intensity, both of which, in turn, depend on fire frequency. Less frequent fires can result in greater episodic N losses because there is a greater accumulation of detrital N to burn. However, when averaged over multiple years, the mean loss of N due to fire in infrequently burned tallgrass prairie is usually less than that from annually burned areas. Ojima et al. (1990) reported fire-induced N losses ranging from 1.25 to 3.0 g N m^{-2} in ungrazed prairie. Assuming that, on average, 90% of

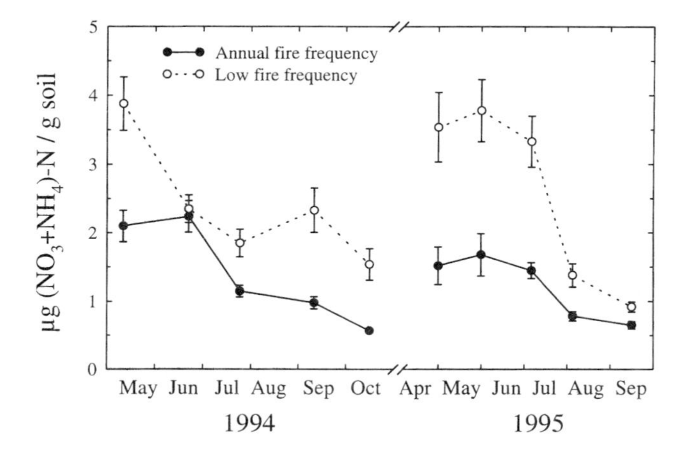

Figure 13.6. Mean concentration of KCl-extractable inorganic N in soils (0- to 14-cm depth) in annually burned watersheds and those subjected to a low fire frequency on Konza Prairie.

aboveground N is volatilized in a fire, we calculate that fire-induced N loss in ungrazed areas of Konza Prairie ranges from about 1 to 4 g N m^{-2}. Thus, annual losses of N in fire can approximate, or exceed, precipitation inputs of new N in frequently burned areas. This observation, along with simulation modeling of the effects of fire in tallgrass prairie, led to the hypothesis that annual burning eventually will deplete reserves of soil organic N (Risser and Parton 1982; Ojima 1987; Ojima et al. 1990, 1994).

Storage of Nutrients in Tallgrass Prairie

The balance between nutrient inputs and outputs discussed in the preceding sections affects the long-term accumulation and storage of nutrients in tallgrass prairie. The bulk of the C and nutrients in tallgrass prairie ecosystems are stored in the soil. Some general characteristics of soils at Konza Prairie are presented in Chapter 4 (this volume). Tallgrass prairie soils contain more C and nutrients per unit volume than comparable forest soils (Seastedt and Knapp 1993) because of the relatively high allocation of C belowground by the dominant grasses and the relatively slow decay rates characteristic of grasslands (Seastedt 1995). On average, about 625 g N m^{-2} are stored in the upper 25 cm of soil at Konza Prairie, which is about 40 times the amount of total N sequestered in plants at peak biomass (Seastedt and Ramundo 1990) and about 125 times greater than the annual N requirements for new plant production. The actual amounts of N (and P) stored in plant biomass vary with fire frequency and topographic position and are affected by differences in site productivity, nutrient availability, and species composition. Of course, only a small fraction of the total soil N (typically much less than 1%) is available for plant uptake at any particular time. Mean concentrations of KCl-extractable inorganic N (as $NH_4^+ + NO_3^-$) on selected annually

Table 13.2. Concentrations of N in bulk precipitation, throughfall, soil water, and streams of annually burned and unburned tallgrass prairie

	Nitrogen Concentration (μg L⁻¹)					
	Ammonium-N		Nitrate-N		Organic-N	
	Burned	Unburned	Burned	Unburned	Burned	Unburned
Bulk precipitation	456	456	530	530	420	420
Throughfall	344	196	345	258	1,669	2,155
Soil water						
20 cm deep	<2	<2	11	32	358	389
80 cm deep	<2	<2	13	12	213	182
Stream water	—	—	3	14	186	259

Source: From Seastedt and Ramundo (1990).

burned and unburned sites are presented in Figure 13.6. Concentrations of extractable N are usually greatest in the spring and are generally greater on unburned watersheds relative to comparable annually burned watersheds. When converted to a per unit area basis using an average bulk density value of 1 g cm^{-3}, mean amounts of extractable inorganic N in ungrazed sites range from 0.2 g N m^{-2} (0–15 cm) on annually burned sites to 0.4 g N m^{-2} on unburned sites. Likewise, measurements of soil solution chemistry at Konza Prairie, obtained with porous cup lysimeters, show very low concentrations of NH_4^+ and NO_3^- (Table 13.2), which tend to become attenuated as soil solution moves deeper into the soil profile (Seastedt and Hayes 1988; Hayes and Seastedt 1989). The remaining soil N exists as chemically fixed inorganic N; as part of the microbial and microfaunal biomass; or as organic forms that vary in their availability for biological transformation, based on chemical recalcitrance or on physical protection from decomposers (van Veen and Paul 1981; Parton et al. 1987). Rice and Garcia (1994) partitioned total soil N in surface soils (0–5 cm) at Konza Prairie into stable organic forms (81.5%), active organic forms (7.4%), and microbial biomass N (11.1%). Obviously, microbial immobilization and storage of soil N in stable organic forms play important roles in N retention in these ecosystems.

Roots and rhizomes also are important nutrient pools in tallgrass prairie (Hayes and Seastedt 1987; Ojima et al. 1994; Heckathorn and DeLucia 1996) and provide storage sites for nutrients that can be utilized to support new plant growth. Benning and Seastedt (unpublished data) summarized root and rhizome N dynamics and the effects of burning, mowing, and N fertilization during the first 4 years of a long-term fire-fertilizer-mowing experiment at Konza Prairie (Chapter 14, this volume). As expected, roots and rhizomes in burned sites contained more total N than those in unburned plots. However, this was due largely to increased root biomass on the burned treatments because tissue N concentration is typically higher in roots and rhizomes on unburned than on burned sites (Ojima et al. 1994; Blair 1997; Benning and Seastedt unpublished data). In this experiment, fire also mediated the ability of the prairie to retain N added in fertilizer. Belowground productivity did increase with N additions, but this increase was modest relative to increases in aboveground productivity. Perhaps the most interesting finding was that the N concentrations of live roots did not increase nearly as much as values for dead roots. This implies that plants may indeed control ecosystem N availability by mediating the immobilization potential of the detrital subsystem.

Appreciable quantities of nutrients can be immobilized in the standing detritus and surface litter that accumulates in unburned tallgrass prairie (Knapp and Seastedt 1986). Senescent tallgrass litter has some of the lowest foliage N concentrations of any litter type. For example, senescent leaf tissue of *Andropogon gerardii* may have N concentrations as low as 0.38%, resulting in an initial C:N ratio of 110. Stems on the tall flowering culms of the dominant C_4 grasses have even higher C:N ratios (up to 250). Hence, the surface litter and standing dead of tallgrass prairie act as a "N sink," immobilizing significant quantities of N during the first few years of decomposition (Seastedt 1988). As a result, accumulated surface detritus significantly modifies throughfall N dynamics in tallgrass prairie ecosystems (Seastedt 1985a). Once this material has decomposed sufficiently, however, it likely liberates large amounts of N upon further decay. These dynamics explain, in part, the higher inorganic N concentrations of soils and soil water in unburned prairie. In contrast to surface litter, root and rhizome litter does not appear to immobilize significant quantities of N during the decay process, in spite of the fact that root litter is only slightly higher in N concentrations than foliage (Seastedt et al. 1992; Blair unpublished data). These litter N dynamics also were observed using buried wood dowels (O'Lear et al. 1996). This material, which initially had a very low N concentration (0.32% N, C:N ratio of 156), also failed to immobilize substantial amounts of N (O'Lear et al. 1996). Thus, belowground litter in tallgrass prairie appears to have a lower immobilization potential than would be predicted on the basis on initial litter quality alone (Seastedt et al. 1992). This is an apparent contradiction to the fact that increased belowground C inputs, via root production, increase the overall N immobilization potential of prairie soils (Ojima et al. 1994). One explanation for this phenomenon is that immobilization processes in grassland soils may be separated physically from the actual decomposing substrate through fungal export of N. Conversely, chronic grazing of decomposing substrates in soil simply may remove microbial N from the substrates, thereby preventing the detection of net immobilization. These questions deserve further study. Given the differences in above- and belowground litter quality and decomposition processes, any factors that alter litter inputs, litter quality or plant root:shoot ratios should significantly affect N cycling in tallgrass prairie ecosystems (Wedin and Pastor 1993).

Intrasystem Nutrient Dynamics

Estimates of nutrient inputs, outputs, and standing stocks, such as those presented previously, are necessary for developing nutrient budgets for tallgrass prairie and predicting the impact of changing land use on nutrient storage, but they reveal very little about the dynamics of intrasystem nutrient fluxes, the controls of nutrient availability, and the linkages between nutrient cycling and plant and animal responses. In order to address these issues, we must examine soil and litter nutrient transformations and plant responses to temporally and spatially variable patterns of nutrient availability.

Litter and Soil Processes

With the exception of newly burned prairie, nutrients in incoming precipitation are transformed almost immediately within the tallgrass prairie canopy. Precipitation inputs are intercepted by live and standing dead vegetation, where evaporation, leaching, and uptake of nutrients alter the amount and chemical composition of the solution, referred to as *throughfall*, as it moves through the canopy. Seastedt (1985a) provided detailed information on throughfall chemistry and deposition rates at Konza Prairie. Over a 2-year study period, he found that throughfall volumes averaged about 76% of precipitation volumes in burned prairie versus 58% of precipitation volumes in unburned prairie because of greater interception and evaporation in the presence of accumulated detritus. He also noted significant reductions in the concentrations of inorganic N and increases in concentrations of organic N in throughfall across both fire treatments, with the greatest changes occurring in unburned prairie. Solution chemistry is altered substantially as it moves through the prairie canopy, which acts as a sink for inorganic N, presumably because of microbial immobilization, and is a major source of dissolved organic N (DON) in throughfall reaching the soil surface (Seastedt 1985a). Organic N concentrations in throughfall are generally four to five times greater than those in bulk precipitation (Table 13.2). Occasionally, large amounts of NH_4^+ and NO_3^- can be transmitted in canopy throughfall to the soil surface, but soil solution concentrations are seldom within an order of magnitude of these values (Seastedt and Hayes 1988; Hayes and Seastedt 1989). Immobilization and uptake processes in the uppermost regions of the soil very quickly reduce the concentrations of both inorganic and organic N in solution, leaving DON the dominant form of N in soil water and stream water (Seastedt and Ramundo 1990). For example, at 20 cm NO_3^- concentrations sampled by porous cup lysimeters averaged 27 μg N L^{-1} versus 13 μg N L^{-1} at 80 cm, about a 50% reduction in NO_3-N concentrations as water moved to the lower depth. Characterization of DON at various stages of movement through tallgrass prairie ecosystems, in terms of both chemical composition and biological availability, remains to be done.

Mineralization is the microbially mediated process whereby organic N is converted to an inorganic form (initially NH_4^+), which can be taken up by microbes (immobilized) or plants (assimilated) or nitrified (oxidized to NO_3^-) by chemoautotrophic bacteria. Although both NH_4^+ and NO_3^- are "available" forms of N, differences exist in the relative mobility of these ions in soils and the degree to which plants and microbes compete for these nutrients. Ammonium is much less mobile in soils than nitrate and therefore is a much smaller component of soil solution, groundwater, and stream water (Hayes and Seastedt 1989; Dodds et al. 1996b), although exchangeable (KCl-extractable) concentrations of

NH_4^+ and NO_3^- in the soil are approximately equal (Turner et al. 1997). Ammonium often is taken up preferentially by microbes, and microbial immobilization of N plays a major role in N availability in the C-rich soils of tallgrass prairie. In fact, soil from Konza Prairie supported greater amounts of microbial biomass C (134 g C m^{-2}, 0–10 cm) and N (33 g N m^{-2}) than soils from a variety of other ecosystem types across North America (Zak et al. 1994). These values are very similar to those reported by Rice and Garcia (1994). Zak et al. (1994) also reported that potential net N mineralization rates during a 32-week lab study averaged 29 g N m^{-2}. Not surprisingly, actual estimates of net N mineralization rates under field conditions at Konza Prairie, where water and energy limitations become important, are an order of magnitude lower than this, ranging from about 1 to 4 g N m^{-2} (0-15 cm) over a growing season (Ojima et al. 1994; Turner et al. 1997; Blair 1997).

Plant Uptake and Nutrient Use Efficiency

Accumulation of N and P in aboveground biomass and litter is an important aspect of nutrient movement through tallgrass prairie ecosystems and can represent a sizable potential loss of nutrients during burning. Therefore, quantifying amounts and patterns of N and P accumulation in aboveground live and dead plant biomass has been part of the Konza Prairie LTER Program. The accumulation of N in total aboveground plant biomass (both live and dead material) varies approximately threefold under different extremes of fire frequency (Fig. 13.7). Hayes (1985; Hayes and Seastedt 1987) provided more detailed information on the seasonal patterns of N concentrations in tallgrass vegetation on Konza Prairie, particularly in relation to seasonal movement of nutrients within plants. These studies indicated that about 46% of the N contained in rhizomes and 58% of the N in roots were transported into aboveground biomass during the spring. As the plants senesced, up to 58% of the N contained in *A. gerardii* leaves was translocated back to the roots and rhizomes for storage, most of it apparently being transported in the form of amino acids. These studies, as well as those done by Seastedt et al. (1989) in an effort to understand invertebrate impacts on plant N dynamics, suggest high rates of N retranslocation during senescence. However, the amount of N retranslocated during senescence (about 50% of maximum aboveground N) is not different than that reported for woody species in low N environments (e.g., Chapin and Kedrowski 1983). One interesting finding of Hayes's work and later studies by Heckathorn and DeLucia (1995, 1996) was that N retranslocation to rhizomes and roots, in terms of both the amount and the percentage of aboveground N, was maximized under drought conditions. Hayes hypothesized several reasons for this. First, the amount of leaf area (and the cost of retranslocation) is minimized under water limitations. Second, flowering, a large drain on N reserves, does not occur during drought years. These two features combine to maximize internal recycling and provide a mechanism for conserving N in low productivity years. These findings reinforce the notion that productivity in "good" years may negatively affect productivity in subsequent years because of the immobilization of N into senescent plant tissue,

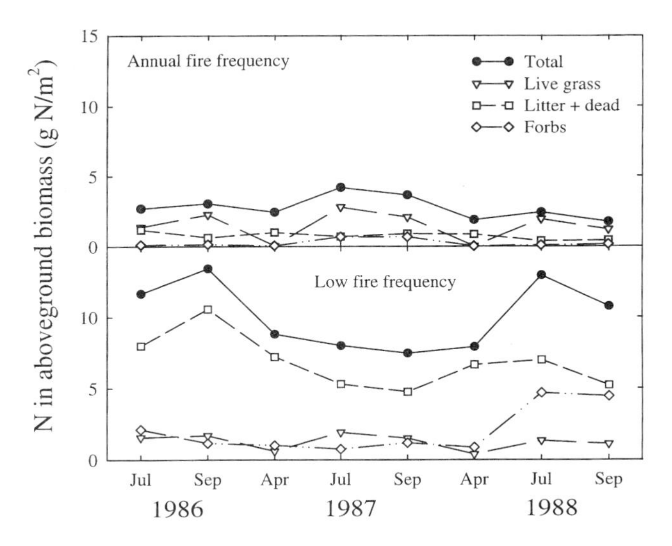

Figure 13.7. Accumulations of N in live and dead aboveground plant biomass in an annually burned watershed and a watershed subjected to a low fire frequency at Konza Prairie.

and they provide a possible mechanism to explain observations of high ANPP following drought years in tallgrass prairie (Chapter 12, this volume).

Effects of Fire on N Cycling

As noted earlier, fire represents a major source of N loss from tallgrass prairie ecosystems, and the potential imbalance in the N cycle of frequently burned prairie has been pointed out by several researchers. This was incorporated into simulation models of tallgrass prairie C and N cycles (Ojima et al. 1990, 1994; Seastedt et al. 1994), leading to the hypothesis that annual burning will result in an eventual decline in soil C and N pools and presumably lower NPP. Testing this hypothesis and documenting the effects of fire extremes (annual burning versus fire exclusion) on soil and plant processes were major foci of our nutrient cycling research during LTER II (1986–1990) and III (1991–1996), and the results of studies done prior to LTER III were summarized by Seastedt and Ramundo (1990). Results from these and other subsequent studies clearly demonstrate the importance of fire regimes in altering N availability and plant response in tallgrass prairie (Knapp and Seastedt 1986; Ojima 1987; Seastedt et al. 1991; Benning and Seastedt 1995). Additional studies at Konza Prairie have provided some mechanistic explanations for the effects of fire by documenting changes in biologically active soil N pools (Garcia and Rice 1994; Rice and Garcia 1994; Chapter 14, this volume) and N mineralization rates (Ojima et al. 1994; Blair 1997; Turner et al. 1997).

Fire results in a series of changes in tallgrass prairie ecosystems that alter N cycling and soil processes. The immediate effects of fire are the removal of accumulated detritus and the loss of some nutrients, including N. The removal of surface plant litter also alters the energy environment and microclimate of the soil, resulting in greater solar inputs and warmer soil temperatures. These conditions promote the rapid phenological development and growth of the dominant C_4 grasses and eventually can lead to changes in plant C allocation and nitrogen use efficiency (NUE). That is, the grasses tend to accumulate more root mass belowground and produce plant tissue that is lower in N concentration. These changes, in turn, increase microbial and plant demand for N, leading to even greater N demand under frequent burning. These changes were incorporated into the CENTURY simulation model, which predicted that repeated annual burning should lead to an eventual loss of soil C and N and presumably an eventual decline in site productivity. However, the evidence for decreased N availability with frequent burning has been mostly indirect and based on responses of plants to N fertilization. Until very recently, few studies had considered the amounts and seasonal changes in soil inorganic N or rates of N mineralization under different fire regimes. Turner et al. (1997) documented net N mineralization rates and seasonal changes in concentrations of inorganic soil N in annually burned and unburned sites at Konza Prairie. Concentrations of inorganic N were significantly greater in unburned than in annually burned sites. Net N mineralization, averaged across different topographic positions, also tended to be greater in unburned sites, although the differences were not statistically significant. However, other studies on Konza have indicated significantly greater net N mineralization in unburned than in annually burned prairie. For example, Blair (1997) found that soils in unburned prairie mineralized two to three times more N over the growing season than did soils in annually burned sites. These results are similar to those reported by Ojima et al. (1994) for a long-term fire experiment located near Konza Prairie.

We should note that some fundamental differences exist between the short-term and long-term effects of fire, which can explain some of the disparate results of an infrequent fire (a pulse of very high productivity, lack of response to added N) compared with those associated with long-term annual burning (sustained high productivity, large response to N additions). The immediate effects of fire include changes in the energy environment and microclimate of prairie soils (Hulbert 1969; Knapp 1984b; Seastedt and Briggs 1991; Knapp et al. 1993a). Following removal of detritus and standing dead, soil temperatures from the surface to 10 cm deep are substantially greater (about 10°C or more) in burned sites than in unburned soils. Elevated soil temperatures have been postulated to increase soil biological processes, including respiration, nutrient mineralization, and plant growth, according to a Q_{10} relationship. This does appear to be true with respect to plant phenological development and total soil CO_2 flux (Knapp et al. 1998). However, soil water availability also can limit soil activity in grasslands, and soils under burned prairie tend to be drier than soils protected by surface detritus (Chapter 12, this volume). Recent studies of N mineralization on Konza Prairie indicate that net N mineralization rates in unburned plots that were burned for the first time are lower than those in comparable plots that re-

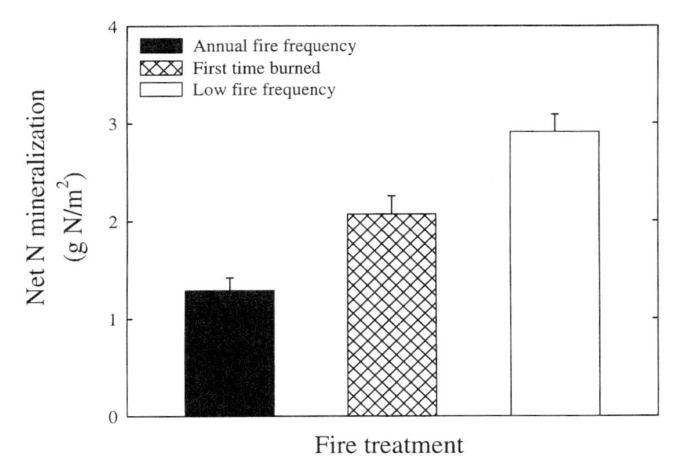

Figure 13.8. The effects of fire frequency on net N mineralization. Mean net N mineralization over the 1994 and 1995 growing season (May–October) was calculated by summing together values from 30-day in situ incubations in plots that were annually burned, unburned (low fire frequency), or infrequently burned (unburned prior to spring 1994 or 1995 fires).

main unburned, in spite of higher temperatures and similar organic N pools (Fig. 13.8). Therefore, some processes such as N mineralization may be affected negatively by burning.

Although frequent fires have been hypothesized to result in long-term reductions in the N capital of tallgrass prairie ecosystems, we have not yet detected any significant reductions in productivity of long-term annually burned plots (Chapter 12, this volume), nor have we been able to detect reductions in total soil C or N concentrations to date (Chapter 14, this volume). Several mechanisms have been proposed to explain these observations. As noted earlier, some evidence indicates that N fixation may be increased by burning (Eisele et al. 1990). Frequent fires also may decrease potential N losses via leaching and denitrification. Although leaching losses generally are considered to be small in both burned and unburned prairie, Seastedt and Hayes (1988) found that concentrations of nitrate in soil solution (porous cup lysimeters) were significantly lower under annually burned prairie. The combination of greater soil moisture and increased nitrate concentrations could lead to some increase in N loss via leaching in unburned prairie. Similarly, greater soil moisture and higher nitrate availability could contribute to potentially higher denitrification rates in unburned prairie (Groffman et al. 1993), although the effects of lower temperatures may mitigate some of this effect.

The relationship between fire frequency and N limitation has been well documented in a variety of studies done at Konza Prairie. Fertilizer experiments (10 g N m^{-2}) have shown that annually burned prairie is most responsive to N additions, whereas unburned prairie is unresponsive (Seastedt et al. 1991; Chapter

12, this volume). This could be a simple result of energy (light/temperature) limitations in unburned prairie (Hulbert 1969; Knapp and Seastedt 1986). However, when previously unburned prairie is subjected to fire, it remains only moderately responsive to N additions (Chapter 12, this volume), suggesting that available N accumulates in the absence of fire. Thus, fire frequency and history appear to control the production response of tallgrass prairies to periodic fires in a way that is mediated by N availability. When fire burns an area of prairie that has not been burned for several years, a "pulse" in ANPP often occurs that exceeds that of either annually burned or unburned prairie (Chapter 12, this volume). This phenomenon has been hypothesized to be a consequence of nonequilibrium conditions that result from a switch in limiting resources following an infrequent fire from energy limitation toward N limitation. During this time, a transient, nonequilibrium maximum rate of productivity is achieved (Seastedt and Knapp 1993). Recently, we were able to document greater in situ net N mineralization rates in infrequently burned prairie than in annually burned prairie (Fig. 13.8), providing further support for the "transient maxima" hypothesis (Seastedt and Knapp 1993). In summary, the "pulse" in ANPP following an infrequent fire apparently is due to an accumulation of inorganic and labile soil N in the absence of fire and its subsequent utilization during the growing season following the fire (Blair 1997). However, this response is nonsustainable because as energy limitations are removed, other factors such as N or water availability will begin to limit productivity. Based on the annual decrease in seasonal net N mineralization rates with sequential fires, N availability probably will drop back within 2 to 3 years to the low levels characteristic of annually burned grasslands.

Effects of Grazing on N Cycling

Grazing by large ungulate herbivores (i.e., *B. bison*) was historically important in tallgrass prairie ecosystems, and management for domestic herbivores (i.e., cattle) is a dominant land-use practice in the Flint Hills region (Hartnett et al. 1997). However, the effects of grazing by large ungulates on N fluxes are perhaps the least studied (and least understood) aspects of nutrient cycling in North American tallgrass prairie. To date, much of the nutrient cycling research at Konza Prairie has focused on the effects of fire, and fewer studies have addressed the impact of grazers. However, based on studies in other grassland ecosystems, there is little doubt that grazers increase rates of N cycling in grasslands (e.g., Huntly 1991; Holland and Detling 1990; Holland et al. 1992; Biondini and Manske 1996; Frank and Evans 1997), and they are likely to affect exports as well. As with fire, grazing by ungulates has both short- and long-term effects (Turner et al. 1993) that have yet to be quantified adequately in tallgrass prairies. Studies at Konza Prairie indicate that removal of aboveground biomass (clipping, mowing, or grazing) can lead to substantially higher N and P concentrations in plant shoots (Turner et al. 1993) and delayed translocation of N belowground at the end of the growing season (Seastedt et al. 1988b). However, in the latter study, total aboveground plant N remained lower in the mowed plots than in unmowed controls because of reduced aboveground biomass. Seastedt

(1995) suggested that the potential for grazers to increase N availability in tallgrass prairie has some particularly significant consequences if the dominant grasses in fact do use N limitation to their competitive advantage. Certainly a comparison of Konza Prairie with surrounding, heavily grazed prairies suggests some very large differences in species composition of the vegetation. This has been supported by recent comparisons of *B. bison*–grazed areas with paired grazing exclosures on Konza Prairie, which indicate that grazing by bison increases plant species richness and diversity, as well as plant community heterogeneity (Hartnett et al. 1996; Chapter 9, this volume). The extent to which these plant community changes are mediated by nutrient availability, as opposed to other effects of grazers, remains largely unknown. However, the rapid alterations in species composition of tallgrass prairie associated with fertilizer additions (e.g., Gibson et al. 1993; Wedin and Tilman 1996) suggest that grazing-induced changes in nutrient availability are at least partially responsible. It also is important to note that grazing interacts with fire in tallgrass prairie, and many of the effects of fire are likely to vary depending on the presence or absence of grazers (Hobbs et al. 1991).

Topographic Variation in Nutrient Cycling Processes

A large degree of topographic relief and associated variability in soil resources are major landscape-level features at Konza Prairie and at other tallgrass prairie ecosystems in the Flint Hills (Chapter 3, this volume). These topoedaphic gradients, together with land-use practices that include different burning regimes, result in varied landscape-level patterns of resource availability (Schimel et al. 1991; Turner et al. 1997) and plant responses (Knapp et al. 1993a; Benning and Seastedt 1995). In particular, the relative importance of water, light, and N as potentially limiting factors varies with topography in ways that are highly dependent on fire frequency (Schimel et al. 1991; Knapp et al. 1993a). For reasons already discussed, N tends to be more limiting in burned than in unburned prairie. However, soil moisture may become more limiting than N availability at shallower upland sites, and this condition is exacerbated by the removal of surface detritus (Knapp et al. 1993a). Therefore, N availability can be expected to be most limiting to plant productivity in annually burned lowland sites (Schimel et al. 1991) and should be least limiting at unburned upland sites. These generalizations are supported by measurements of net N mineralization at upland and lowland sites on burned and unburned watersheds (Turner et al. 1997) and by plant productivity responses to added N (Turner et al. 1997; Benning and Seastedt 1995). However, this also means that a general inverse relationship occurs between ANPP and N availability across the landscape at Konza Prairie (Fig. 13.9) because ANPP tends to be highest in burned lowland sites and lowest in unburned uplands (Briggs and Knapp 1995). This is somewhat surprising for an ecosystem that generally is thought to be N-limited (Risser and Parton 1982; Seastedt et al. 1991). As we have continued to examine soil and plant responses in more spatially explicit ways, some additional unexpected landscape-level patterns in soil processes have begun to emerge. For example, O'Lear et al. (1996)

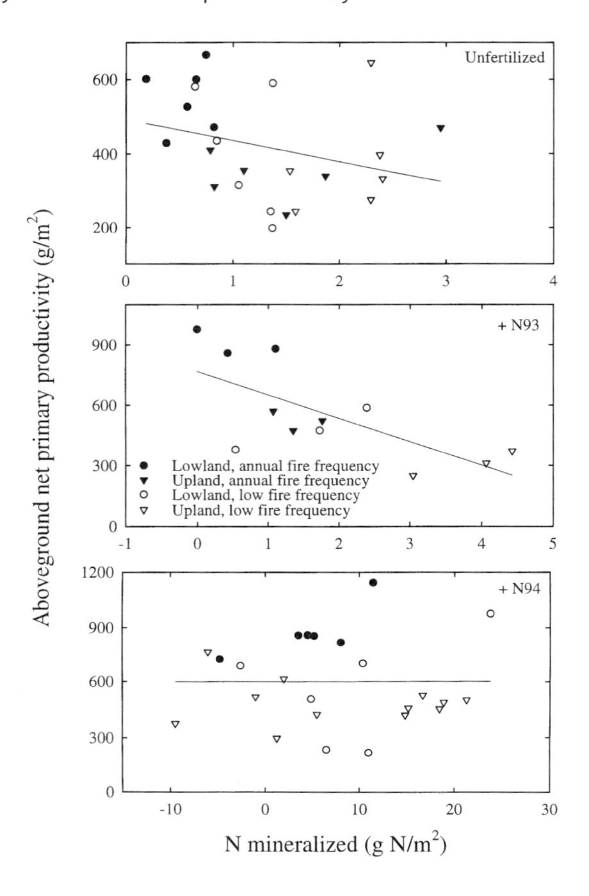

Figure 13.9. The relationship between aboveground annual net primary productivity and cumulative net N mineralization in burned (solid symbols) and unburned (open symbols), upland (triangles) and lowland (circles) prairie that was unfertilized, fertilized in 1993 (+N93), or fertilized in 1994 (+N94). The general lack of relationship suggests that the controls of plant productivity and soil N availability operate somewhat independently across the tallgrass prairie landscape. From Turner et al. (1997), with permission of the Ecological Society of America.

found that decay rates of buried wood dowels were faster in shallow-soil, upland sites and slope sites than in deep-soil, lowland sites. As with soil N mineralization, this pattern is the opposite of that generally observed for plant productivity (i.e., greater at lowland sites compared with uplands). Based on these studies, we hypothesize that different factors, or different responses to the same factors, control aboveground (ANPP) and belowground (decomposition, N mineralization) processes, resulting in a "decoupling" of these processes across topo-edaphic gradients. We expect that these differences are driven by quantity and quality of belowground plant inputs, as well as by abiotic factors (soil moisture and temperature), although this has not yet been studied adequately. The implications of different patterns and controls of productivity and decomposition for soil organic matter dynamics and C storage across topographic gradients remain to be investigated in these grasslands.

Summary

Considerable progress has been made in developing nutrient budgets for tallgrass prairie and in refining our understanding of the factors controlling N availability and losses in these ecosystems. In particular, our understanding of the nutrient-related mechanisms underlying the responses of tallgrass prairie to fires and fire frequency has increased tremendously, based largely on research done at Konza Prairie. Although many of our studies have confirmed the importance of N as the nutrient with the greatest potential for affecting the structure and function of tallgrass prairie ecosystems, they also have shown that N limitation is not a universal feature of tallgrass prairie and that its relative importance varies with fire, grazing, topography, and precipitation. Perhaps one of the most surprising findings was the general inverse relationship between patterns of soil N availability and plant productivity across the tallgrass prairie landscape. This lack of relationship is due to the potential for multiple resources (i.e., water, light, and N) to alternately limit plant productivity under different environmental conditions and management regimes. This supports our emerging conceptual view of tallgrass prairie as a nonequilibrium system in which the relative importance of N varies in both space and time (Chapter 1, this volume).

To date, the long-term effect of annual burning has been a major focus of our nutrient cycling studies. These studies have been invaluable in elucidating the mechanisms by which the tallgrass prairie is able to maintain relatively high rates of productivity in spite of the losses of N associated with annual fires. Indeed, the empirical studies done at Konza Prairie and elsewhere (Ojima et al. 1994) have indicated that plasticity in nutrient use efficiency by the dominant C_4 grasses plays an important role in maintaining the productivity of these grasslands. Our empirical studies have not detected any decrease in soil C and N pools in annually burned relative to unburned prairie, which was predicted by earlier models. Some of the mechanisms acting to prevent decreases in soil C and N are reduced potential for leaching and denitrification losses of N, increased N immobilization potential related to greater belowground C inputs, and reduced soil C and N mineralization rates because of changes in resource quality (higher C:N ratios) and drier soil conditions under burned prairie. These mechanisms were identified only as a result of long-term studies on the effects of fire in tallgrass prairie; these studies have provided insights into the function of tallgrass prairie ecosystems that could not have been obtained solely through modeling efforts. This is an important example of the crucial need for long-term data in studies of ecosystem dynamics.

Acknowledgments This is contribution no. 97-193-B from the Kansas Agricultural Experiment Station, Kansas State University, Manhattan.

14

Belowground Biology and Processes

Charles W. Rice
Timothy C. Todd
John M. Blair
Timothy R. Seastedt
Rosemary A. Ramundo
Gail W. T. Wilson

A large proportion of the net primary productivity in grassland ecosystems is allocated belowground, and much of the C fixed by plants is processed by heterotrophic soil organisms (Elliott et al. 1988). This pattern of C allocation and processing by the soil biota leads to the accumulation of the large stores of soil organic matter and nutrients characteristic of the grasslands of the central United States. Soils of tallgrass prairie, in particular, are characterized by high belowground productivity and biomass, large accumulations of organic matter and nutrients, and a large and diverse assemblage of soil biota (Chapters 1, 4, this volume). Although soil processes involved in the cycling of nutrients are discussed in Chapter 13 (this volume), this chapter focuses on belowground productivity, soil biota, and the processing and storage of soil C, with emphasis on some key factors influencing belowground populations and processes. A conceptual model of tallgrass prairie ecosystem function (Fig. 14.1) illustrates the interconnectedness of above- and belowground processes in these ecosystems, as well as the importance of the soil biota in affecting the flow of energy (C) and nutrients through soil.

In spite of their importance, studies on belowground biota and processes in tallgrass prairie have been relatively limited, especially when compared with their aboveground analogs. For example, very few estimates have been made of belowground biomass or productivity in tallgrass prairie (e.g., Kucera et al. 1967; Sims and Singh 1978), largely because of the difficulty and destructive nature of the sampling required to quantify root biomass and turnover. Most of the studies available prior to the initiation of the LTER Program were summarized by Risser et al. (1981). Based on these earlier studies, belowground biomass usually has been estimated to be about twice that of aboveground biomass, but little information has been available on spatial and temporal patterns of belowground productivity or responses to prevalent management practices or natural variability. Although fire, grazing, nutrient limitations, and drought all have been shown to significantly affect aboveground productivity (Chapter 12, this volume), their effects on belowground processes have not been well documented. Likewise, very few data were available on the abundance and composition of the soil biota or the fate of C allocated belowground. Therefore, quantifying patterns of belowground biomass and productivity, detailing the structure and function of soil communities and elucidating the responses of belowground communities and processes to key experimental manipulations, have been areas of emphasis in the Konza Prairie research program.

In order to focus our research efforts in soil ecology, a long-term Belowground Plot experiment (Fig. 14.2) was established in 1986 as part of the Konza Prairie LTER Program. The goal of this experiment is to examine the soil-related mechanisms tied to tallgrass prairie responses to fire, grazing, and nutrient limitations. The experimental treatments include burning (annually burned versus unburned); aboveground biomass removal (annually mowed versus unmowed); and annual fertilization (N only, P only, N + P, and no fertilizer). Results from this experiment and other relevant studies done on Konza Prairie are summarized in this chapter. Our objectives are to build on the earlier studies in tallgrass prairie and to pool data from the Belowground Plot experiment into a synthetic presentation of our current understanding of patterns and controls of belowground biota and processes in tallgrass prairie ecosystems (Fig. 14.1).

Roots

Plant roots and rhizomes provide a direct link between above- and belowground processes and biota; act as important energy and nutrient reserves for plants (Chapter 13, this volume); and provide most of the inputs of energy and organic C to the soil biota (Fig. 14.3). However, relatively few estimates have been made of root and rhizome biomass and productivity in tallgrass prairie ecosystems, especially when compared with studies of aboveground plant productivity (Chapter 12, this volume). Indeed, Weaver's (1954, 1958) treatises on North American prairie remain the benchmarks for studies of root morphology and distribution in tallgrass prairie.

Available studies of root biomass indicate that from two to four times more plant biomass occurs belowground than aboveground in tallgrass prairie, with estimates of total belowground peak biomass (0–90 cm) ranging from about 700–2100 g m^{-2} throughout a range of tallgrass prairie sites (Risser et al. 1981; Seastedt and Ramundo 1990). On Konza Prairie, root biomass (excluding rhizomes) to a depth of 30 cm has been estimated to be 859 to 1086 g m^{-2} (Seastedt and Ramundo 1990), which is comparable to values from other sites in the Flint Hills (Ojima 1987). Recent estimates of root and rhizome biomass and depth distribution were obtained in an annually burned, deep-soil site on Konza Prairie as part of a cross-site, soil core transplant experiment (Blair unpublished data).

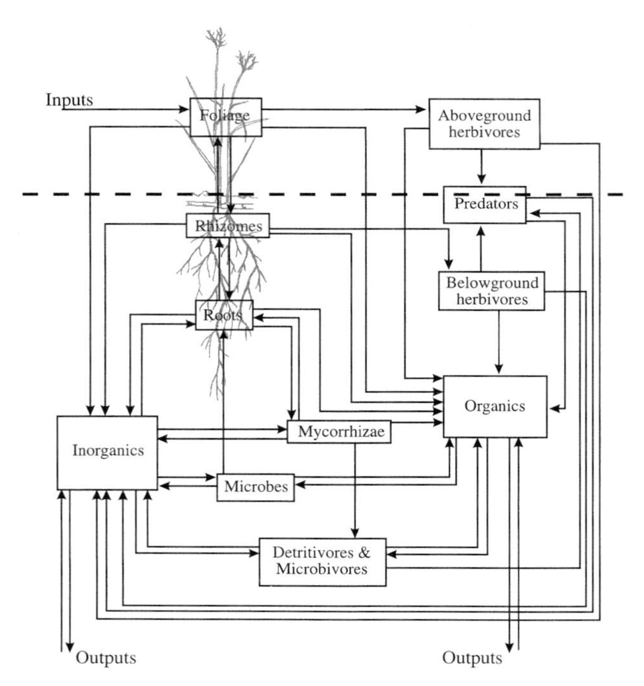

Figure 14.1. A conceptual model of energy (C) flow and nutrient cycling in tallgrass prairie ecosystems, emphasizing the importance of both belowground biotic and abiotic components, as well as the linkages between above- and belowground. From Seastedt and Ramundo (1990).

Total belowground plant biomass (roots + rhizomes) in Konza Prairie cores (25 cm diameter × 60 cm deep) averaged 1914 g m^{-2}. Peak aboveground biomass averaged 629 g m^{-2}, resulting in a mean root:shoot ratio of 3.2. Rhizomes averaged 487 g m^{-2} and constituted about 25% of total belowground biomass. Of the total root mass recovered, averages of 44% occurred at 0 to 10 cm, 25% at 10 to 20 cm, 21% at 20 to 40 cm, and 10% at 40 to 60 cm. These values can be compared with data from Kucera and Dahlman (1968), which indicated that approximately 80% of the total root mass occurred in the upper 25 cm of soil. Variability in root biomass and depth distribution in tallgrass prairie is likely to be high given the effects of topography, climate, edaphic factors, and various management practices (i.e., fire and grazing) on aboveground productivity (Chapter 12, this volume).

Root biomass, production, and disappearance all fluctuate seasonally, but root biomass generally increases throughout the growing season, provided adequate precipitation and soil water are available (Hayes and Seastedt 1987; Benning 1993). The influence of soil water content on root production and turnover was documented at Konza Prairie using a root window approach described by Hayes and Seastedt (1987). Total root length, new root production, and disappearance of old roots all were affected negatively by a drought year. Standing crop of root bio-

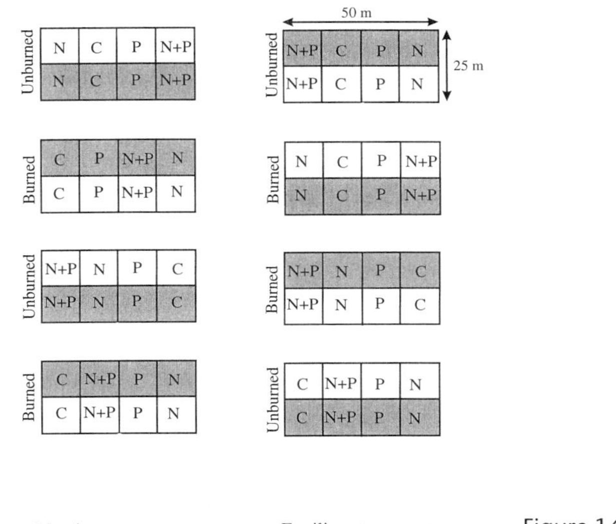

Mowing treatment

Unmowed

Mowed

Fertilizer treatment

C = Unfertilized
N = 10 g N/m^2
P = 1 g P/m^2
N+P = 10 g N + 1 g P/m^2

Figure 14.2. Experimental design of the Belowground Plot experiment initiated on Konza Prairie in 1986.

Figure 14.3. The greatest proportion of plant biomass in tallgrass prairie occurs belowground as roots and rhizomes. Both live and dead plant biomass provide the energy to support a wide array of belowground consumers across several trophic levels. (A. Knapp)

mass, or total root length, is the net result of the production of new roots and the disappearance of old ones (i.e., root turnover). The decrease in root production (56%) associated with an extended dry period was greater than the decrease in rates of disappearance of old roots (32%), resulting in a net decrease in total root length (Hayes and Seastedt 1987). Changes in turnover rates of root length in response to soil water content also have been documented at Konza Prairie. Turnover rates during the drought year were estimated to be 564%, as opposed to 389% in the following year when precipitation amounts were near average. The rapid turnover of root length during the growing season reflects the dynamic nature of the plant root system in tallgrass prairie, as well as the potential for plant roots to respond to spatial and temporal environmental variability. Not surprisingly, root dynamics in the upper 10 cm were most sensitive to changes in soil water content, with pulses of new root growth corresponding to periods of high rainfall during the growing season. However, time lags in plant responses and phenological constraints may confound the relationship between precipitation and root dynamics.

Patterns of root production and disappearance observed in the root windows were correlated with changes in root biomass, although estimates of the turnover rate of root biomass (25–40%) have been much lower (Dahlman and Kucera 1965, 1969; Sims and Singh 1978; Hayes and Seastedt 1987). Precipitation patterns also influence the status of roots in tallgrass prairie, with drought conditions resulting in reduced live root biomass but possibly increased dead root biomass (Hayes and Seastedt 1987). Short-term increases in total root biomass and root:shoot ratios in response to reduced precipitation have been observed in Konza Prairie cores transplanted to a more arid site (Blair unpublished data), but whether this was the result of altered patterns of C allocation or changes in root turnover remains to be resolved.

Responses to Fire

Fire frequency significantly alters root growth dynamics, root biomass, and root nutrient content. Root biomass appears to be greater in frequently burned tallgrass prairie than in unburned prairie (Table 14.1; Fig. 14.4), and this response has been documented in several studies on Konza Prairie (Garcia 1992; Benning 1993). Hayes and Seastedt (1987) reported similar rates of root disappearance in burned and unburned plots, suggesting that the increase in root biomass associated with frequent burning is a function of increased rates of root production, as opposed to decreased root senescence (Seastedt and Ramundo 1990). Perhaps one of the most significant belowground responses to burning is a decrease in the tissue N content of rhizomes and roots (Benning 1993; Ojima et al. 1994), resulting in increased C:N ratios of belowground plant biomass in frequently burned prairie. These fire-mediated increases in root productivity and decreases in root tissue quality have important effects on N cycling processes (Chapter 13, this volume), as well as on belowground food webs, as discussed later in this chapter. The combination of increased root productivity and slower decomposition rates associated with frequent burning also is likely to play an important role in the accrual of soil organic matter in these grasslands (Seastedt 1995).

Table 14.1. Comparisons of root and rhizome biomass of burned and unburned tallgrass prairie

Depth of Sample	Study Site	Burned (g m^{-2})	Unburned (g m^{-2})	Reference
35 cm	Illinois	1,064	839	Hadley and Kieckhefer 1963
5 cm	Missouri	956	669	Kucera and Dahlman 1968
100 cm	Illinois	2,107	1,908	Old 1969
30 cm	Kansas	1,002	790	Ojima et al. 1994[a, b]
30 cm	Kansas	1,086	859	Seastedt and Ramundo 1990[a, b]
20 cm	Kansas	1,618	1,362	Garcia 1992
20 cm	Kansas	960	838	Benning 1993[b]

Source: From Seastedt and Ramundo (1990).

[a]Calculated from C data, assuming mass=2.5 times the amount of C.
[b]Rhizomes not included.

Responses to Grazing

Belowground plant responses to large ungulate grazers appear to be variable in different grassland ecosystems (Milchunas and Lauenroth 1993). To date few studies have considered root responses to the presence of aboveground grazers at Konza Prairie, although data reported by Vinton and Hartnett (1992) suggested reduced root biomass in response to *Bos bison* grazing. Belowground plant responses to grazers are likely to result from both removal of aboveground plant biomass and redeposition of more labile forms of nutrients in the form of dung and urine (Detling 1988; Frank et al. 1994). Studies on Konza Prairie in which mowing has been used to remove aboveground plant biomass have yielded variable results. In most cases, root biomass has been reduced by foliage removal (Garcia 1992; Todd et al. 1992; Turner et al. 1993), although increases also have been observed, particularly in the absence of burning (Benning 1993). Like burning, repeated annual mowing also reduces the N content of roots (Turner et al. 1993; Blair, unpublished data), and additions of supplemental N appear to increase root biomass under mowed, but not unmowed, tallgrass prairie (Garcia 1992; Benning 1993).

Responses to Nutrients

Belowground plant responses to N additions in tallgrass prairie are not interpreted as easily as the increases in aboveground plant productivity often reported in response to supplemental N (Owensby and Anderson 1969; Seastedt et al. 1991). In an early analysis of root data from a long-term, fire-mowing-fertilization study on Konza Prairie (the Belowground Plot experiment), Benning (1993) reported increases in total belowground biomass of approximately 15% in response to N fertilization. However, subsequent sampling and analysis of root data from these same plots in 1994 indicated reductions in live root biomass following 10 years of chronic N fertilization (Rice unpublished data). This suggests

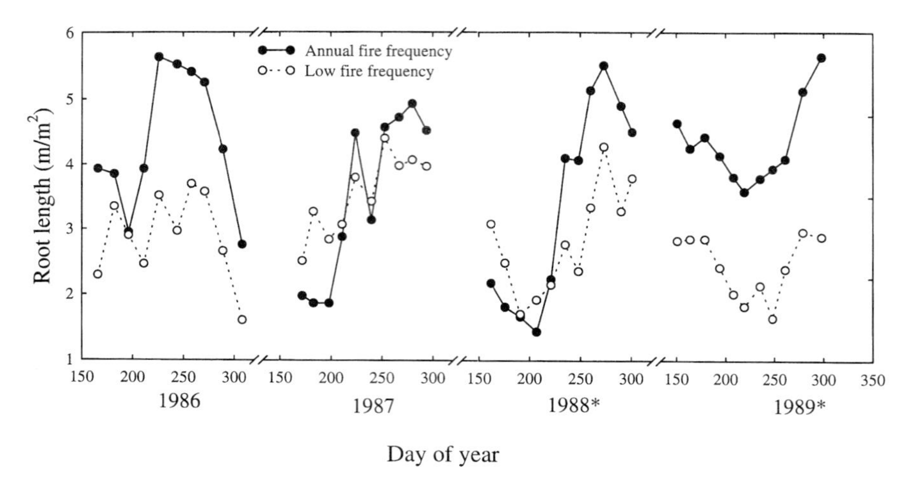

Figure 14.4. Root lengths measure from root window in unburned (low fire frequency) and annually burned tallgrass prairie. From Benning (1993). * beside year on the x-axis denotes drought years.

differences in short- and long-term responses of plants to nutrient enrichment, although these contradictory responses may be confounded with different precipitation patterns, reflecting dry and wet conditions, respectively. This points out the need for long-term studies to evaluate the responses of belowground systems to experimental manipulations and natural variability. In contrast to root biomass, N content of roots exhibits consistently large (about 50%) increases in response to N fertilization.

Mycorrhizal Fungi

Arbuscular mycorrhizae (AM) fungi are ubiquitous in tallgrass prairie and form associations with the roots of most prairie plant species. These fungal symbionts and their extraradical hyphal networks have strong influences on plant uptake of P, soil structure (aggregation), and rhizosphere interactions. The significance of these AM fungi and the processes they mediate is evidenced by their effects on aboveground host plant growth, competitive relationships, grazing responses, and overall plant community structure (Chapter 6, this volume). Here we focus on the effects of fire, grazing, nutrient additions, and topography on the AM fungal community.

Fire, grazing, and topoedaphic conditions influence both the distribution and abundance of mycorrhizal fungi in tallgrass prairie and the development and function of the symbiosis. Ten to 20 species of AM fungi are isolated commonly from Konza Prairie soils (Chapter 4, this volume). Spatial variation in fungal species composition is related primarily to fire frequency, topographic position, and grazing (Gibson and Hetrick 1988; Eom et al. unpublished data); however, the strengths of these relationships vary considerably among years (Bentivenga and Hetrick 1992a, b).

The effects of annual mowing, burning, and N and P fertilization regimes on AM fungi were studied over a 10-year period after initiation of these treatments in the belowground plot experiments. Spring burning significantly reduced AM fungal species diversity and evenness but increased spore abundance (Eom et al. unpublished data). This appears to be a general response because most of the 17 fungal species present increased in abundance with repeated annual burning. Although fire altered fungal community structure, neither fire nor mowing significantly affected AM fungal colonization of roots or the development of extramatrical hyphae (EMH) (Eom et al. unpublished data). In contrast, N additions greatly increased both root colonization and EMH, whereas P amendment decreased EMH development. Unlike fire, nutrient addition had no significant effects on AM species composition or diversity.

Mycorrhizal community structure also can be altered by cattle grazing. Studies on Konza Prairie indicate that moderate grazing decreases AM fungal spore diversity, evenness, and mycorrhizal root colonization. Sites subjected to higher grazing intensity had lower species diversity and root colonization, although AM species evenness was unaffected. Alternatively, EMH was increased significantly under high grazing intensities.

Effects of these different treatments or management practices on AM fungal communities and function may be mediated through changes in soil resources or microclimate or through changes in their host plants. Effects of fire may be related to both direct and indirect effects of altered soil temperatures, soil water potential, and host plant species composition. Evidence suggests that the direct effects of burning or grazing on mycorrhizal fungal composition and abundance may be short-lived, whereas significant long-term effects of these practices may occur indirectly through changes in host plant species composition (Bentivenga and Hetrick 1992a). The resulting changes in the composition, development, and functioning of AM fungi are extremely important not only because they are critical to belowground processes but also because of their strong links to and influences on plant demography, competitive relationships, and diversity of plants in the tallgrass prairie community.

Soil Organic Matter

As mentioned earlier, one characteristic feature of tallgrass prairie is the relatively large accumulation of soil organic matter (Chapter 1, this volume). Organic matter is both a source of and a temporary sink for several major plant nutrients. Indeed, one reason for the high productivity of these soils when placed under cultivated agriculture is the release of these stored nutrients. The amount of C stored as organic matter in grassland ecosystems is an important consideration in the global C balance, and much recent work has focused on the assessment of C stored in soils (Lugo 1992; Lal et al. 1995). Storage of soil organic matter integrates the effects of factors such as vegetation and soil organisms, climate, parent material, time, and disturbances (Jenny 1941). He portrayed the integration of these factors by constructing soil organic matter isoclines along longitudinal and latitudinal transects in North America. As mean annual temperature increased, soil organic matter levels decreased along a latitudinal transect. Along a longitudinal transect, increasing precipitation increased soil organic matter levels. Vegetation type also greatly affects soil organic matter amounts. In a given area, grasslands produce a greater accumulation of soil organic matter than forests under similar climatic conditions (Seastedt and Knapp 1993). The greater soil organic matter content in prairie is due partly to the greater allocation of C belowground compared with forests. Soil organic matter is not just a reservoir of energy and nutrients; it influences other soil properties that affect soil quality, including structure, cation exchange capacity, water infiltration, and adsorption of organics (Doran and Parkin 1994). Thus, soil organic matter represents a key index of many aspects of ecosystem function.

Soil organic matter includes a continuum of components, or fractions, ranging from recently added plant material (litter) to "aged" compounds such as fulvic and humic acids that have turnover times on the order of centuries. For ease of definition and modeling, soil organic matter often is divided into three pools: active, slow, and passive (Parton et al. 1993; Paul et al. 1995). The more labile pools (active and slow) should be more sensitive to driving environmental variables and changes in plant production, whereas the recalcitrant pool represents long-term storage of C and nutrients. Work on Konza Prairie shows that the more labile pools (active and slow) represent 3% and 27%, respectively, of the total organic C in these soils (Rice and Garcia 1994). For N, 11% and 7 % of total soil N is distributed in the active and slow pools of organic matter, respectively.

Measuring changes in total soil organic matter over time or in response to experimental manipulations is often difficult because of the large quantities of soil organic matter and the large proportion of C residing in the more recalcitrant (passive) fraction. This fraction has turnover times on the order of centuries. Short-term studies conducted on Konza Prairie (under 10 years) often indicate little change in total soil organic matter levels in response to aboveground manipulations. For example, we have not detected a significant change in organic matter levels after 10 years of annual burning. Simulation modeling of the effects of fire in tallgrass prairie predict a depletion of soil organic matter over a period of 20 to 100 years (Risser and Parton 1982; Ojima 1987; Ojima et al. 1990, 1994), but validation of these results is difficult because of the long time scale required to measure these changes. The reduction in soil organic matter levels predicted with frequent burning in spite of higher aboveground net primary production (ANPP) compared with unburned prairie apparently is due to enhanced soil organic matter turnover because of increased soil temperatures after fire (Ojima et al. 1990). However, the major source of soil organic matter in tallgrass prairie is root inputs, and increased root production and decreased root tissue quality (i.e., C:N ratio) may act to offset potential increases in soil organic matter mineralization rates. In fact, because roots and rhizomes constitute a large proportion of plant production, we would not expect to see dramatic changes in soil organic matter levels with aboveground detritus removal by either fire, mowing, or grazing. The hypothesis that changes in soil organic matter pools are driven largely by changes in root production and turnover will be evaluated with continuation of the Belowground Plot experiment.

Given the difficulty in detecting short-term changes in total soil organic matter in tallgrass prairie, we may be able to anticipate long-term changes in total soil organic matter by measuring changes in the more labile fractions. Annual burning has increased the proportion of soil organic C in the slow organic C fraction at the expense of the passive fraction relative to proportions in unburned prairie (Rice and Garcia 1994). The increase in the slow fraction is probably the result of increased belowground plant production, providing a greater input of new C to the soil. The higher ratio of slow C to total organic C may indicate long-term changes in total soil organic C (Anderson and Domsch 1989; Insam et al. 1989; Insam 1990). In particular, the loss of C in the passive fraction may be an early indication of long-term losses in soil organic matter as predicted by the CENTURY model (Risser and Parton 1982; Ojima 1987; Ojima et al. 1990, 1994); this also will be evaluated with continued measurement in these long-term treatments.

An additional anthropogenic change that has the potential to alter C source-sink relationships in tallgrass prairie is increased concentrations of atmospheric CO_2. In a study of tallgrass prairie responses to elevated CO_2, we have measured a slight, nonsignificant increase in soil organic matter after 3 years (Rice et al. 1994), although both above- and belowground plant production increased under a doubled CO_2 environment. We also have measured an increase in the more dynamic microbial and active fractions of soil organic matter in response to elevated CO_2 (Chapter 16, this volume). These changes suggest that tallgrass prairie soils have the potential to sequester additional C under an elevated CO_2 scenario.

Microorganisms

Microorganisms constitute the living component of soil organic matter and are thought to contain 1 to 4% of the total organic C (Anderson and Domsch 1989; Sparling 1992) and 2 to 6% of the organic N (Jenkinson 1988). Soil microbes also are considered the primary agents responsible for transforming nutrients contained in litter and soil organic matter to forms available for plant uptake (Jenkinson and Ladd 1981). Clark (1977) indicated that N requirements in the shortgrass prairie ecosystems were met by internal translocation by plants and mineralization of soil organic matter by microorganisms. Data from Konza Prairie suggest that microbial mineralization of N must supply about half of the annual N requirements of tallgrass prairie vegetation (Seastedt and Ramundo 1990).

The quantity of the soil microbial biomass occurring in tallgrass prairie soils is regulated by substrate and water availability, protection by soil particles, and temperature (van Veen et al. 1984; McGill et al. 1986). Thus, fire, grazing, and nutrient availability should directly and indirectly regulate microbial biomass dynamics. Because microbial biomass represents the base for many other soil biota, the dynamics of the microbial biomass have effects that carry throughout the soil food web. That is, changes in microbial biomass and composition may affect invertebrate populations that feed on the microorganisms as discussed later in this chapter. We have not performed an extensive survey of the composition of the microbial community on Konza Prairie, although colony counts suggest 10^5 to 10^6 colony-forming units (CFUs) of bacteria and actinomycetes per gram of soil and 10^4 CFUs of fungi per gram of soil (Garcia 1992). However, colony counts of microorganisms do not necessarily reflect the composition of the entire microbial community, nor do they represent the activity of different microorganisms. Using selective inhibitors (bactericides and fungicides), we have shown that microbial activity in tallgrass prairie is fungal-dominated (Garcia 1992). In fact, fungi appear to be responsible for approximately 74% of total microbial respiration under standard laboratory conditions. Studies on Konza Prairie indicated that frequent burning did not affect the relative contributions of bacteria and fungi to total soil microbial activity. However, the addition of fertilizer N did increase the bacteria:fungal ratio, which may have been the result of a lower C:N ratio of the plant detritus that favored bacterial activity.

The temporal dynamics of soil microbial biomass in tallgrass prairie follow a seasonal pattern of low biomass in the spring and recovery by late summer and early fall (Garcia and Rice 1994). This late-season increase in microbial biomass presumably is tied to retranslocation and turnover of roots (Hayes and Seastedt 1987; Seastedt 1988). Interannual variability in microbial biomass C is related positively to plant production. Thus, microbial biomass C is generally greater in annually burned prairie unless production is limited by drought. In contrast, microbial biomass N appears to be correlated negatively to plant production and drought. That is, when ANPP is high in normal to wet years, N demand by the plant increases, and retranslocation and root turnover of N may be reduced (Hayes and Seastedt 1987; Seastedt 1988). The net result is an increase in the C:N ratio of the microbial biomass during wet years and a lower C:N ratio during drought years (Fig. 14.5). This pattern has been assessed only for 3 years and needs to be evaluated further on longer-term data sets. If this pattern holds true, it would suggest that N is conserved in the living biomass, either plant or microorganisms. This temporary storage of N in the more active phase of the nutrient cycle may provide increased N availability and support a pulse in plant production following a drought (Chapter 12, this volume).

Grazing in most grasslands generally decreases the input of C and the C:N ratio of above- and belowground plant biomass, resulting in reduced microbial biomass and potential for N immobilization (Holland and Detling 1990). In our belowground plot experiments, mowing also decreased microbial biomass by reducing plant C inputs (Seastedt 1988; Holland and Detling 1990). In tallgrass prairie, the amount of C and N in the microbial biomass reflects its importance in the cycling of both elements. Although a large portion of microbial biomass could be considered "dormant," the turnover of microbial biomass in the surface 5 cm is equivalent to one-third of the N annually accumulated by plants. Therefore, microbial biomass may act as a regulator of N dynamics, allowing for con-

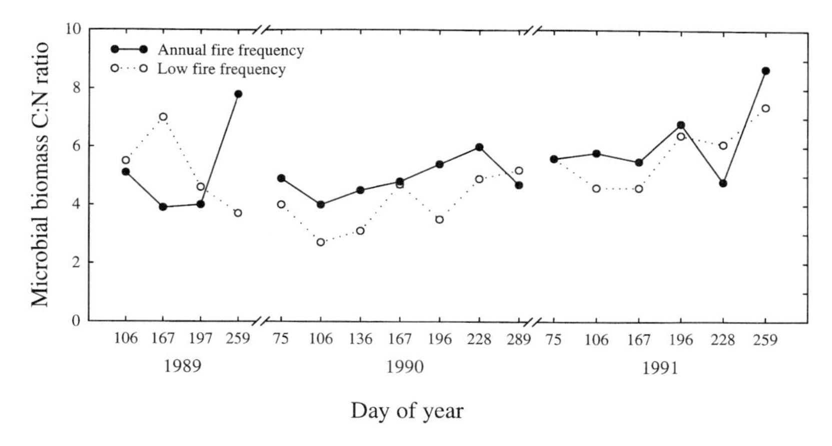

Figure 14.5. C:N ratio of microbial biomass in unburned (low fire frequency) and annually burned tallgrass prairie.

servation of the nutrients in tallgrass prairie. In a study of soils from several ecosystems across the United States, soils from Konza Prairie respired the greatest amounts of C per unit of N mineralized (Zak et al. 1994), providing further support for the concept of N limitation in tallgrass prairie ecosystems (Chapter 13, this volume). The percentage of total soil organic C contributed by microbial biomass C was also high relative to that of other native ecosystems (Zak et al. 1994). This suggests that a greater percentage of the organic matter is labile in tallgrass prairie relative to other ecosystems.

Microbial Activity

Microbial activity is ultimately responsible for the decomposition of organic matter and subsequent release of nutrients and is regulated partly by soil water content, temperature, and interactions with other soil biota (Insam 1990; Wardle and Parkinson 1990). Patterns of microbial activity in tallgrass prairie closely followed changes in soil water content (Garcia 1992); when soil temperatures were over 14°C, microbial activity was regulated primarily by soil water content (Garcia 1992). Other studies also have demonstrated that abiotic factors can regulate microbial activity in tallgrass prairie soils (Kucera and Kirkham 1971; Grahmner 1990). This may explain the short-term negative effect of burning on microbial respiration (Fig. 14.6). Removal of surface litter, increased evaporation, and lower soil water content in burned prairie (Chapter 12, this volume) may cause a reduction in microbial activity following spring fires (Garcia 1992). Under optimal water and temperature conditions, microbial activity in tallgrass prairie soils from Konza Prairie was among the highest measured in soils from 13 different native ecosystems in the United States (Zak et al. 1994). This is further evidence that these prairie soils have large stores of labile C capable of supporting microbial metabolism, and that abiotic factors limit microbial activity

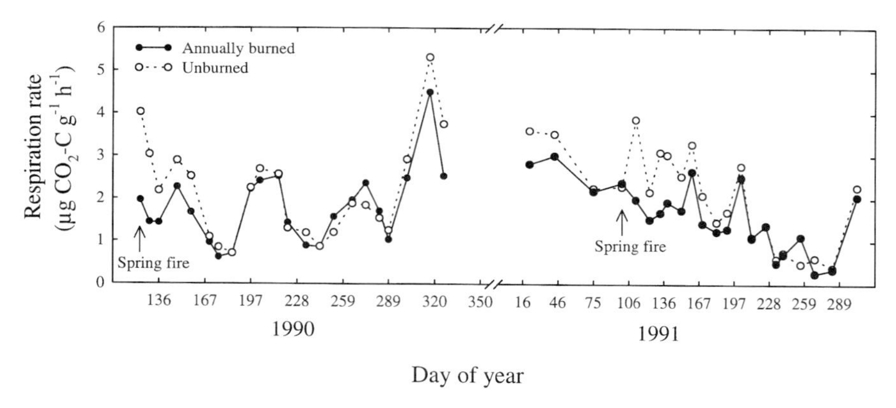

Figure 14.6. Short-term reductions in microbial respiration in response to spring fires in annually burned tallgrass prairie (Garcia 1992). Also shown are data from unburned sites.

under natural conditions and contribute to the accrual of soil organic matter in these ecosystems.

Invertebrates

Prior to the LTER Program on Konza Prairie, studies of the relationships between soil invertebrates and belowground processes in tallgrass prairie were limited largely to research conducted at the Osage Site in Oklahoma as part of the International Biological Program (Risser et al. 1981). In their discussion of invertebrate consumers, these authors reported that no information on energy flow and nutrient relations could be found for this grassland biome. Therefore, comparisons were made with the more xeric shortgrass prairie, where 90% of the invertebrate energy transfer is belowground, with the detrital food web, in particular, being dominated by nematodes (Coleman et al. 1976; Hunt et al. 1987). Similarly, herbivorous nematodes are postulated to have large impacts on plant production in shortgrass and mixed-grass prairies (Smolik 1977; Stanton et al. 1981; Ingham and Detling 1990). Based on data collected at the Osage Prairie Site, Risser et al. (1981) concluded that energy transfer by soil invertebrates in the tallgrass prairie was similar to that in shortgrass prairie but that detrital activity was dominated by groups other than nematodes.

Regulatory mechanisms for soil invertebrate populations also were described poorly for tallgrass prairie prior to the Konza Prairie LTER Program. Fire and grazing, two primary factors affecting productivity and belowground processes in tallgrass prairie, have received considerable attention, but few studies have related population changes to habitat or resource modifications. Warren et al. (1987) reviewed the extensive literature available on grassland arthropod responses to burning. These responses were highly variable, but several studies indicated that soil-inhabiting arthropods of tallgrass prairie were not affected directly by fire but ultimately tended to be favored by the greater productivity associated with frequent burning (Lussenhop 1976, 1981). Invertebrate responses to grazing were best documented for herbivorous nematode densities, which are typically lower in grazed than ungrazed tallgrass and mixed grass prairie (Norton and Schmitt 1978; Risser et al. 1981; Smolik and Lewis 1982). This pattern is radically different from that observed in shortgrass prairie, where grazing tends to stimulate herbivorous nematode densities (Smolik and Dodd 1983; Ingham and Detling 1984), as does supplemental water and N (Smolik and Dodd 1983).

Influence on Ecosystem Processes

Significant plant growth responses to experimental manipulations of soil invertebrate densities have not been observed on Konza to date (Seastedt et al. 1987, 1988a; Todd et al. 1992). This is in sharp contrast to reports of large increases in plant production following similar induced reductions in invertebrate densities in shortgrass and mixed-grass prairies (Smolik 1977; Stanton et al. 1981; Ingham and Detling 1990). Although the effects of soil fauna on plant productivity in tallgrass prairie remain unresolved, root decomposition and N mineralization rates have been demonstrated to be affected measurably by soil macroinvertebrates (James and Seastedt 1986; Seastedt et al. 1987, 1988a). James and Cunningham (1989) reported fine root fragments and approximately twice the total organic matter content of bulk soil in the gut contents of seven species of earthworms found on Konza Prairie, suggesting that earthworms in tallgrass prairie may be feeding preferentially in organically enriched sites in the rhizosphere. In a laboratory study of both native and introduced earthworm species from Konza Prairie, James and Seastedt (1986) presented evidence for an interactive effect of earthworms and plant roots on soil N dynamics. Root growth of *Andropogon gerardii* in pots was stimulated in the presence of native earthworms (*Diplocardia* spp.), whereas nitrate leaching was decreased or increased depending on the presence or absence of plants, respectively. This suggests that earthworms can play a significant role in affecting rates and timing of N mineralization in tallgrass prairie soils, and that plant roots can utilize this N in ways that limit potential leaching losses in these soils. Perhaps the most significant observation was that this effect was not apparent in pots containing the introduced species, *Aporrectodea turgida*, also common to tallgrass prairie soils. Nutrient deposition in earthworm casts exhibits seasonal and spatial patterns (Fig. 14.7) that are likely to further alter nutrient availability and stimulate plant productivity at the ecosystem scale, although the magnitude of these effects has not been determined. Seastedt et al. (1988a) reported higher rates of both root decomposition and N mineralization in the presence of earthworms and macroarthropods, with nearly twice as much N mineralized during a 2-year period. Based on these observations and the conspicuous absence of consistent measurable effects of soil invertebrates on plant productivity, the authors speculated that plant-invertebrate interactions in tallgrass prairie involve opposing processes of similar magnitude, with positive effects on nutrient availability and negative effects caused by root herbivory.

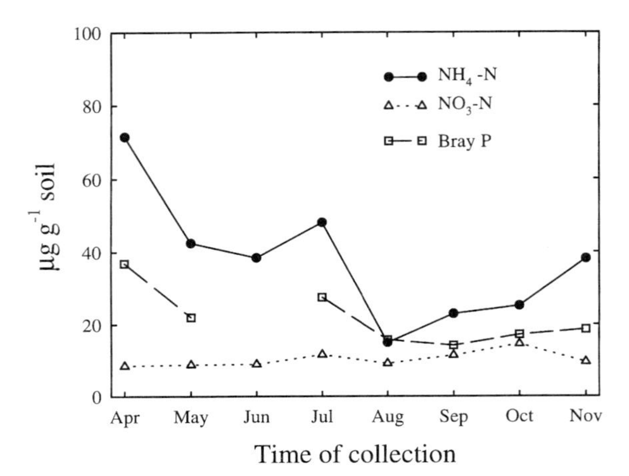

Figure 14.7. Mean monthly concentrations of NH_4-N, NO_3-N and Bray P in surface-collected earthworm casts from Konza Prairie soils. From James (1991).

The effects of other soil invertebrates (nematodes and microarthropods) on N mineralization in tallgrass prairie have yet to be investigated, but studies of food webs in a shortgrass prairie indicate that bacterial-feeding amoebas and nematodes account for most (over 80%) of the N mineralized by soil fauna (Ingham et al. 1986a, 1986b; Hunt et al. 1987). This scenario is unlikely to be true for the tallgrass prairie ecosystem, however, because bacterial-feeding nematodes constitute a relatively small proportion of the nematode community (see Chapter 4, this volume). Instead, the primary invertebrate contribution to the detrital food web and nutrient cycling in tallgrass prairie appears to be through the actions of the macrofauna, particularly earthworms.

Regulatory Mechanisms for Invertebrate Populations

The abundance and diversity of invertebrate populations in tallgrass prairie soil are influenced strongly by climatic factors such as precipitation patterns and by disturbances resulting from management practices such as burning, grazing, and supplemental N. Timing and amounts of precipitation can affect soil invertebrates directly by altering soil water content and indirectly by affecting plant productivity and organic inputs to the soil. Burning, grazing, and N fertilization result in changes in both quantity and quality (primarily N content) of belowground resources (Seastedt et al. 1988a; Todd et al. 1992; Benning 1993; Ojima et al. 1994). These resource responses have been manipulated in various studies on Konza Prairie during the LTER Program to investigate the complex food webs that occur belowground and to identify factors regulating the population dynamics of soil invertebrates. A summary of invertebrate responses to these experimental manipulations is included in Table 14.2.

Precipitation inputs have large but variable effects on tallgrass prairie soil invertebrate populations. Supplemental irrigation applied during the growing season resulted in twice the number of total nematodes during periods with typical

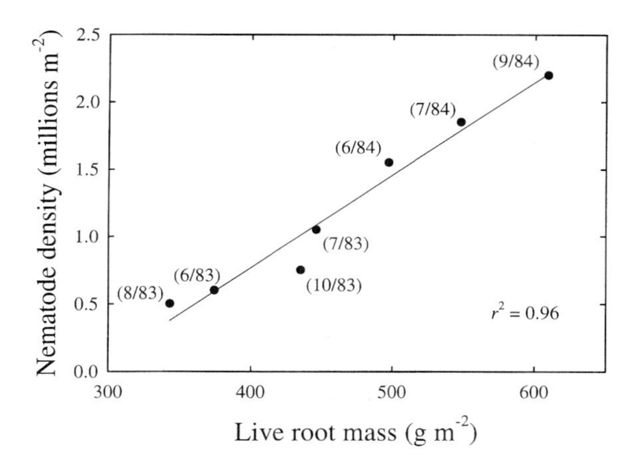

Figure 14.8. Relationship between herbivorous nematode densities and live root mass in annually burned tallgrass prairie. Numbers in parentheses represent sampling dates. From Hayes and Seastedt (1987); Seastedt et al. (1987).

growing season moisture deficits, with increases in herbivorous populations approaching 300% (Todd, unpublished data). James (1992b) reported increases in earthworm numbers and biomass in the presence of irrigation, although individual species responses varied. Similarly, autumn nematode densities during periods of drought were as much as 60% lower than those observed in years with normal precipitation (Todd 1996), and both earthworm activity and nematode numbers were depressed during the drier summer months (Seastedt et al. 1987; James 1991). Although soil water content directly affects invertebrate activity (e.g., earthworms display periods of dormancy during dry periods; James 1992b), some responses appear to better reflect patterns in belowground primary production, such as suppression of new root growth under drought conditions, even when total root biomass remains stable (Hayes and Seastedt 1987; Benning 1993). Consequently, the status of existing roots (live versus dead biomass) may provide the strongest interpretation of nematode responses to precipitation, as illustrated by the relationship between seasonal variation in herbivorous nematode densities and amounts of live roots (Fig. 14.8). Similar responses to soil moisture and belowground production have not been observed for all soil invertebrate populations on Konza Prairie. Microarthropods, for instance, appear to be favored by drier soil conditions, such as those common to upland sites (O'Lear 1996), although the mechanisms underlying this response have not been investigated. Comparisons of long-term patterns in precipitation and invertebrate population densities suggest that such responses are nonlinear and that even those invertebrate groups that respond favorably to increased precipitation are affected adversely by excessive moisture levels. This relationship may be due to direct alteration of the soil environment (e.g., water-filled pore spaces, aerobic status) or may reflect patterns of root production (e.g., cessation of new root growth under high soil water levels; Risser et al. 1981).

The responses of soil invertebrates to fire are similarly variable. The nematode community is not strongly affected overall, but populations of the dominant herbivores (*Helicotylenchus* spp.) and microbivores are increased by burning (Todd 1996). Numbers and biomass of macroarthropod herbivores (Scarabaeidae) and native earthworms (*Diplocardia* spp.) also respond favorably to fire (James 1982, 1988; Seastedt 1984a; Seastedt et al. 1986). These positive invertebrate responses to fire likely reflect increases in belowground plant productivity and microbial biomass, which are common with frequent burning (James 1988; Garcia 1992; Garcia and Rice 1994; Ojima et al. 1994). Microbivorous (primarily bacterivorous) nematodes, for instance, responded similarly to C supplements, which also stimulated production in tallgrass prairie (Seastedt et al. 1988a). In contrast, some microarthropods (mainly oribatid mites and collembolans) and the introduced earthworm, *A. turgida,* exhibited negative responses to fire, ostensibly because these fauna exploit plant litter as a food resource (Seastedt 1984b; James 1988; James and Cunningham 1989). Evidence indicates that burned prairie may represent an unfavorable environment for introduced earthworms because of higher soil temperatures or decreased soil water content (James 1988; James and Cunningham 1989).

Grazing effects have been simulated by mowing in most invertebrate studies to date on Konza Prairie. Once again, invertebrate responses have been inconsistent, with herbivorous nematodes consistently exhibiting 30 to 40% decreases in population densities (Todd et al. 1992; Todd 1996), whereas arthropods and native earthworm species often were more abundant in mowed prairie (Seastedt et al. 1986; Todd et al. 1992). Seastedt et al. (1988b) interpreted arthropod herbivore increases as a response to the increased N content of the reduced root resources in grazed or mowed prairie. Although short-term effects on root N content from mowing may be positive, long-term effects are negative (Turner et al. 1993; Blair unpublished data), which may explain the variable responses by soil invertebrate populations. Microbivorous nematodes, for instance, displayed short-term increases and long-term decreases in response to mowing, whereas fungivores displayed the opposite trend (Todd et al. 1992; Todd 1996). Reductions in herbivore densities and corresponding increases in microbivore or fungivore densities in response to mowing or grazing appeared to be typical responses of nematode communities in tallgrass and mixed-grass prairies (Risser et al. 1981; Smolik and Lewis 1982; Todd et al. 1992). The herbivore response, in particular, correlates well with observed reductions in plant cover and root biomass or N content in mowed prairie (Todd et al. 1992; Gibson et al. 1993; Turner et al. 1993).

In contrast to mowing, ungulate grazing effects in grasslands are characterized by their spatial heterogeneity. Thus, although plant responses to defoliation may be similar, the distribution of those responses in grazed prairie will be affected by herbivore feeding preferences. Dung and urine deposition and other nongrazing activities of herbivores (e.g., wallowing and trampling) further increase spatial heterogeneity in grazed systems by providing a patchy distribution of resources for soil microbial and faunal communities. To date, only the effect of dung deposition on spatial patterns of soil invertebrates has been investigated on Konza Prairie. Using *B. bison* dung, James (1992a) demonstrated that *A. turgida* and certain, but not all, *Diplocardia* spp. migrated to dung pats and fed upon fecal material. On one sampling date, biomass of *A. turgida* under dung pats was 10 times that in the surrounding area. This emphasizes the extent to which

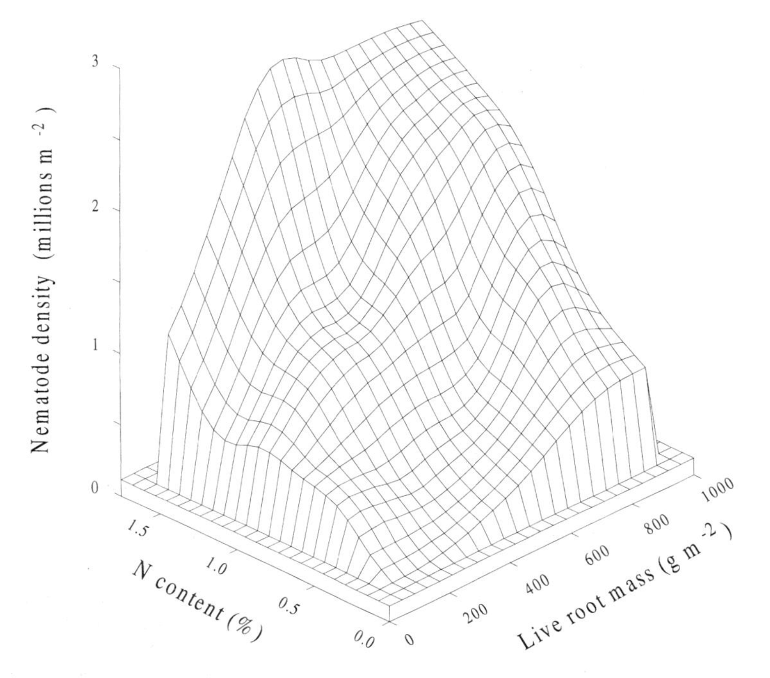

Figure 14.9. Relationship between herbivorous nematode density and live root mass or N content in live roots in tallgrass prairie. Data are from several burning, mowing, and supplemental N treatments in wet and dry years. From Todd (1996).

grazing can contribute to the aggregation of highly mobile groups of soil fauna and, given the differences in attraction observed among earthworm species, suggests that ungulate grazers may affect not only the spatial distribution but also the species composition of soil invertebrate communities.

The most consistent and predictable population responses by soil invertebrates to experimental manipulations in tallgrass prairie have been related to nutrient inputs, especially inorganic N. Nitrogen fertilization stimulates densities of herbivorous and microbivorous nematodes and both macro- and microarthropods (Seastedt et al. 1988a; Todd 1996; Seastedt unpublished data). However, root biomass typically does not mimic invertebrate population responses to N fertilization (Benning 1993; Blair unpublished data); consequently, herbivore densities are related more closely to total amounts of N in roots than to root biomass both within and across years. Herbivorous nematodes, for instance, display dramatic increases in density only in the presence of concomitant increases in both the amount and content of N in live plant roots (Fig. 14.9). Furthermore, this relationship is strongest for arthropod and nematode herbivores in the presence of burning, which is likely a reflection of the greater N limitation of annually burned tallgrass prairie (Seastedt et al. 1991). These observations support the hypothesis by Seastedt et al. (1988a) that the C:N ratio of roots is a primary limiting factor for soil invertebrate herbivores in tallgrass prairie. In contrast, increases

Table 14.2. Summary of belowground responses to experimental manipulations in Konza Prairie studies

Treatment	Positive Response	Negative Response
Irrigation	Native earthworms (James 1992b)	Microarthropods (O'Lear 1996)
	Herbivorous nematodes (Todd, unpublished)	
Fire	Root biomass (Garcia 1992; Benning 1993)	Root N content (Benning 1993; Ojima et al. 1994)
	Passive soil organic matter (Rice and Garcia 1994)	Soil organic matter (Ojima et al. 1990)
	Microbial C (Garcia and Rice 1994)	Microbial N (Rice and Garcia 1994)
	Native earthworms (James 1982 1988)	Microbial activity (Garcia 1992)
	Herbivorous macroarthropods (Seastedt 1984a)	Introduced earthworms (James 1988)
	Herbivorous nematodes (Todd 1996)	Microarthropods (Seastedt 1984b)
	Microbivorous nematodes (Todd 1996)	Predaceous nematodes (Todd 1996)
Mowing	Root biomass (Benning 1993)	Root biomass (Garcia 1992; Todd et al. 1992; Turner et al. 1993)
	Native earthworms (Todd et al. 1992)	Root N content (Turner et al. 1993)
	Herbivorous macroarthropods (Seastedt et al. 1986)	Microbial C and N (Garcia and Rice 1994)
	Microbivorous nematodes (Todd et al. 1992)	Bacteria (Garcia 1992)
	Fungivorous nematodes (Todd 1996)	Herbivorous nematodes (Todd et al. 1992; Todd 1996)
Supplemental N	Total root/rhizome biomass (Benning 1993)	Live root biomass (Unpublished data)
	Root N content (Benning 1993)	Predaceous nematodes (Todd 1996)
	Microbial N (Garcia and Rice 1994)	
	Macroarthropods (Seastedt unpublished data)	
	Microarthropods (Seastedt et al. 1988a)	
	Herbivorous/Microbivorous nematodes (Todd 1996)	

in microbivorous nematode densities and concomitant decreases in the ratio of fungivores to microbivores in the presence of N fertilization simply may reflect similar changes in the microbial community (Garcia and Rice 1994). In this case, the soil invertebrates appear to be responding only to changes in the quantity of food resources (bacterial densities), which, in turn, are regulated by root quality.

Summary

Belowground communities are important components of tallgrass prairie ecosystems. Long-term research at Konza Prairie has yielded many new insights into the temporal dynamics of belowground communities and processes and how these are influenced by climatic variability and various land management practices (e.g., fire and grazing). Generalized responses of plants, soil organic matter, and specific groups of soil biota to several experimental treatments related to management practices are summarized in Table 14.2. These belowground responses are linked in a variety of ways to aboveground community dynamics and whole-ecosystem responses. For example, the effects of fire on aboveground plant productivity (Chapter 12, this volume) and N cycling (Chapter 13, this volume) are mediated partly by changes in belowground processes. Increased allocation of C belowground and changes in soil microclimate produce short-term changes in soil microbial biomass and activity, as well as longer-term changes in the amounts and quality of organic matter stored belowground. The soil invertebrate community, in turn, is affected by changes in plant root inputs (both quantity and quality) and changes in the soil microbial community. Although a number of studies have documented responses in soil invertebrate communities to experimental manipulations of fire, grazing, nutrient additions, and water availability, fewer studies have demonstrated a direct effect of invertebrate activity on plant responses. The potential for these responses has been shown (e.g., the effects of earthworms on soil N dynamics and the response of root-feeding nematodes to root N content), but carefully designed studies are needed to separate the beneficial and detrimental effects of different components of the soil invertebrate community.

Soil microbes regulate rates and patterns of nutrient availability to tallgrass prairie plants in a variety of fundamental ways. The importance of mycorrhizal associations for P uptake by tallgrass prairie plant communities also has been demonstrated in several studies done on Konza Prairie. Clearly, this relationship is important in the low-P soils typical of our region (Chapter 4, this volume). Further studies are needed to clarify the relative importance of mycorrhizal symbiosis for tallgrass prairie on more P-rich soils. Studies at Konza Prairie have confirmed the limiting nature of N availability for both plants and soil microbes and have suggested the importance of microbial immobilization of N as a mechanism for N conservation in these grasslands. The potential linkages between seasonal and periodic droughts and changes in microbial immobilization of N warrant further study.

Finally, belowground measurements collected during the Konza Prairie LTER Program to date emphasize that ecosystem models developed for other North American grasslands (Parker et al. 1984; Hunt et al. 1987) are inadequate to describe the tallgrass prairie. Higher productivity and moisture availability, greater N limitation, fungal dominance of microbial activity, the sizable contribution of earthworms, and higher ratios of plant- and fungal-feeding to bacterial-feeding nematodes are unique characteristics of tallgrass prairie that require more detailed attention before comprehensive food web and nutrient cycling models can be developed for this ecosystem. Additionally, management regimes, particularly burning frequency and grazing intensity, are integral components of the tallgrass prairie ecosystem, and research on Konza Prairie has demonstrated that such practices result in complex changes in soil biota and belowground processes that warrant further elucidation.

Acknowledgments This is contribution no. 97-202-B from the Kansas Agricultural Experiment Station, Kansas State University, Manhattan. Partial financial support was provided by the Department of Agronomy, Kansas State University. We thank Judy Adams for her careful editing and typing of the manuscript.

15

A Landscape Perspective of Patterns and Processes in Tallgrass Prairie

John M. Briggs
M. Duane Nellis
Clarence L. Turner
Geoffrey M. Henebry
Haiping Su

In earlier chapters, landscape position was identified as an important characteristic for understanding many of the spatial and temporal patterns measured at Konza Prairie. For example, annual spring burning consistently increases total aboveground net primary production (ANPP) only in lowlands sites (Chapter 12, this volume). In addition, landscape position may affect the movement and availability of nutrients (Chapter 13, this volume). This chapter will focus on broader issues in landscape ecology, specifically, how a landscape-level perspective can help evaluate the relationship between ecosystem patterns and processes in tallgrass prairie.

Landscape ecology as a discipline was not discussed in Risser et al. (1981). This is not surprising considering the recent development of this field. In a later paper, Risser et al. (1984) presented a working definition of landscape ecology as "the development and dynamics of spatial heterogeneity, spatial and temporal interactions and exchange across heterogeneous landscapes, influences of spatial heterogeneity on biotic and abiotic processes, and management of spatial heterogeneity." The concept of landscape ecology was further defined and broadened in scope by Forman and Godron (1986), Turner (1989, 1990), and Wiens (1997) and has evolved to emphasize the structure and dynamics of landscape mosaics and their effect on ecological phenomena. We will build upon this theme and, more specifically, will consider the development and dynamics of spatial heterogeneity in tallgrass prairie, as well as interactions and exchanges across this heterogeneous landscape. Assessing and evaluating spatial heterogeneity in tallgrass prairie has been the focus of most of the landscape ecology work on Konza Prairie.

Traditionally, ecological studies have been conducted primarily in plots of 1

m^2 or less (Kareiva and Anderson 1988; Brown and Roughgarden 1990). Increased interest in large-scale issues such as global change, loss of biodiversity, rapid changes in land cover, and a growing human population have caused scientists to become increasingly aware of how ecological patterns and processes change in time and space, especially with regard to how anthropogenic pressures impact natural ecological communities. Landscape ecology, with an emphasis on spatial heterogeneity and large spatial scales, can help address these concerns. However, to do so, it is imperative to develop the database, tools, and techniques for these important tasks. Over the last 15 years, our research at Konza Prairie has produced a database that, when combined with the proper tools and techniques, can help ecologists understand complex ecosystems.

Importance of Techniques

As with many emerging disciplines, the development of appropriate tools and technologies is critical to the success of question-driven research in landscape ecology. Indeed, the development of tools specific to the tallgrass prairie represented a substantial portion of our early research efforts (Nellis and Briggs 1987; Su et al. 1990; Briggs and Nellis 1991; Henebry 1993; Henebry and Su 1993). In order to facilitate quantitative measures of the tallgrass prairie landscape at a variety of spatial and temporal scales, we have used two primary tools, remote sensing and geographic information systems (GIS), for capturing, manipulating, processing, and analyzing spatial or geo-referenced data. These systems address both geometric data (coordinates and topologic information) and attribute data (e.g., information describing the properties of geometric spatial objects such as points, lines, and areas or electromagnetic energy responses to surface or canopy conditions). With these tools, ecological processes can be measured at spatial scales beyond the typical plot measurements. For example, the interactions between land management practices (e.g., burning) and seasonal changes in the Flint Hills are quite complex, yet they can be visualized and quantified across the Konza Prairie landscape with remote sensing and GIS techniques (Plate I). In addition, interannual variability in ecosystem patterns and processes (especially peak aboveground biomass) can be quite dramatic in tallgrass prairie (Plate II; Chapter 12, this volume). Contrast the differences between a near-record wet year (1993) and a year of average precipitation (1987). In 1987, watersheds that were burned in the spring are easily distinguished, while in 1993, few differences among watersheds were detectable (Plate II). Over the last 10 years, we have used remote sensing data coupled with a GIS to monitor entire watersheds and assess the impacts of different management practices on landscape patterns over time and space (Briggs and Nellis 1991; Groffman and Turner 1995; Su et al. 1996). Such analyses typically combine traditional point-level measurements with large spatial data sets to conduct landscape-level measurements on Konza Prairie. In the following we review specific examples of landscape-level analyses made possible because of the development of tools and techniques for measuring and analyzing spatial data.

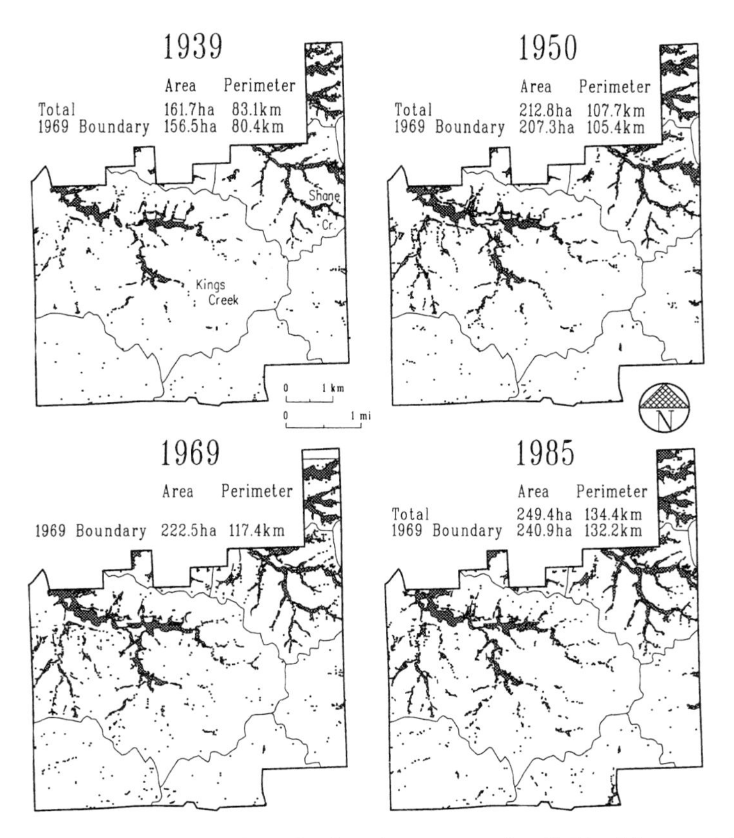

Figure 15.1. GIS representation of gallery forest expansion digitized from aerial photos from 1939, 1950, 1969, and 1985. Note how the area of forest increased from 161.7 ha in 1939 to 249.4 ha in 1985. Extent of major drainages on Konza Prairie are outlined. From Knight et al. 1994; reprinted with permission of SPB Academic Publishing.

Historical Changes in the Konza Prairie Landscape

Large-scale changes in land cover are ideal foci for a landscape approach. One question of fundamental importance for our long-term studies is: How has the tallgrass prairie landscape changed within the past 100 years? Our analysis indicated that the most extensive change that has occurred on the Konza Prairie landscape within the past 100 years has been an increase in forested areas. Based on analysis of the original land office surveys for Konza Prairie, Abrams (1986) reported a dramatic increase in gallery forest between 1859 and 1939. In 1859, only two areas of about 5 ha of continuous forest were noted. Because these surveys were conducted only along section lines (i.e., N-S and E-W), the forested area probably was underestimated. However, in general, the Flint Hills region in the 1800s was described as rolling prairie devoid of any woody vegetation (Abrams 1986).

Knight et al. (1994) conducted a detailed spatial analysis of the gallery forest of Konza Prairie by digitizing aerial photographs taken over a 46-year period (Fig. 15.1). Because this information was placed into a GIS database along with other GIS coverages such as soils, a digital elevation model, and fire history of Konza Prairie, a rigorous analysis of the expansion of the gallery forest was possible. In contrast to the 5 ha of forest recorded in 1859 (Abrams 1986), over 159 ha of forest was present on Konza Prairie in 1939. Since then, the area of forest has increased in each sampling period, such that a total of 250 ha of forest was noted in 1985. More important, Knight et al. (1994) concluded that the expansion of gallery forest on Konza Prairie was not limited by the landscape and, in effect, only 10 to 15% of the alluvial-colluvial deposits along stream channels (the best habitat for forest to exist on Konza Prairie) were forested. Instead, other factors such as fire still may be limiting the expansion of forest (Chapter 6, this volume).

Spatial Patterns in the Present Konza Prairie Landscape

Satellite-borne remote sensing data have allowed us to examine present-day landscapes at larger spatial scales than is possible with conventional methods. For example, estimates of ANPP from watersheds on Konza Prairie are derived from sampling 20 to 40 0.1-m^2 quadrats (Chapter 12, this volume). In contrast, satellite-obtained data encompass the entire watershed at a scale unobtainable with other data collection systems. When used in conjunction with point data on ecosystem processes, satellite data can be exceptionally valuable for estimating patterns and processes across watersheds and/or even larger areas (Plates I, II).

NDVI and ANPP

The most commonly used vegetation index of "greenness" (the Normalized Difference Vegetation Index; NDVI) contains satellite-derived information on both the amount and the condition of vegetation. In general, NDVI can be considered an index of photosynthetic capacity, as well as an estimator of aboveground vegetative biomass. The NDVI is a difference ratio of the radiances computed from the formula: NDVI = (brightness value from infrared band – brightness value from red band)/(brightness value from infrared band + brightness value from red band). The NDVI is a bounded ratio that varies from – 1.0 to +1.0, with actively growing vegetation having positive values. Although the NDVI shows some sensitivity to other parameters (such as atmospheric conditions), the dominant signal appears to be related primarily to surface conditions (Box et al. 1989). Estimation of aboveground biomass in tallgrass prairie is facilitated by the fact that the canopies are simple in structure, with much of the aboveground biomass contributing directly to reflectance. In addition, biomass production is a cumulative process within a single season, with peak green biomass roughly equivalent to ANPP (Briggs and Knapp 1995). In most years, peak biomass in tallgrass prairie occurs in August or early September. For these reasons, satellite data derived from reflectance characteristics of the vegetation have proven to be valuable tools in the study of tallgrass prairie processes at the landscape scale on

Figure 15.2. An NDVI image of 15 August 1987 (from Landsat Thematic mapper data). Brighter areas designate higher NDVI values and, thus, more green vegetation. Note interstate I-70 along the southern boundary (bottom of figure) of Konza Prairie, the agricultural fields to the north (top of figure), and the dendritic pattern of the gallery forests along stream channels. Burned watersheds are characterized by their uniform brighter shades.

Konza Prairie (Nellis and Briggs 1987; Briggs and Nellis 1989; Nellis and Briggs 1989; Benning and Seastedt 1995; Briggs et al. 1997).

Tallgrass prairie vegetation produces an NDVI signal that exhibits strong seasonal patterns associated with climate-driven phenological development and strong spatial patterns related primarily to topography and land use (Fig. 15.2). In most years, greenup begins in late April or early May, peaks in June or early July, and declines slowly thereafter. Despite increases in aboveground biomass, greenness decreases until August or early September, as senescent material accumulates in the canopy. Areas dominated by C_3 vegetation exhibit two temporally distinct pulses of activity (and greenness); areas dominated by C_4 vegetation exhibit a single pulse of activity and maximum greenness at midseason (Fig. 15.3). Such differences in the temporal dynamics of C_3 and C_4 systems may be useful in determining spatial patterns of plant life-form distributions.

In grasslands, spatial patterns in greenness are determined largely by the effects of topographic position and land use on plant species composition and productivity. Greenup occurs more rapidly and to a greater extent in lowlands than in

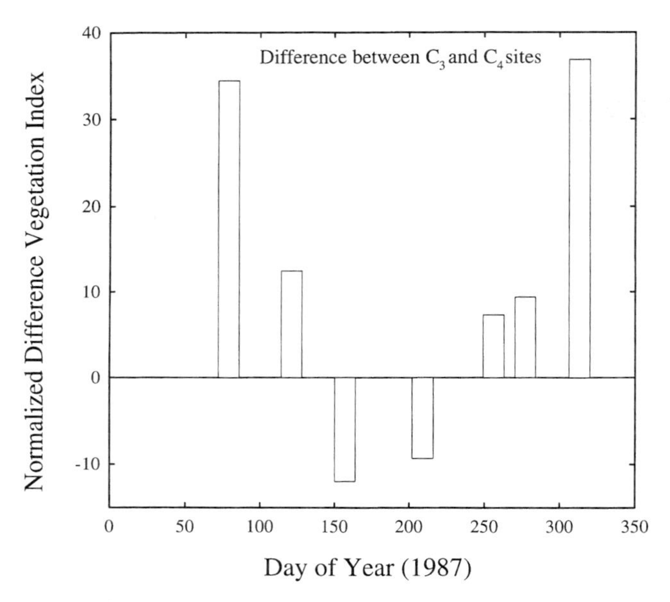

Figure 15.3. Scaled NDVI data that illustrate the differences between C_3 and C_4 grasslands in eastern Kansas. The C_3 grasses have a bimodal distribution with peaks occurring in spring and fall, whereas C_4 grasses have a single peak in the summer.

uplands. Greenup also is more rapid in areas burned in the spring, partly because fire removes senescent material from the canopy and promotes rapid growth of the dominant C_4 vegetation (Plate I; Chapter 12, this volume). The relative magnitude of greenness also is affected by grazing and burning. Because burning removes senescent material produced in previous growing seasons, NDVI values generally are higher on burned areas than on unburned areas. Grazing by large ungulates removes substantial quantities of aboveground biomass. Values of NDVI are generally lower on grazed areas than on ungrazed areas, reflecting differences in standing crop at the time of data acquisition rather than differences in biomass produced (Turner et al. 1992). Grazing also introduces increased temporal variation in reflectance characteristics that is a function of the frequency and intensity of biomass removal and regrowth responses of the vegetation.

At Konza Prairie, we have accumulated an extensive database relating ground-based measures of primary production and satellite spectral reflectance measurements. In order to evaluate the validity of using NDVI as a surrogate for aboveground biomass in tallgrass prairie, we obtained thematic mapper (TM) data for the years 1984 to the present. The dates of the TM scenes were within 2 to 3 weeks of the time biomass was harvested to estimate ANPP (Chapter 12, this volume). All scenes were rectified geometrically, and NDVI was calculated for each of the scenes. Because the locations of the ground reference plots for clipping are in a GIS, the NDVI values for each of the harvested plots were obtained. A regression model then was used to describe the relationship between

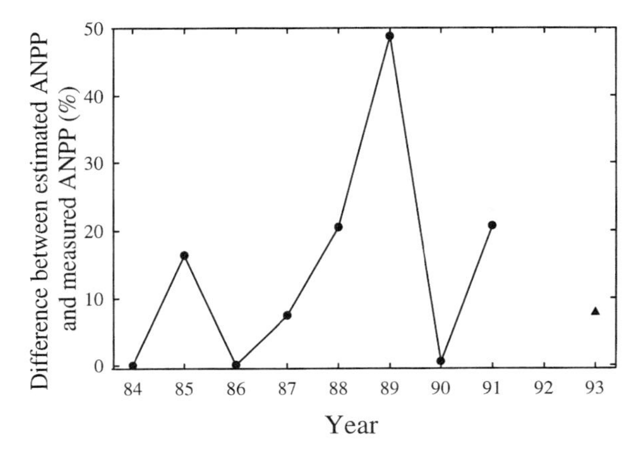

Figure 15.4. Difference between ANPP estimated using an NDVI regression model derived from TM data for the years 1984 to 1991 and ANPP measured from LTER plots. Data from 1993 (triangle) also are shown.

NDVI and ANPP using half of the data set (randomly selected). A significant relationship was found between NDVI and ANPP for the time period of 1984 to 1991 ($P < 0.001$). However, only a small amount of the variance was explained ($r^2 = 0.11$). Using this regression model for each of the years on the other half of the data set, comparisons of ANPP estimated using the regression model of NDVI with actual ANPP yielded interesting results (Fig. 15.4). In four of the years (1984, 1986, 1987, and 1990), little difference (less than 10%) occurred between estimated and measured ANPP. In the other years, however, differences were as high as 49%, and overall the difference was 14.3% $\pm$ 5.65. In addition, data from a 1993 TM scene showed a difference of 8% (Fig. 15.4).

Thus, although ANPPs determined from ground harvest data and NDVI estimates were very similar in some years, overall, we suggest that NDVI values in tallgrass prairie should not be used as a substitute for measured ANPP across all years, especially if the relationship in one year is used to predict ANPP in other.

This is also true if predictive capability is the goal of using NDVI and ANPP over a landscape that has different land cover types. For example, we have found that, within a year, separate regression models relating reflectance and biomass on burned and unburned areas may be necessary. The reason for this is due to the buildup of litter. Data summarized over a 10-year period show that a large amount of dead material (litter) is present (382 g/m^2 $\pm$ 19) at most sites after a few years (2–5). This litter layer affects reflectance readings and thus dictates that separate relationships between remote sensing data (i.e., NDVI) and ANPP be determined for burned and unburned tallgrass prairie. Year-to-year differences in the relationship between ANPP and reflectance can be large. In fact, for the least complex vegetation community on Konza Prairie (ungrazed, annually burned watersheds that are dominated by C_4 grasses; Chapters 6, 9, this volume), different and significant relationships are found for each year (Fig. 15.5). Furthermore, a general relationship derived from data obtained over multiple years would result in less accurate ANPP estimates than could be obtained from relationships limited to a single year's observations (Fig 15.5).

Given this limitation, we calculated mean ANPP for an entire watershed by

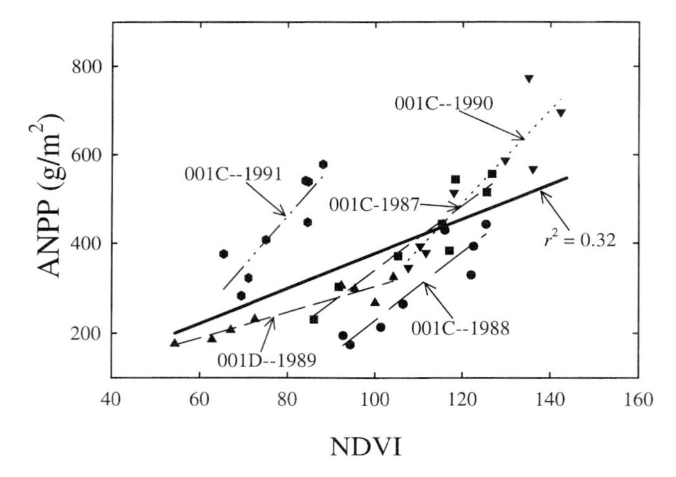

Figure 15.5. Relationships between scaled NDVI derived from TM data and ANPP measured in annually burned watersheds over a 5-year period (1987–1991). In any single year/site, the relationships are strong ($r^2 > 0.80$), but the slopes and the intercepts of the regression lines differ markedly between watersheds and years. When all data are combined into a general regression model, the r^2 decreases to 0.32 (solid line). This suggests that caution is needed when using general NDVI relationships to predict ANPP.

applying the regression model of NDVI to the mean NDVI for each of the watersheds that are sampled for ANPP. Comparing the estimated ANPP calculated from the NDVI regression model with the ground-based data showed that ANPP was underestimated from the NDVI regression model on watersheds when production was high but overestimated when production was low (Fig. 15.6). The underestimation of ANPP at the watershed level when production was high might be due to half of our ground ANPP data coming from lowland soils, which have the highest ANPP potential (Chapter 12, this volume) but are not the dominant soil type on Konza Prairie (Chapter 4, this volume). Thus, we may have overemphasized the lowland soils and, thus, overestimated ANPP on a watershed level when scaling up from plot data. One explanation for the NDVI overestimate of ANPP when productivity was low is that NDVI is calculated from the entire watershed, so other plant life-forms that contribute to high NDVI values (e.g., trees and shrubs; Fig. 15.2) are included in the reflectance data but not in our ground-based plots. Both of these speculations require further study.

Other Applications of NDVI

Briggs and Collins (1994) assessed the relationship between satellite-derived NDVI measurements from transects used to determine vegetative community composition on Konza Prairie. From these ground-level measurements of vegetative composition, a measure of spatial heterogeneity along each transect was determined as the mean dissimilarity in plot species composition among samples (Collins 1992). Using a series of texture algorithms on the NDVI measurements,

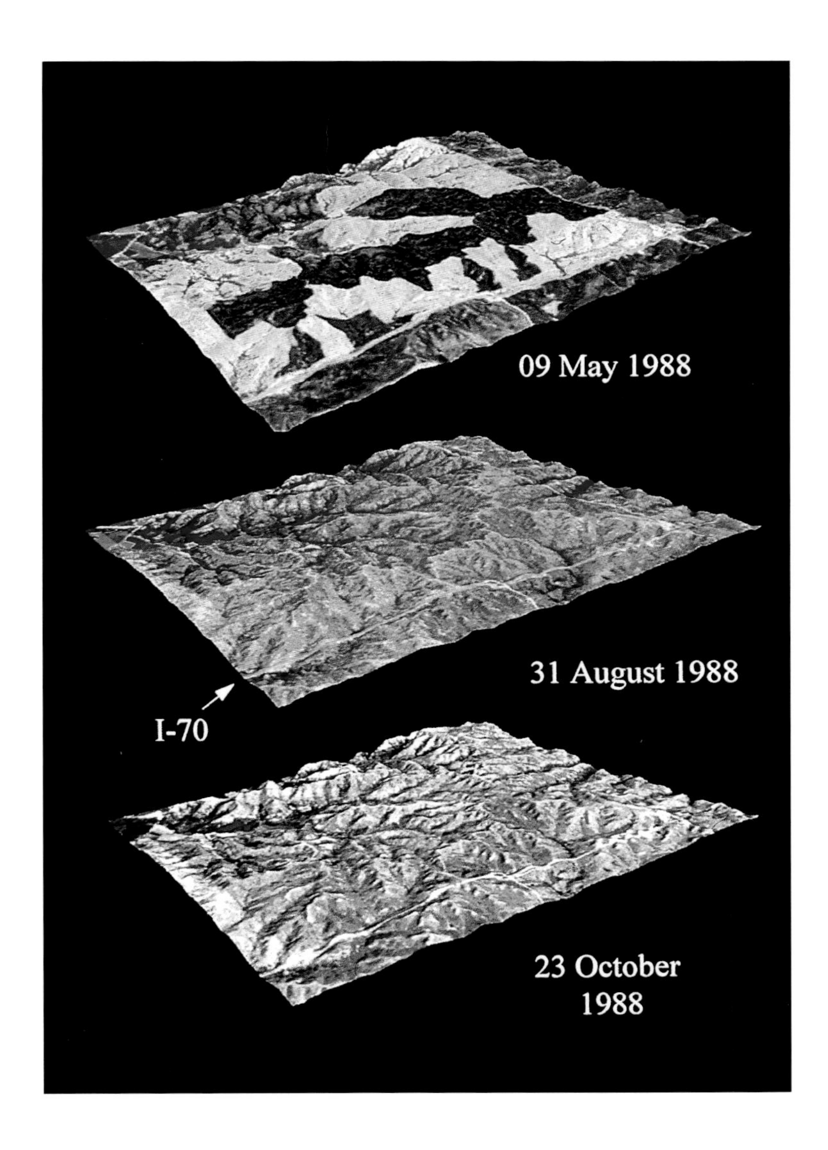

Plate I. Striking seasonal changes in reflectance caused by differences in vegetative phenology and land management. Red areas represent areas of high vegetative cover. Contrast, for example, the burned watersheds (black areas) in May with the unburned watersheds in October. Also note the western boundary (left) of the Konza Prairie with private lands that are grazed more heavily. These are Systeme Pour l'observation de la Terre (SPOT) satellite images (resolution of 20 meters) draped over a digital elevation model of Konza Prairie, and are false-color composites (Band 3 [0.79–0.89 μm], Band 2 [0.61–0.68 μm], and Band 1 [0.50–0.59 μm], displayed in red, green, and blue, respectively).

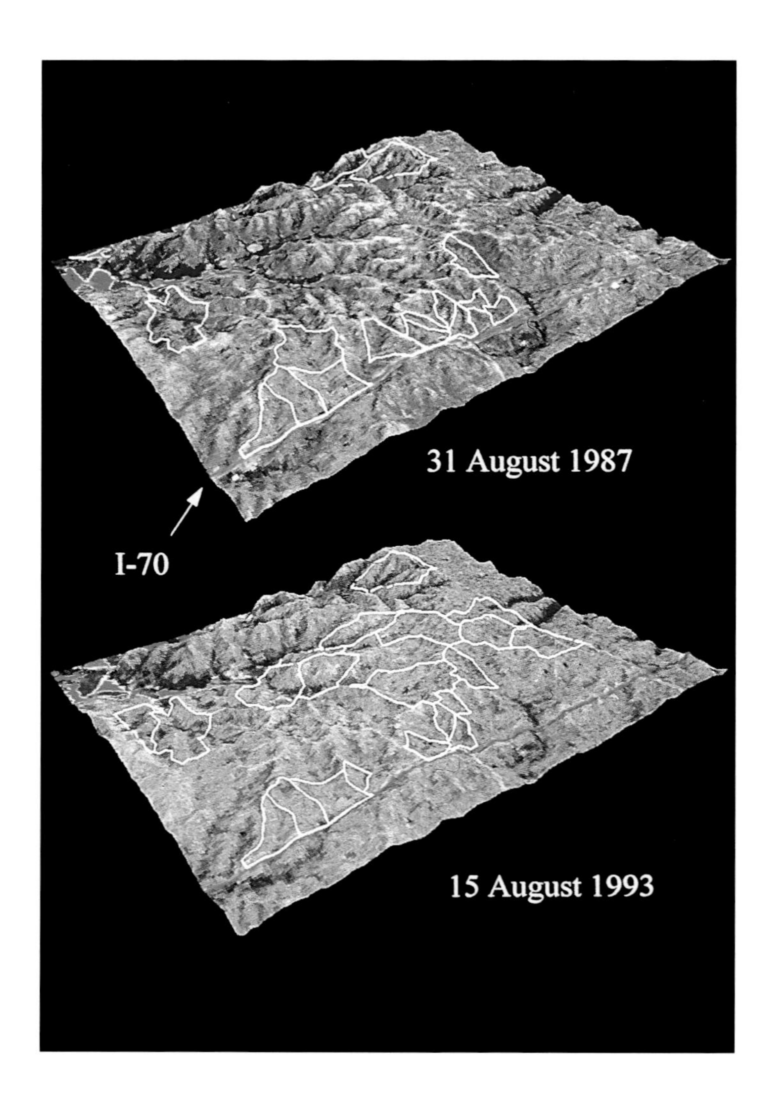

Plate II. The interannual variation in reflectance that occurs in the Konza Prairie. Red colors represent areas of high vegetative cover (note the dendretic pattern of the gallery forests along drainages). In 1993 (a year with near-record precipitation) little visual impact of burning can be noted; however, in 1987 (a year with average precipitation) burned watersheds are clearly distinguished from sites protected by fire. Boundaries of watersheds that were burned each of the years are outlined in white. These are Landsat Thematic Mapper (TM) satellite images (resolution of 30 meters) draped over a digital elevation model of Konza Prairie, and are false-color composites (Band 4 [0.76–0.90 μm], Band 2 [0.52–0.60 μm], and Band 1 [0.45–0.52 μm], displayed in red, green, and blue, respectively).

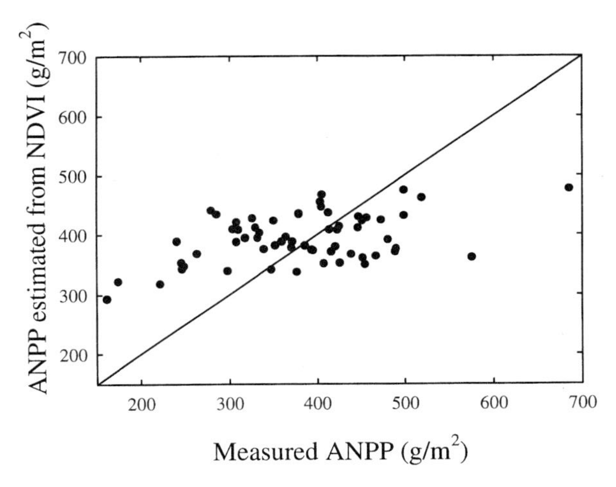

Figure 15.6. Comparison of mean ANPP measured in LTER plots and ANPP estimated from NDVI for entire watersheds. The line represents the 1:1 relationship.

which previously had been shown to be useful in quantifying spatial heterogeneity in tallgrass prairie (Nellis and Briggs 1989; Briggs and Nellis 1991), the relationship between the satellite-derived measurements and the ground-level measurements was evaluated over several years. Overall, these researchers reported that remote sensing can be used to provide information about the spatial heterogeneity of tallgrass prairie plant communities at larger spatial scales. However, the ability to detect the spatial patterns was affected by management practices; relationships were strongest in infrequently burned sites, which have the greatest spatial heterogeneity in plant species communities and biomass (Chapters 6, 12, this volume).

These studies illustrate the inherent complexity in tallgrass prairie. Because multiple factors limit productivity or affect vegetative species composition in tallgrass prairie ecosystems (Chapters 6, 9, 12, this volume), even if a remotely sensed measurement is found to be correlated with a ground-level pattern or process, the relationship likely will be altered by interannual variation in climate or management. This is a serious problem with remotely sensed measurements (even those as powerful as NDVI) because they are often not a product of the process with which they are correlated, especially as spatial scales and management vary.

Despite this caveat, remote sensing has helped us to scale up ecosystem processes over larger spatial scales. For example, inherent relationships occur between plant productivity and N gas fluxes in tallgrass prairie because both are controlled by water and N availability. Although temporal variability in soil moisture, N availability, and gas flux within a single growing season can be relatively high, an annual time scale integrates this variability. Variation in ANPP across the landscape is also an integrated product of variation in water and N availability over the entire growing season. Thus, strong relationships between point estimates of ANPP and N cycling processes can provide a means of estimating processes at the scale of landscapes or larger.

Groffman and Turner (1995) used this information to examine relationships between plant productivity and fluxes of N over a large (15 × 15-km) study area. This study site (known as the FIFE site; Sellers et al. 1988), includes all of Konza Prairie. By understanding the relationship between fire frequency and NDVI, these researchers were able to obtain landscape-level measurements of N_2 and nitrous oxide (N_2O) fluxes over a 2-year period. Furthermore, by coupling point measurements with remote sensing data in a GIS, they were able to account for the key factors (topographic position, fire frequency, and soil types) that were responsible for most of the variance in their measurements and compare their values with modeled output and data from other studies. Thus, when the limits of the data are recognized, remotely sensed data can be used to scale up from point measurements to landscape-level measurements.

Development of a General Model to Examine Landscape Dynamics

As discussed in earlier chapters (Chapters 6, 8, 12, this volume), land-use changes affect almost all ground-level processes in the tallgrass prairie. How do these ground-level processes translate into changes in landscape images that are derived from remotely sensed data? If we are going to use remote sensing data to describe these ground processes, it is important to understand the relationship between them.

Remote sensing can help in understanding and inferring process from pattern. An important part of these procedures is knowing the relationship between sensor resolution and scene object (i.e., what is on the ground that is contributing to the pixel value that the sensor is recording). For example, if the sensor's spatial resolving power or pixel is larger than that of the scene object, then many scene objects contribute to a single pixel value; following the nomenclature of Strahler et al. (1986), this is termed an *L-resolution.* Conversely, if multiple contiguous pixels are required to display a scene object, it is referred to as having an *H-resolution.* A similar situation exists for the sensor's temporal and spectral design aspects. Henebry (1993) and Henebry and Su (1993) developed a working model for examining Konza Prairie landscape dynamics. By focusing on the spatial structure of the imagery, they "ordinated" images along axes of spatial heterogeneity (SH) and spatial dependence (SD) (Henebry and Su 1993). Heterogeneity here denotes a spatially delimited numerical measure of variation or range; it also can define patchiness or the H-resolution aspect of spatial pattern. Dependence is a spatially delimited measure of the correlational or covariational structure within the image; it quantifies homogeneity or the L-resolution aspect of spatial pattern. If heterogeneity and dependence are plotted orthogonally (as ordinate and abscissa, respectively), then the resulting coordinate system, SH-SD space, can be divided into general regions of pattern. When this SH-SD space is used to map remotely sensed images, a series of images can describe a trajectory

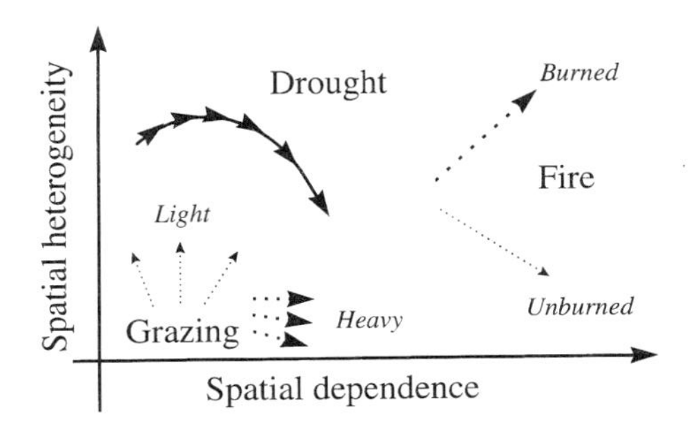

Figure 15.7. Effects of primary forcings (drought, fire, and grazing) on spatial patterns of NDVI in tallgrass prairie at spatial resolutions of over 100 m.

of a landscape (i.e., how the landscape as measured using remote sensing changes over time). The result is a model that can make testable predictions (Fig. 15.7).

Use of this model provides insight into the possible effect of measurement or scale on landscape analysis. For example, at sensor spatial resolutions under 100 m, drought and fire affect multiple contiguous pixels. Grazing, on the other hand, occurs at subpixel scales. Moreover, the landscape consequences of grazing are tied strongly to interactions of stocking rate and landscape characteristics (Senft et al. 1987) and, on a broader spatiotemporal scale, to the interactive effects of climate and coevolutionary history on the plant community (Milchunas et al. 1988). In the tallgrass prairie, lighter stocking of domestic livestock increases species diversity and spatial patchiness, whereas heavier stocking can reduce species diversity and "homogenize" the landscape (Collins 1987; Glenn et al. 1992). Similarly, low grazing intensity can increase SH with little effect on SD, but heavier grazing intensity can decrease SH while increasing SD (Fig. 15.7)

This model has had a variety of applications at Konza Prairie. Henebry (1993) found evidence of a drought-grazing interaction affecting spatial reflectance patterns in a 9-year series of Konza TM images; specifically, during a drought, patch size (correlation area) decreased and intrapatch homogeneity (first-order spatial autocorrelation) increased in an area grazed by *Bos bison*. In addition, Su et al. (1996) used this technique on a series of nine Landsat TM images from 1983 to 1991 to examine the effects of disturbance (fire and grazing), topography, and precipitation on spatial pattern in tallgrass prairie. They found that TM data were useful in detecting the spatial structure of tallgrass prairie and could be used as a metric to monitor and assess changes in ecological conditions as affected by natural (climatic) and/or human impacts.

Measuring the Impact of Grazing on Spatial Patterns

As alluded to earlier, grazing as a landscape-level disturbance is more complex than fire. Litter accumulation in ungrazed, unburned prairie can lower ANPP and basal cover of plants and reduce species diversity (Towne and Owensby 1984; Chapters 6, 9, 12, this volume). In contrast, disturbances created by grazing or occasional burning can increase plant species diversity (Collins 1987) and community heterogeneity (Collins 1992). Moderate grazing increases the number of plant species compared with ungrazed tallgrass prairie (Kelting 1954; Penfound

1964). "Spot" grazing creates and maintains patches that provide niches for plants that otherwise could not become established, such as some C_3 grasses and annuals. Large herbivores remove and trample plant cover during grazing and can create significant bare patches (i.e., *B. bison* wallows). Changes in soil exposure and structure lead to changes in surface hydrology and sediment transport, thereby affecting drainage and deposition networks (Laycock and Conrad 1967; Blackburn 1984). By reducing the litter layer and exposing more soil, grazing makes the vegetation more susceptible to climatic extremes, especially drought (Albertson and Weaver 1944; Frank and McNaughton 1992). The patchiness induced by grazing, trampling, and wallowing leads to an uneven distribution of fuel loads and thus burn patterns across the watershed, creating a positive feedback loop that reinforces patchiness.

In order to assess these complex patterns associated with the reintroduction of *B. bison* on Konza Prairie, we have used both GIS and remote sensing to model to the impact of herbivores and to quantify the spatial patterns induced by their activities. Since 1991, locations of *B. bison* on Konza Prairie have been mapped twice per week from March through October. Large-scale photographs overlaid with a 30-m grid system were used in the field as a reference for recording locations. Within each 30-m cell, the number of *B. bison* occupying that cell at the time of field observation was noted and then entered into a GIS. The *B. bison* data then were summarized by month and served as the dependent variable for modeling *B. bison*–landscape relationships. The independent variables (cell attributes) used were fire frequency, soil type, distance to permanent water sites, vegetation types, percent slope, and slope aspect. The vegetation layer was derived using NDVI applied to Landsat TM data for each of the observation years. Selectivity indices, ranging from strong avoidance to random use to strong selection, were calculated for each cell and used to relate spatial and temporal *B. bison* distribution patterns with the assigned cell attributes.

From this GIS modeling approach, we were able to determine that *B. bison* preferred burned watersheds during April, May, and June but were observed less commonly in these watersheds later in the growing season (Fig. 15.8). Soil type, which has been shown to influence productivity of vegetation (Chapter 12, this volume), was a significant parameter in *B. bison* grazing pattern only from July through October. Surprisingly, slope and access to permanent water sites had no impact on *B. bison* grazing preference. These results indicate that the grazing behavior of *B. bison* is not random, nor is it equivalent to that of domestic cattle (Hartnett et al. 1996b). Because the response of *B. bison* grazing activities to fire regime has both a temporal and a spatial component, adjacent watersheds that differ in fire frequency and/or soil type are likely to be impacted differently by a *B. bison* herd with equal access to both areas.

Regional Characteristics Scaling, Future Directions

As noted earlier (Chapter 1, this volume), the native tallgrass prairie within the Flint Hills of Kansas constitutes the largest remnant of what was once one of the

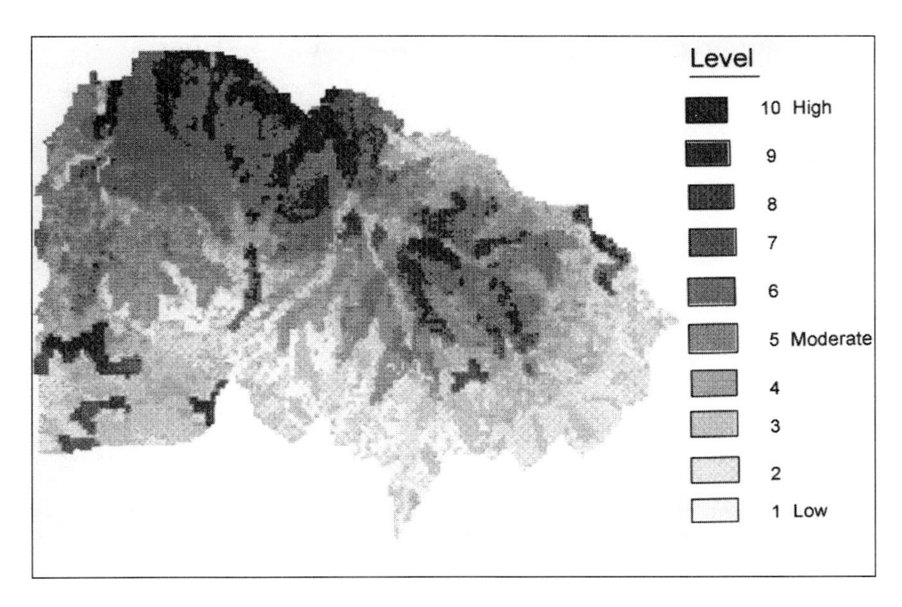

Figure 15.8. Predicted *Bos bison* distribution patterns for the fall of 1992 as determined from a GIS model. Highest densities of *B. bison* are predicted to occur in areas indicated by the darkest shading.

largest biomes in the conterminous United States. Composed of shale and limestone layers, the terrain of the Flint Hills has sheltered this tallgrass prairie from extensive cultivation. Instead, grazing has been the principal component of the region's agricultural economy since the early 1880s (Kollmorgen and Simonett 1965). Fourteen Kansas counties currently are designated as "bluestem pasture" for the purposes of agricultural statistics. Within this region, approximately 97% of the 1.46 million ha of rangeland were leased in 1992, generating between $35 and $60 million in rental fees (Kansas Agricultural Statistics 1992). Over the last 14 years, an average of 88% of pastures have been leased by early spring (Kansas Agricultural Statistics 1992), suggesting that the grazing pressure in the Flint Hills is fairly constant.

Unlike many grassland systems that have been altered significantly by cattle grazing (Archer et al. 1995; Schlesinger et al. 1990), the Flint Hills appear to have been affected only moderately by over 100 years of grazing by cattle. In fact, this area may be unique in that cattle grazing has helped maintain this particular ecosystem. As noted earlier, woody plant invasion in the Flint Hills is increasing, and the trend is more prevalent east of this area. For example, Spencer et al. (1984; pers. comm.) estimated that in eastern Kansas, from 1981 to 1994, the area of woody vegetation occupied from grasslands nearly doubled. This increase was not limited to gallery forest or riparian areas because upland areas have experienced a 100% increase in woodlands. Only within the Flint Hills, where spring burning is common, do large expanses of tallgrass prairie still exist.

Table 15.1. Comparison of landcover types of the Flint Hills and Konza Prairie.

Land Cover Class	Percentage of Area in Flint Hills	Percentage of Area at Konza Prairie
Residential	0.66	0.0
Commercial/industrial	0.29	0.0
Urban grasslands	0.26	0.0
Urban woodlands	0.01	0.0
Urban water	0.01	0.0
Cropland	24.33	6.21
Grasslands	65.30	84.12
Woodlands	6.72	7.81
Water	2.07	0.9
Other (mostly highways)	0.36	0.2

Source: Data for the Flint Hills are from the Kansas landcover mapping project, Kansas Applied Remote Sensing Program, and are based on Landsat thematic mapper data in 1990.

This spring burning has been promoted by the livestock grazing industry. Without this industry, and thus with a decrease in burning, invasion of the grassland by woody species would increase and endanger the tallgrass prairie ecosystem.

A comparison of the rest of the Flint Hills can reveal how representative of the region Konza Prairie is and will aid us in extrapolating beyond the boundaries of our site. Data from the Kansas Land Cover Project, which used TM data classified into 10 land cover types (with verification from aerial photography), revealed that over 67% of the Flint Hills is classified as grassland, 20% as cropland, and 7% as woodland (Table 15.1). Konza Prairie is similar in most classes but has an underrepresentation of cropland. Thus, Konza Prairie is representative of the Flint Hills in a general sense. However, less than 40% of Konza is grazed by either cattle or *B. bison*, whereas most of the grasslands within the Flint Hills probably are grazed on an annual basis.

As we continue to expand beyond the boundaries of Konza Prairie to develop predictive models, it is imperative that we remind ourselves that the tallgrass prairie is a unique ecosystem. Knapp and Seastedt (Chapter 1, this volume) point out that this system represents an area where water, energy, or nutrients all may be the primary limiting resource constraining system response at any point in time or space. The nonequilibrium nature of this system has been illustrated within this chapter and explains why single variables (even those as powerful as NDVI) fail to accurately predict a relatively simple response such as peak aboveground biomass.

Summary

Research in landscape ecology on Konza Prairie has enabled us to quantify large-scale patterns and processes on the tallgrass prairie. As a result, we have developed an extremely valuable spatial database. For many of our analyses, we have applied standard techniques (i.e., NDVI to examine spatial patterns of

ANPP, GIS to model movements of *B. bison*, remote sensing and GIS to examine soil patterns; Su et al. 1990), but we also have developed new tools (using SH-SD space to map remotely sensed images; Henebry 1993) that should aid ecologists in understanding complex ecosystems. One of the most important lessons we have learned in our research on Konza Prairie is the value of a long-term data set for assessing landscape-level phenomena. This was evident from our work using NDVI to predict ANPP (Figs. 15.4, 15.5) and from the limitations associated with scaling-up procedures that are based on short-term data sets. Working within these limitations, researchers at the landscape scale can provide insights into how large-scale spatial patterns affect ecosystem structure and function.

Part V

Toward the Future

16

Climate Change, Elevated CO_2, and Predictive Modeling

Past and Future Climate Change Scenarios for the Tallgrass Prairie

Timothy R. Seastedt
Bruce P. Hayden
Clenton E. Owensby
Alan K. Knapp

Understanding the response of tallgrass prairie to natural and anthropogenic changes in climate requires the same hierarchical perspective used to study prairie responses to fire and grazing. Changes in the physical and chemical environment are expressed at the physiological level of the organism. However, these physiological responses may not be expressed at the community level because they can be masked or obscured by biotic interactions such as competition and predation. These community interactions are themselves constrained by soil characteristics and natural variation in climate, storms, and other events, as described earlier (Chapter 1, this volume). Therefore, extrapolations of responses of the tallgrass biota observed from microcosm, greenhouse, or small plot studies must be treated with some healthy skepticism. In similar fashion, extrapolations from validated simulation models also must be viewed with caution. At most these products are testable predictions of complex hypotheses, and failures of such models (rather than successes, which may be spurious) are likely to provide the greatest potential for insights into key processes controlling the structure and function of the grassland.

This chapter considers the future of the tallgrass prairie in three ways. First, we assess the potential consequences of atmospheric CO_2 enrichment; second, we evaluate past climate, climatic variation, and climatic trends in the tallgrass prairie region; and third, we review ecosystem modeling efforts that have attempted to mimic climate, fire, and grazing modifications to the prairie. These models can provide insight into tallgrass prairie responses to future climates;

thus, they are important tools to guide management strategies that favor the native flora and fauna under a variety of future environmental conditions.

Overview

Previous chapters have identified mechanisms responsible for the maintenance of the dominant tallgrass species (Chapter 6, this volume) and those factors that govern patterns of productivity (Chapter 12, this volume). Many of the unique features of the tallgrass prairie (Table 1.1) developed from the intensive plant competition for energy, nutrients, and water. The area is a grassland only because of a high fire frequency and the absence of abiotic and biotic factors that could destroy the integrity of the prairie sod. Historical grazing intensity by large ungulates is unknown, but grazing pressures on the system must have been at a level that allowed for some litter accumulation and frequent fires to occur. This combination of physical and biological conditions coupled with a complex set of plant characteristics to produce a nonequilibrium system that is an overachiever in terms of its plant productivity and soil fertility characteristics (Seastedt and Knapp 1993; Seastedt 1995).

The life zone classification by Holdridge (1947) used precipitation and the ratio of actual evaporation to precipitation to demonstrate a global pattern of vegetation types. This system has been used to predict future ecosystem types under changing climatic regimes; thus, it can suggest the ecosystem transformations that are possible under shifting temperatures and aridity. However, the original Holdridge system did not include humid grasslands, despite their widespread occurrence. Perhaps this is because humid to semiarid grasslands occur across a wide range of temperature zones and, consequently, their delineation was difficult based on simple climatic factors (Chapter 1, this volume).

However, the climate of the tallgrass prairie region is responsible for many of its unique characteristics. As noted earlier, the midcontinental United States is characterized by cold winters and hot summers, with most precipitation occurring in late spring and early summer (Chapter 2, this volume). Potential evapotranspiration (PET) approximates actual evapotranspiration (AET) in the eastern portions of this region (Budyko 1974; Risser et al. 1981). This relationship establishes a strong link among hydrologic cycles, the net radiation balance, and the vegetation. Although the absolute amount of evapotranspiration may be much larger in wetter areas of North America, the ability of the vegetation to control the relative amount of water transpired rather than lost to groundwater or surface water may be maximized in this region. Because rainfall tends to exceed PET in the nongrowing season, plant roots must pump this fraction of the rainfall input out of the soil during the growing season if annual AET is to equal PET. Such activity can cool the site via two important mechanisms. First, the use of solar radiation to evaporate water prevents surface warming. Second, enhanced cloud formation as a result of the transfer of the latent heat to thermal energy in the atmosphere reduces peak solar input. The accumulation of soil organic matter improves the water-holding capacity of the soil; hence, a positive feedback loop to plant production is generated by the soil-building properties of the vegetation. Studies during long-term droughts indicated that soil water storage could enhance the survivorship of certain prairie plants for several growing seasons (Weaver and Albertson 1944). Thus, plants help construct soils that subsequently allow for persistence of components of the flora through time periods that otherwise might not sustain these species. Restored grasslands on soils mined by traditional agriculture do not have as large a buffering system and therefore may exhibit reduced stability to drought and increased responses to climate change.

The sum of these attributes helps define tallgrass prairie. Using plant, ecosystem, regional climate, and modeling perspectives, we will discuss how these features may interact with future climates and climatic changes.

Tallgrass Prairie Responses to Elevated CO_2

Debate is ongoing about the magnitude of potential climatic changes in the Great Plains in response to increases in the atmospheric concentrations of "greenhouse gases" (cf. Ojima et al. 1993a, b). There is little debate, however, that one of the most important greenhouse gases, CO_2, has been increasing rapidly in the last 100 years and will continue to increase in the foreseeable future (Keeling et al. 1995). Field studies of the effects of elevated CO_2 on an intact tallgrass prairie ecosystem have been conducted for 8 years in large open-top chambers (Owensby et al. 1996). These chambers were located within 15 km of Konza Prairie and thus provide a excellent database from which to make predictions about the future of tallgrass prairie.

Responses of Water Relations to Elevated CO_2

Textbook comparisons of C_3 versus C_4 photosynthetic pathways suggest that the C_3 species should be more responsive to enhanced atmospheric CO_2 concentrations. If this were indeed the case, CO_2 enrichment would pose a major threat to the dominant species of the tallgrass prairie and could reduce the overall productivity of this region via conversion of C_4 grasslands to C_3 grasslands. As noted previously (Chapter 1, this volume), periods of water stress are common seasonally, and more severe droughts have occurred frequently in this biome in the past. The key to understanding and modeling tallgrass prairie responses to elevated CO_2 is knowledge of the interactive effect of CO_2 and plant and ecosystem water relations.

Results from several years of study indicate that plant water relations are impacted strongly by elevated CO_2 in this and other grasslands (Field et al. 1997; Owensby et al. 1997). Measurements of midday leaf xylem pressure potentials in *Andropogon gerardii* in chambers with ambient and elevated (twice ambient) CO_2 suggest that leaf water status will be improved at elevated CO_2 in both wet and dry years (Knapp et al. 1993b, 1994b; Fig. 16.1). Moreover, stomatal conductance in *A. gerardii* and most other species was always significantly reduced

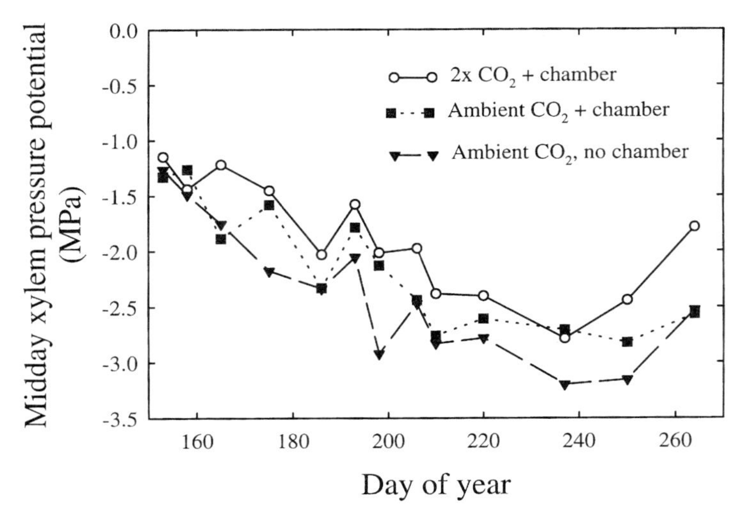

Figure 16.1. Seasonal course (1991) of midday leaf xylem pressure potential in *Andropogon gerardii* growing in native tallgrass prairie. Plants were exposed to elevated (double ambient) CO_2 concentrations and ambient CO_2 concentrations in open-top chambers. Also shown are data from field plants adjacent to the chambers. The seasonal mean leaf xylem pressure potential was significantly higher ($P < .05$) in plants at elevated CO_2 compared with those in ambient CO_2 chambers, and field plants had significantly lower xylem pressure potentials than those in ambient CO_2 chambers. From Owensby et al. (1993c), with permission of the Ecological Society of America.

(20–50%) at elevated CO_2 (Knapp et al. 1993b, 1996). In addition, stomatal responses to rapid fluctuations in sunlight similar to those resulting from intermittent clouds or within-canopy shading were also more rapid (Knapp et al. 1994a). Thus, improved plant water status at elevated CO_2 may result from an overall reduction in stomatal conductance and more rapid stomatal adjustments to variations in sunlight.

With no direct photosynthetic response to elevated CO_2, as predicted for C_4 species (Strain and Cure 1985), these alterations in stomatal conductance still would result in increased water use efficiency. Indeed, studies of the photosynthetic responses of *A. gerardii* in open-top chambers confirmed that no direct enhancement of photosynthesis occurred at elevated CO_2 during years with adequate rainfall (Nie et al. 1992; Knapp et al. 1993b). In dry years, however, photosynthetic rates of plants exposed to ambient CO_2 (and more severe levels of water stress) were significantly lower than those of plants at elevated CO_2. Thus, although photosynthetic capacity may not be enhanced at elevated CO_2, water stress–related reductions in photosynthesis are less likely to occur under elevated CO_2 for this C_4 grass.

Whole-chamber (ecosystem) water vapor and C fluxes also have been measured in tallgrass prairie (Ham et al. 1995). In a year dominated by periods of high

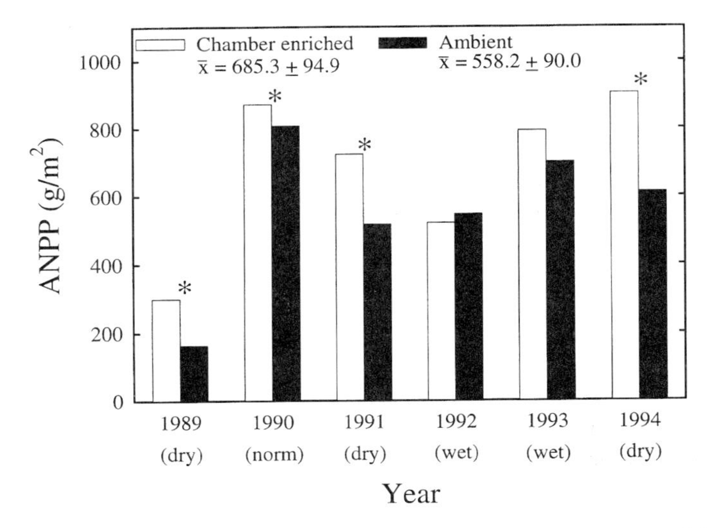

Figure 16.2. Aboveground net primary production (ANPP) in tallgrass prairie measured as peak standing biomass for plots exposed to double ambient CO_2 (enriched) and ambient CO_2 concentrations. Dry, norm, and wet refer to growing-season precipitation. Significantly different means are designated by an asterisk ($P < .05$). Updated from Owensby et al. (1996).

soil water availability, evapotranspiration was reduced by 22% at elevated CO_2, but net C exchange was not affected, except late in the season by delayed senescence (Ham et al. 1995). As a result, ecosystem-level water use efficiency was increased by over 50%. However, during afternoon periods of high evaporative demand, when differences in plant water status between plants at elevated and ambient CO_2 were maximal (Owensby et al. 1993a), ecosystem quantum yield (mmol CO_2/mmol PAR) was significantly higher at elevated CO_2 (Ham et al. 1995). Thus, data from the leaf to the ecosystem level suggest that, at elevated levels of CO_2, the C_4-dominated tallgrass prairie will have significantly higher water use efficiency and will maintain more favorable water status when subjected to periodic moisture stress. In the event that climate changes result in decreased precipitation and/or increased temperatures, the increased water use efficiency at high CO_2 may serve as a buffer against substantial ecosystem change.

Responses of Biomass and Species Composition to Elevated CO_2

Owensby et al. (1993b) measured above- and belowground biomass production, leaf area and plant community species composition in tallgrass prairie for an 8-year period. Compared with ambient CO_2 levels, elevated CO_2 increased the production of C_4 grass species but not of C_3 species in years with normal or below-normal precipitation (Fig. 16.2). Belowground biomass production, estimated in root ingrowth bags, responded similarly, but the relative increase at elevated CO_2 was greater than aboveground responses. At elevated CO_2, species composition of the C_4 grasses did not change, but the C_3 grass *Poa pratensis* declined and some C_3 forbs increased. The reduction in C_3 grass cover may have been due to the lack of grazing in these open-top chambers, which allowed the taller C_4 grasses to overtop the shorter C_3 grasses. The taller C_3 forbs increased the most at elevated CO_2, supporting the hypothesis that the canopy response (i.e., competition for light) associated with elevated CO_2 affected interspecific competition. The common prediction that C_3 species in mixed C_3/C_4 plant communities will gain a competitive advantage at elevated CO_2 was not supported during the 8 years of this open-top chamber study (Owensby et al. 1993a). The combination of increased biomass and leaf area of the C_4 grasses, differential drought responses, and a lack of selective grazing on the dominant C_4 grasses may have precluded any increase in C_3 grass production. The critical role that large ungulates may play in reducing the competitive dominance of the C_4 grasses also was shown in Chapter 9 (this volume).

Biomass Responses to Elevated CO_2 and Nitrogen Limitations

At current levels of atmospheric CO_2, biomass production in frequently burned tallgrass prairie is N-limited (Chapters 12, 13, this volume). Under elevated CO_2, N limitations may be more severe, although increased nutrient use efficiency may partly offset this effect (Owensby et al. 1996). Elevated CO_2 has both short-term and long-term implications for nutrient limitations. The short-term response reflects the ability of the ecosystem to increase productivity with increased C availability. The long-term response is related to reduced N availability because of slower decomposition of low-quality litter. In a 2-year study of the interactions of N fertilizer and elevated CO_2, Owensby et al. (1993c) measured much greater responses of aboveground and belowground biomass to elevated CO_2 in N-fertilized versus control chambers. These increases were maximal in dry years. Tissue N concentrations were lower for plants at elevated CO_2 in fertilized plots versus plants at ambient CO_2 in fertilized plots, but total standing crop of N was greater at elevated CO_2. Increased root biomass at elevated CO_2 likely increased N uptake. Overall, the proportional increase in biomass at elevated versus ambient CO_2 was the same when more N was added to both treatments, but the absolute responses were much greater. Owensby et al. (1993a) concluded that the response to elevated CO_2 in tallgrass prairie was suppressed by chronic N limitation.

Longer-term, nutrient cycling studies also have been conduced at elevated CO_2 in tallgrass prairie (Owensby et al. 1993b). In general, although total N in aboveground and belowground biomass may be increased at elevated CO_2 (because of biomass increases), tissue N concentrations in roots and shoots are usually lower. This could alter rates of decomposition, but studies of the effects of elevated CO_2 on leaf litter decomposition suggest that such an effect is minor (Kemp et al. 1994). This is because tallgrass prairie species translocated a large proportion of leaf N belowground late in the season (Chapter 13, this volume). Thus, surface litter quality was little altered by elevated CO_2.

Similarly, Rice et al. (1994) assessed the effect of elevated CO_2 in unfertilized and N-fertilized plots on the C and N stored in soil organic matter and microbial biomass. Increased plant production under elevated CO_2 may increase soil C and thus increase N limitations. Soil C and N were not affected measurably by elevated CO_2, except when N was added. The response of microbial biomass to elevated CO_2 was dependent on soil moisture levels. Elevated CO_2 increased microbial biomass C and N in a dry year but not in a wet year. Alterations in ecosystem water relations at elevated CO_2 were suggested as the primary mechanism for these microbial responses. Adding N at elevated CO_2 significantly enhanced microbial activity and is consistent with the view that N limitations will constrain tallgrass prairie responses to elevated CO_2 (Owensby et al. 1996).

These studies suggest that the nonequilibrium, multiple limiting-resource nature of tallgrass prairie likely will continue to predominate as grasslands respond to a high CO_2 world. Shifts from greater water limitation to greater N limitation are possible, but they depend on the magnitude of any concurrent alterations in temperature or precipitation patterns for the Great Plains.

Climate Change

Changes in ambient temperatures and precipitation will likely accompany CO_2 enrichment; however, substantial uncertainties exist regarding the magnitude and even the direction of these changes. Global circulation models (GCMs) do not exhibit consistent predictions for rainfall in the tallgrass prairie region; these models are somewhat more consistent in predicting modest temperature increases (Ojima et al. 1993a, b). However, ample data suggest that patterns observed in the past century will be continued in the next. Thus, as analysis of past and current climatic trends may be useful in predicting future climatic regimes for the tallgrass prairie region.

As reported in Chapter 2 (this volume), several authors have noted the association of ecotones between biomes with long-term average frontal boundaries and climatic complexes (Borchert 1950; Bryson 1966; Mitchell 1969). Moreover, Bryson's (1966) climatic complexes and species assemblage patterns for flora and fauna (Chapter 2, this volume) have steep gradients in Kansas. Considering that and taking a climatic determinism perspective, we can conclude that, if climates change, an associated change in species is likely. Konza Prairie is an ideal long-term observation site for recording biotic changes in response to climate change. These changes could vary from simple shifts in the relative composition of the biota, as reported by Weaver and Albertson (1944) for the dust bowl era, to wholesale community changes as suggested by the Holdridge system (1947). The latter are unlikely, however, assuming that fire remains a management option.

Bryson (1966) interpreted pollen and plant macrofossil evidence for late glacial times and proposed a model for the position of climatic frontal boundaries during that time (Fig. 16.3). In this model, the Prairie Peninsula and associated fronts are south and west of their current positions. Wendland (1980) studied the position of the boreal forest–prairie ecotone at 1,000-year intervals

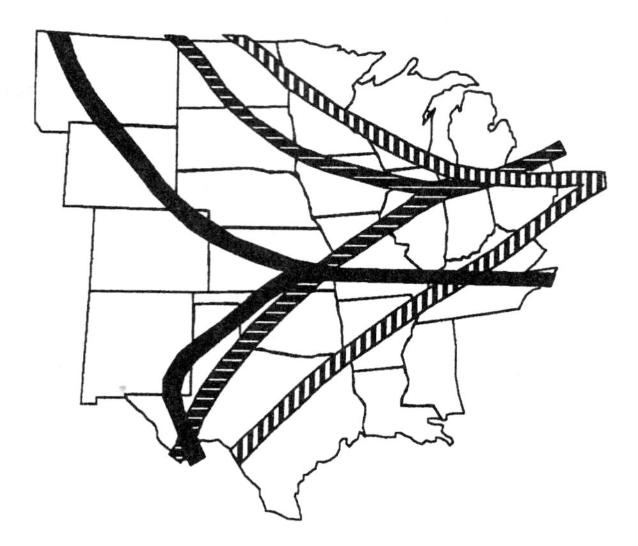

Figure 16.3. Frontal boundaries outlining the Prairie Peninsula: Black frontal boundaries represent late-glacial times (ca. 18,000 yr B.P.); darker striped frontal boundaries represent Cockburn-Cochrane time (ca. 8,000 yr B.P.); and lighter striped frontal boundaries represent Sub-Boreal time (ca. 5,000 to 3,500 yr B.P.). Adapted from Bryson et al. (1966).

for the last 10,000 years and found that it was extended, relative to its current position, both to the east and to the west. Bryson and Baerreis (1968) and Bryson et al. (1970) found that around A.D. 1200 an eastward expansion of steppe species occurred into the northwest corner of Iowa in association with the location of frontal boundaries at that time. These studies suggest that tracking long-term alterations in the locations of frontal boundaries can provide insight into potential vegetation changes.

Historical Analysis of Storms across the Prairies

Storms on the prairies arise from disturbances (vortices) that come ashore on the West Coast and travel across the mountains. These systems re-form as low-pressure systems or cyclones along fronts. The main sites of re-formation of these storms are the Gulf Coast of Texas, west Texas, central Colorado, southern Wyoming, and southern Alberta. Hayden (1981), in a study of North American storminess, found that the mean annual number of cyclones over the Prairie Peninsula was greatest around 1915 and reached a minimum around 1965. This pattern is evident in most areas of the Prairie Peninsula. The annual frequency of cyclones over the eastern two-thirds of Kansas is shown in Figure 16.4. The change is nearly sinusoidal, with a maximum number of storms (over 25 per year) around 1906 and a minimum near zero around 1965.

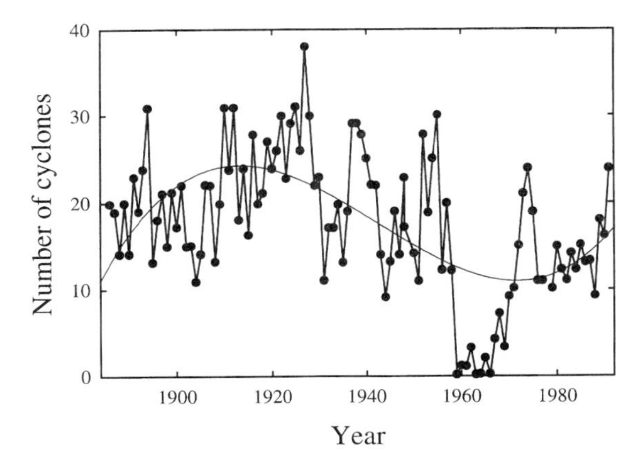

Figure 16.4. The annual numbers of cyclones with one or more closed isobars for at least 24 hours over the eastern two-thirds of Kansas from 1885 to 1990.

To assess the seasonal and century-long patterns and variations of the tracks of frontal cyclones across the Prairie Peninsula region, storm occurrences for each month from 1885 to 1992 were tabulated for 38, 2.5° latitude by 5° longitude grid cells for the region east of the 100th meridian. National Weather Service monthly publications of the tracks of the centers of cyclones were used to develop this data set (Hayden 1981). The resulting data matrix had the dimensions of 38 by 1,272. An S-mode principal components analysis of this data set was run, following the procedures of Hayden (1981, 1982). Six storm tracks were defined by the principal components of storm frequency (Fig. 16.5). Only the first two component storm tracks, labeled 1 and 2 in Figure 16.5, have long-term changes in storm frequencies along them. The first track arises out of west Texas and is aligned northeastward crossing Oklahoma, Missouri, Illinois, Indiana, and Ohio and on to New England. This component explained 24.9% of the

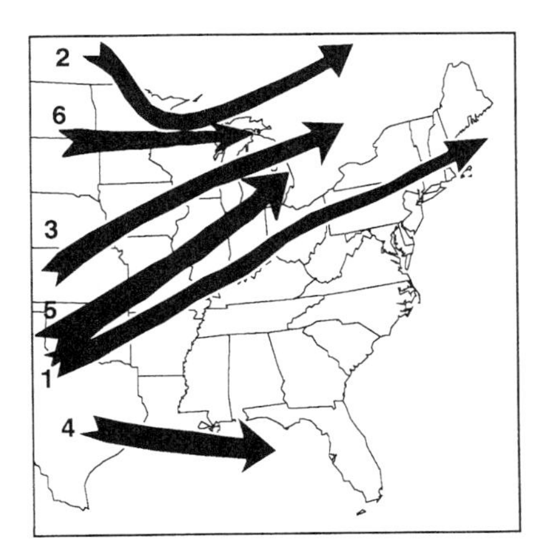

Figure 16.5. Storm tracks out of the Prairie Peninsula. Storm tracks are numbered according the principal component number in the analysis of cyclone frequencies for the period 1885–1990.

total variance in the data set. The second component, the so-called Alberta storm track, represents cyclones that follow the boundary between the boreal forest and the northern prairies. It explained 11.1% of the variance.

Figures 16.6 and 16.7 show the average monthly and average annual factor scores for the first two storm track components. Component 1 has positive factor scores in winter (November through April), indicating that storms along this track are more frequent than average in winter and less frequent than average in summer. The time series of annual factor scores indicates a century-long modulation in the frequency of storms along this storm track, with a maximum frequency around 1915 and a minimum around 1970. The amplitude of the long-term change is approximately the same as the amplitude found on a seasonal basis and must be considered a major climate change.

The second component storm track, the track along the northern margin of the Prairie Peninsula, has positive factor scores in June through November, indicating a maximum frequency of storms in autumn and early winter and a minimum frequency in late winter and early spring. This component storm track also shows a long-term sinusoidal variation in the frequency of storms. Storms were most frequent in the early years of this century and reached a minimum in the late 1960s. As with the first storm track, the amplitude of the long-term trend is of the same order as the seasonal cycle. Although the two storm tracks occur during different months, both exhibit similar long-term climate changes.

Long-Term History of Prairie Peninsula Anticyclones

The prairies region of North America experiences major migratory high-pressure cells, or anticyclones, that are classified into three types based on their origin: Pacific, Canadian, and transpolar. Transpolar anticyclones have their origin in Asia and crosspolar regions; they enter North America and sweep southward. Canadian and transpolar anticyclones are restricted largely to the portion of the year when the high latitudes are snow covered. The end of this period in the spring usually coincides with the last snow on the tundra and, on average, occurs around the second week in June. The autumnal onset of polar anticyclones is less well defined in time but usually occurs during the nongrowing season. Anticyclones from the North Pacific move generally from west to east, through the Prairie Peninsula and up the Ohio Valley; like polar anticyclones, they are restricted to fall, winter, and spring. A principal components analysis of North American high-pressure frequencies was performed to assess the seasonal and century-long variations in the movement of high-pressure cells through the prairies. Monthly occurrences of days with pressures equal to or exceeding 1,020 mb for half-month intervals from 1899 to 1990 were tabulated for 2.5° latitude by 2.5° longitude grid cells from the Northern Hemisphere sea level pressure data (Trenberth and Paolino 1980; Hayden 1984).

The average yearly factor scores indicate a decline in the occurrence of polar anticyclones since 1899. Average factor scores were about 3.0 in the first decade of the 1900s and about −3.0 in 1990. The significance of this climate change is great because the range of variation over the last 90 years is equal to about 30%

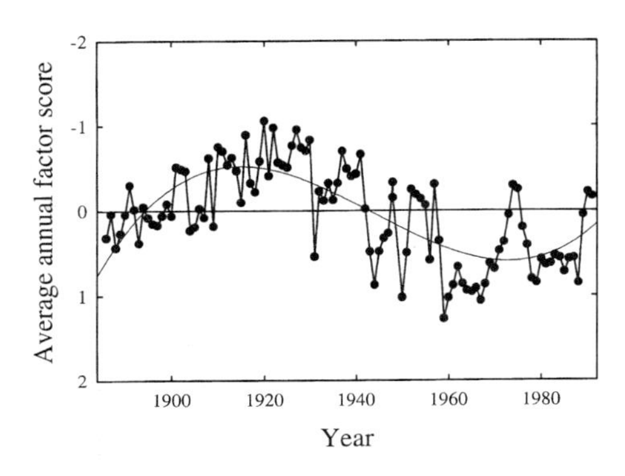

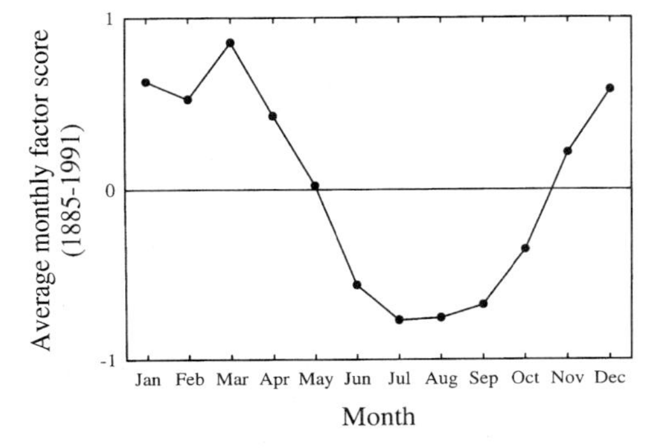

Figure 16.6. Average annual (top) and average monthly (bottom) factor scores for storm track number 1.

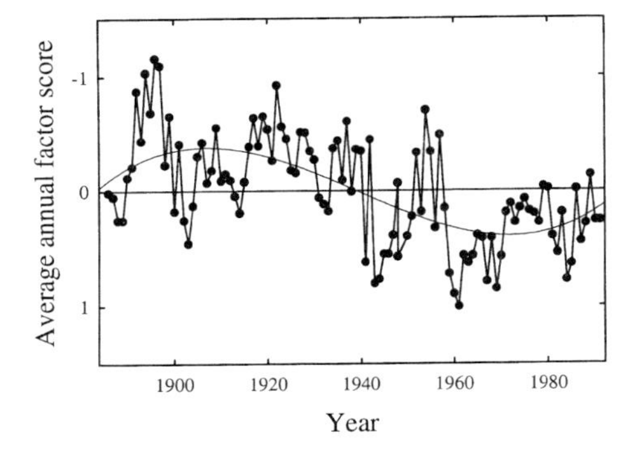

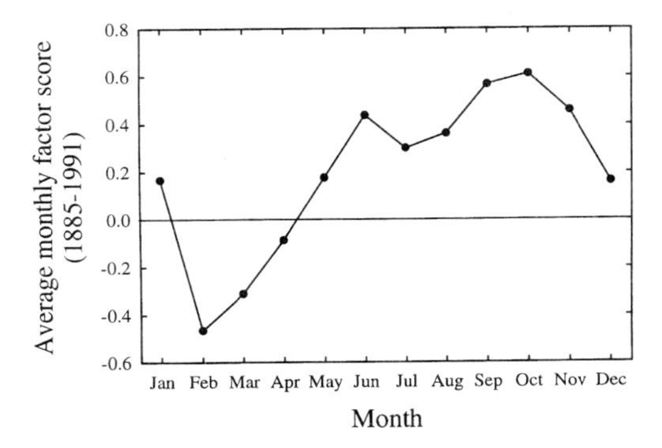

Figure 16.7. Average annual (top) and average monthly (bottom) factor scores for storm track number 2.

of the annual or seasonal amplitude. Thus, the prairies are experiencing fewer polar anticyclones currently than in earlier decades. This does not mean, however, that the intensity of the anticyclones also has changed or that the low temperatures associated with this changed frequency can be deduced from these data. However, over this same period, the polar front has, on average, been farther south and east in recent decades, and temperatures in the southeastern United States have been about 3°C colder than those in the middle of the last century. A reasonable working hypothesis, then, is that fewer but more intense high-pressure cells have come out of Canada in recent decades than earlier in the current century. In contrast, the frequency of Pacific migratory anticyclones across the Prairie Peninsula has increased by about 20% since 1899.

In summary, analyses of the tracking of storms and high-pressure systems across the prairies of North America reveal the inconstancy of weather systems at the century time scale. Storminess over the prairies was maximum shortly after the turn of the last century, reached its nadir during the 1960s, and has begun to increase again. This increased storminess in winter may well have an impact on the prairies through an increase in winter recharge of soil moisture. High-pressure systems likewise have changed over the last century. The transpolar "Siberian Express" and Canadian high-pressure cells have declined in frequency during this century, whereas migratory high-pressure cells moving eastward from the Pacific Ocean have increased. These changes are especially important for the winter season. These movements of high-pressure cells should not be confused with the transport of mass evidenced by wind fields. Other than for the period of the dust bowl in the 1930s, when westerlies and drought in the Prairie Peninsula were pronounced, few studies have focused on temporal changes in wind fields.

Relating the Past to the Future

The region of subsiding, westerly airstreams from the Pacific in winter is bounded on the south and east by its confluence with southerly airstreams from the return polar air mass system of the southeastern United States. To the north, the wedge of Pacific air is bounded by its confluence with northerly airstreams of continental polar and Arctic origins. Eight thousand years ago, the strength of the westerlies in the midlatitudes was greater than it is today, and the wedge of Pacific air extended east to Pennsylvania, making conditions in these eastern regions less suitable for forests but suitable for prairies. At this time the Prairie Peninsula had reached its maximum extent. During late-glacial times (18,000 yr B.P.), the wedge of westerly air was far to the south and extended a short distance into Texas. This period probably represents the minimum extent of the Prairie Peninsula.

The strength of the westerlies and the latitude at which they predominate are determined partly by the north-south temperature contrast and the latitude at which the maximum thermal gradient is found. It is interesting, then, to speculate on the consequences of the current trend in atmospheric CO_2 and the so-called greenhouse effect. Numerical models charged with twice the atmospheric concentration of CO_2 cause the atmosphere in the high latitudes to warm markedly, while the temperature in the low latitudes changes little if at all. This circumstance results in a reduction in the north-south temperature contrast and a corresponding change in the location of this contrast to a higher latitude than today. In addition, this effect would be most noticeable in winter. The reduction in the westerlies that is implied by the reduced temperature contrast should provide for a westward retraction of the Prairie Peninsula. In addition, the models suggest a more vigorous influx of maritime tropical air from the Gulf of Mexico into the continental interior. This would favor rainfall from thunderstorms and a more reliable source of summer soil moisture. This scenario, then, would offer the prospect of the westward encroachment of woody species and perhaps an expansion of forests into the prairies. Indeed, such expansion already has been documented and is an ongoing process (Chapter 15, this volume).

Predictive Models

In this section, we review the research that has employed conceptual or mathematical modeling to predict responses of the tallgrass prairie to global change scenarios. Given a specified fire and grazing regime, predicting change in the structure and functioning of tallgrass prairie requires knowledge of biotic changes (including invasions by alien species); the temporal dynamics of temperature and precipitation patterns; and the temporal dynamics of changes in atmospheric inputs of nutrients such as N and CO_2. We make the optimistic assumption that increases in atmospheric pollutants or mutagens will not occur. At present, no models have adequately incorporated all of the needed elements that affect change. Therefore, all predictions are potentially flawed and are at least incomplete.

Early Analyses and Simple Models

Summaries and synthesis of data resulting from research conducted in conjunction with the International Biosphere Programme (IBP) grassland projects appeared almost two decades ago (French 1979; Breymeyer and Van Dyne 1980; Risser et al. 1981). Findings from the grassland studies also were incorporated into more global views of productivity (Lieth and Whittaker 1975; Webb et al. 1978, 1983) and decomposition (Swift et al. 1979). These early ecosystem efforts focused on the role of water as the key variable controlling productivity of the grasslands (cf. Towne and Owensby 1984; Abrams et al. 1986; Sala et al. 1988).

Analyses of factors affecting soil decomposition and soil characteristics such as C storage, with notable exceptions (Hunt 1977), did not appear until well after the end of the IBP. Melillo et al. (1990) evaluated controls and concluded that litter decomposition and soil organic matter dynamics in mature grasslands were predictable based on climate and soil texture patterns (Brady 1990). These variables were used successfully to develop simulation models of C dynamics (Hunt 1977; McGill et al. 1981; van Veen and Paul 1981; Parton et al. 1987). In dry grassland sites, soil respiration is primarily a function of soil moisture (Hunt

1977; Warembourg and Paul 1977; Orchard and Cook 1983). In mesic grasslands, temperature is the main factor controlling decomposition, and the clay fraction of these soils is an important covariate in determining C storage amounts. Because water is a dominant control of plant productivity, but temperature is the key variable controlling decay processes in mesic grasslands, C storage is determined by the interaction of these variables. Jenny (1941, 1980) demonstrated this pattern for the Great Plains, and some preliminary data (e.g., O'Lear et al. 1996) suggest that a similar pattern might be expected for complex landscapes such as that found on Konza Prairie.

Simulation Models

Konza was the beneficiary of the second generation of simulation models that evolved from IBP efforts. Development and validation of the CENTURY model (e.g., Parton et al. 1987, 1994; Schimel et al. 1990; Ojima 1987; Ojima et al. 1990, 1993a, b, 1994) extensively used Konza Prairie data. Particularly noteworthy about this effort was that early (mid-1980s) versions of the model predicted the "transient maximum" response of infrequently burned prairie, which subsequently was verified by field studies at Konza Prairie (Seastedt et al. 1991; Chapter 12, this volume). The model correctly predicted N accumulation under the energy (light) limiting conditions of the unburned prairie. Holland et al. (1992) demonstrated the usefulness of the model in testing aspects of the "grazing maximization hypothesis" proposed by McNaughton (1979). However, rigorous testing of grazing effects using the model have not been conducted on Konza, and empirical studies (e.g., Turner et al. 1993) suggest that overcompensation may not occur in tallgrass prairie because of shifts in species composition to less productive C_3 species on grazed sites. At present, the model is not capable of accounting for these species shifts and therefore should fail to match empirical data. The model also has been used to predict and test decomposition and N immobilization patterns observed in above- and belowground litter decay at Konza (Seastedt et al. 1992). Results suggested that mechanisms may exist to allow apparent N mineralization from decaying buried litter in an environment that appears to favor N immobilization. A more recent version of the model has allowed investigators to evaluate the significance of life-form (C_3 versus C_4) growth patterns, in conjunction with fire, grazing, and climate change scenarios (Seastedt et al. 1994; Fig. 16.8) and the effects of CO_2 enrichment on plant and soil characteristics (Ojima et al. 1993a, b). Predictions of these models are discussed in the following.

Omissions in Current Models

Although productivity models were forced to deal with the obvious significance of aboveground grazing in grasslands, they have yet to incorporate belowground herbivory. Given the apparent sensitivity of belowground herbivores to C and N concentrations (e.g., Seastedt et al. 1988a), such models will have to be modified to deal with anthropogenic intrusions into C:N ratios (e.g., Fajer et al. 1989; Owensby et al. 1993b).

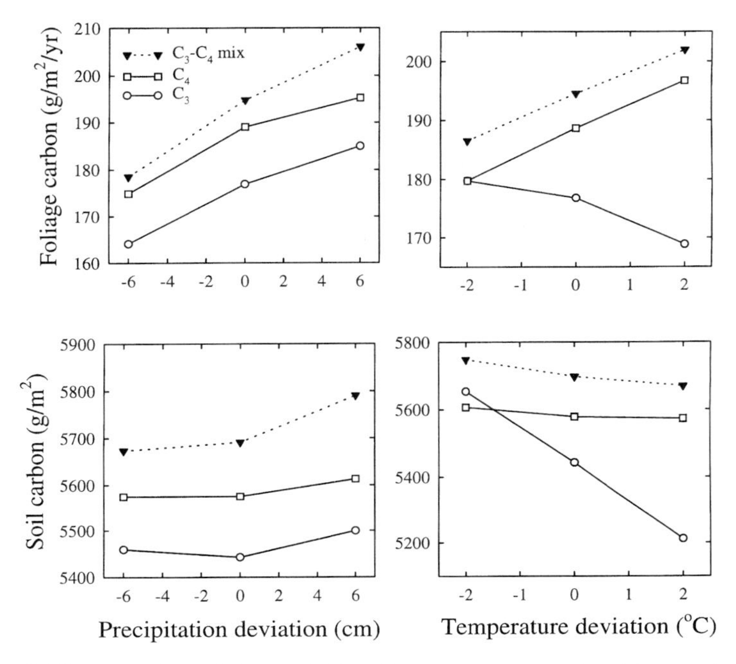

Figure 16.8. Predictions of the sensitivity of long-term unburned prairie to variation in temperature and precipitation. These results show that the composition of the prairie (C_3, C_4, or both) controls both the sensitivity and the direction of the response to climate drivers. From Seastedt et al. (1994), with permission of the Ecological Society of America.

In similar fashion, decomposition models or entire system models have not included detritivory as a control. The success of the previously cited models indicates that this process tends to be under strong temperature and moisture controls, but exceptions have been identified (e.g., Whitford et al. 1981). Moreover, the efficiency of microflora activities also may vary over nutrient gradients (Fisk 1995). Hence, global change may result in the presence or absence of "keystone detritivores" such as earthworms and may modify efficiencies of microflora, so that explicit incorporation of decomposer characteristics may be required to maintain or improve the accuracy of predictive models of decomposition and mineralization.

Requirements for the Next Generation of Models

Consideration of climate change in grasslands certainly must include direct responses of the system to increased atmospheric CO_2 levels, as well as less certain changes in temperature, precipitation, or cloud cover (Schimel et al. 1990). However, the dominant grassland plant species may not respond rapidly to moderate alterations in temperature and precipitation. This is because the dominant grasses have been exposed historically to tremendous interannual and long-term climatic variability (Chapter 1, this volume). For example, plant water stress, assessed as deviations in xylem pressure potential from levels measured in irrigated plants, can be nonexistent in one year and as severe as that experienced in shortgrass prairie or southwestern desert plants the next year (Chapter 12, this volume). A variety of adaptations such as leaf rolling, osmotic adjustment, and the deep rooting habit of these species enables them to withstand this stress (Weaver 1954; Knapp 1984b; Redmann 1985; Heckathorn and DeLucia 1991). Similarly, the photosynthetic response of *A. gerardii* to temperature is very broad (Knapp 1985b), probably because of the variability in temperature, both temporally and geographically, that this widespread species experiences. Indeed, in *A. gerardii*, as well as in *Bouteloua gracilis*, the dominant species of the drier and/or warmer west to southwest, photosynthesis is maintained at 90% of maximum levels over a 15°C range of leaf temperatures. Although little is known about the absolute maximum thermal tolerance of these species (preliminary evidence indicates that quenching of fluorescence from *A. gerardii* photosystems decreases rapidly above 45°C; Knapp unpublished data), an increase of a few degrees in air temperature is unlikely to directly impact leaf-level photosynthesis as strongly in *A. gerardii* as in species with more restrictive temperature-photosynthesis relationships. Conversely, an increase in growing season length may have a substantial effect on C uptake and plant growth (Myneni et al. 1997; Ham and Knapp 1998).

When a variety of directional climate change scenarios are imposed on the grassland ecosystems through CENTURY model simulations, effects are detectable but small relative to those changes induced by management activities (Burke et al. 1989; Seastedt et al. 1994; Copeland et al. 1996). This is not surprising given that key parameters such as soil temperature (Hulbert 1988; DeLucia et al. 1992; Fahnestock and Knapp 1994) can vary by as much as 8 to 10°C between burned versus unburned or grazed versus ungrazed portions of the prairie. Under "average" weather conditions, a specific combination of fire and grazing treatments can have large effects both on plant productivity in the short term and on soil C in the long term (Fig. 16.8). We emphasize that grassland response to climate change will be accomplished by management-climate interaction effects, not by directional climate forcings alone (Westoby et al. 1989; Schlesinger et al. 1990).

The large year-to-year variability in temperatures and precipitation tends to obscure directional climate change effects on plant and soil responses in grasslands. Of interest is the fact that the regression equations evaluating local responses to such variables as precipitation tend to be less sensitive (have a lower slope) than do simulations with validated models such as CENTURY (Fig. 16.9). The local system appears to be less sensitive to variations in precipitation or "buffered"; i.e., positive responses to good years or negative responses to bad years are lower than those predicted by the model developed from regional means (Chapter 12, this volume). Lauenroth and Sala (1992) also observed this buffering in shortgrass prairie and suggested a lag in plant response. Risser (1988) emphasized that the relationship between actual evapotranspiration (AET) and ANPP had been well established by Rosenzweig (1968) and Webb et al. (1978). However, in deep soil, semihumid to semiarid zones, precipitation from one year can contribute to plant productivity in subsequent years (Cable 1975), which is not factored into these regression models. Moreover, such relationships are not always established easily, even with excellent databases

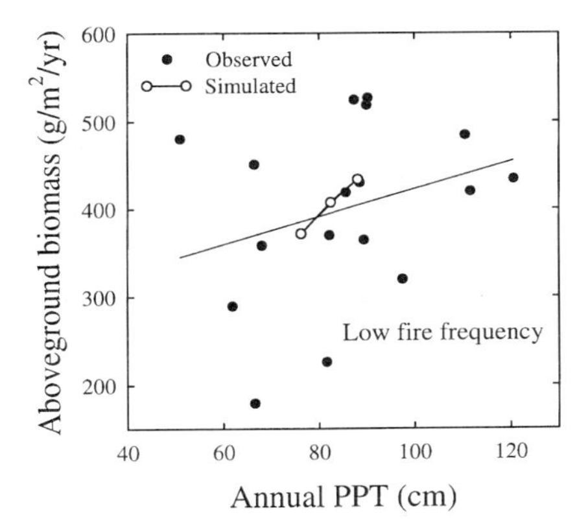

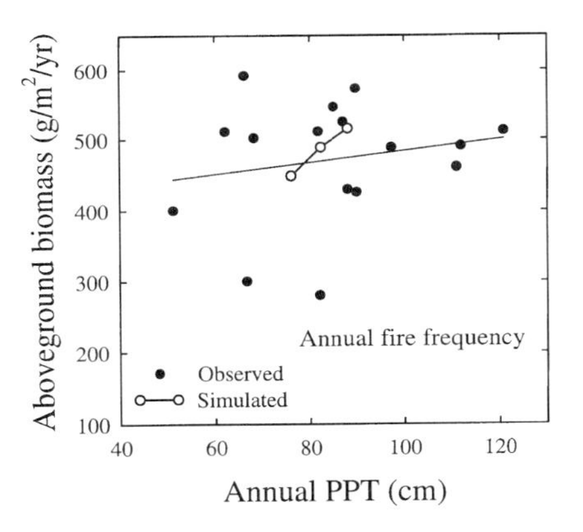

Figure 16.9. The sensitivity of an unburned prairie (top) and burned prairie (bottom) to precipitation changes, as estimated by the CENTURY model (version 3.0, open symbols), compared with 14 years of clipped plot data (solid symbols). The model, which predicts average values generated from 100-year simulations, indicates that the system is more responsive to precipitation than the actual data suggest. "Buffering mechanisms," which apparently limit productivity in wet years but increase productivity in dry years, are suggested.

(Towne and Owensby 1984; Briggs and Knapp 1995). One possible reason for this complication is that competition between relatively low-productive (often C_3) and high-productive (often C_4) life-forms can be influenced by the vagaries of climate and N availability (e.g., Weaver and Albertson 1944; Owensby et al. 1970; Seastedt et al. 1991). Changes in availability of essential resources lead to some nonlinear responses that require more detailed models to enhance predictability within a given year. Finally, the fact that grasslands are responding to multiple resource limitations, not just precipitation, means that C allocation strategies in one year affect growth potentials in subsequent years. Thus, regression models and systems models such as CENTURY lack the required detail to capture this inherent biotic variability. However, CENTURY does depict well the switching of N and energy limitations under infrequent fire (Ojima et al. 1994; Seastedt et al. 1994). Finally, in defense of system models, productivity varies across local landscapes in patterns that are not consistent among years (Schimel et al. 1991; Knapp et al. 1993b; Benning and Seastedt 1995). "Validation data" for the model generally come from localized areas that will not incorporate these landscape patterns. Thus, on average, the local landscapes may exhibit less interannual variability than do small plots.

Climate-Nutrient Interactions

The extent to which vegetation can respond to enhanced CO_2 has been discussed previously. Anthropogenic N inputs could further confound these responses (Owensby et al. 1993b). As this latter variable is a function of human activities largely independent of climate change, its influence is particularly significant. Nitrogen limitation already affects the prairie directly, in terms of its influence on productivity (Chapter 12, this volume), and indirectly, as a mechanism affecting plant species competitive interactions (Chapter 9, this volume). Nitrogen availability is also under the indirect influence of climate. "Good" years of high plant productivity reduce N availability by locking this material into organic matter. "Bad" years (e.g., drought) should increase N availability in the following growing season. Moreover, N enrichment could actually diminish prairie productivity by replacing the dominant C_4 species with less productive C_3 species (Wedin and Tilman 1996). This effect has been observed following extensive droughts (e.g., Albertson and Weaver 1944) and, as described elsewhere, tends to result in increased N availability in soils.

Summary

Direct human manipulations to the tallgrass prairie (e.g., plowing, artificial grazing regimes, introduction of exotic species, or N fertilization) may have more dramatic effects on the native ecosystem than those generated by modest changes in mean annual temperatures or precipitation and CO_2 enrichment. Given enlightened management (e.g., liberal use of fire and nonchronic grazing), changes in most areas likely will not involve either large-scale species extinctions or large changes in grassland productivity. This is due partly to the inherent variability and heterogeneity of the past and present systems. In other words, natural and anthropogenic manipulations already have influenced the localized climate, and the "adapted" species are already here. The large variation in microclimate induced by various natural and anthropogenic manipulations, in conjunction with a highly stochastic continental climate, suggests that the tallgrass prairie of the next century probably will not exhibit the degree of change that has been caused already by overgrazing and tillage of the soil. Persistent drought could progressively reduce the significance of the tallgrass species; N enrichment could have a similar effect. However, CO_2 enrichment should counterbalance impacts of both drought and N amendment.

Acknowledgments The elevated CO_2 research in tallgrass prairie reported here was supported by the U.S. Department of Energy, Office of Health and Environmental Research, Environmental Sciences Division. Patrick Coyne, Neal Adam, Erik Hamerlynck, Jace Fahnestock, Fred Caldwell, and Mike Craft were valuable members of the tallgrass prairie research team. This is contribution no. 97-203-B from the Kansas Agricultural Experiment Station Kansas State University, Manhattan.

17

The Dynamic Tallgrass Prairie
Synthesis and Research Opportunities

Scott L. Collins
Alan K. Knapp
David C. Hartnett
John M. Briggs

The overarching goal of the Konza Prairie Long-Term Ecological Research (LTER) Program is to understand pattern and process in tallgrass prairie ecosystems. The conceptual framework guiding this research program is that fire, grazing, and climatic variability are essential and interacting factors responsible for the structure and function of tallgrass prairie. Collectively, the chapters in this volume describe more than 15 years of research on Konza Prairie devoted to understanding how these essential factors affect this complex, variable, and sometimes baffling ecosystem.

In this concluding chapter, we review the defining characteristics of tallgrass prairie and summarize the essence of past research from our nonequilibrium perspective. Next, we reiterate a few of the unique and intriguing highlights derived from our long-term research program. We highlight these "nuggets" because they are either contrary to prevailing ecological theory, unique to our system, or otherwise counterintuitive. Finally, we turn our attention to future avenues of inquiry at Konza Prairie. Our main thrusts will continue to be fire, grazing, and climate variation, but our questions will be directed more toward global change issues, such as land use and fragmentation, exploitation of grassland systems, and climate change. We end in this fashion because we feel that research, especially long-term research, should be both explanatory and predictive and should integrate all aspects of inquiry from basic ecology to resource conservation and management.

Characteristics of This Nonequilibrium System

As noted in Chapter 1, we have adopted an explicitly nonequilibrium perspective in our LTER Program. This perspective is a logical consequence of the defining characteristics of tallgrass prairie (Table 1.1), which are products of multiple limiting factors and complex interactions. In contrast to many other systems where ecological processes are constrained by chronic limitations of a single resource, organismic to ecosystem processes and dynamics in tallgrass prairie are products of spatial and temporal variability in three primary limiting resources: light, water, and N. Variability in and switching among these primary limiting resources are caused by both extant and historical regimes of fire, grazing, and climate. Moreover, responses to these factors are strongly dependent on topographic and landscape position. As a result of this complexity, tallgrass ecosystems exhibit distinctly nonequilibrium dynamics.

In contrast to the nonequilibrium view of biotic systems, research on Konza has shown that geomorphology and associated soil development are in a dynamic steady state (Chapters 3, 4, this volume). Topographic variation associated with the geomorphology of the northern Flint Hills will continue to gradually erode and change, but these processes occur at time scales that are much longer than those associated with ecological dynamics. Thus, although we view the ecological processes as nonequilibrium, they take place in a geologic time frame that is relatively stable. To some extent, this helps to minimize the effects of one potentially confounding driving variable, geomorphology, in what is already a highly complex biological system.

In describing the factors affecting community organization in the grasslands of eastern Africa, McNaughton (1983b:314) stated, "Undue emphasis on competition or predation or disturbance or any other single, composite factor as a community organizer is highly misleading in the context of Serengeti grasslands, and, I believe, communities in general." We strongly concur with this assessment. Results from our long-term research program demonstrate that system dynamics at Konza Prairie result from multiple factors that are linked by a complex web of ecological interactions. As in the Serengeti, our system is characterized by multiple contingencies in space and time. Biotic interactions such as herbivory, competition, or plant-microbe relationships can be influenced strongly by each other as well as by abiotic variables. An important consequence of this complexity is that indirect effects are pervasive and important. For example, fire effects on animal population dynamics are primarily indirect, through changes in grazing patterns and vegetation structure (Chapter 8, this volume). Large grazers have significant indirect effects on soil invertebrates mediated through their influence on root dynamics (Chapter 14, this volume). Fire indirectly influences plants by changing soil microbial transformations and nutrient availability belowground (Chapter 13, this volume) and the abundance and distribution of herbivores aboveground (Chapter 15, this volume). Plant–mycorrhizal fungi interactions strongly influence prairie plant responses to their competitors and herbivores (Chapter 6, this volume). Thus, the tallgrass prairie is a system in which biotic interactions cannot be understood when studied in isolation.

The emphasis on complex interactions, multiple limiting resources, and non-equilibrium dynamics gives the impression that we can only stand back and articulate post hoc descriptors and explanations of population, community, and ecosystem dynamics. Clearly, this is not the impression we want to perpetuate. Indeed, many ecological responses in tallgrass prairie are well known and predictable (Fig. 17.1). Fire, climatic variability, and grazing are known to influence organismic and ecosystem responses independently and interactively. We also know that these responses are dependent on topographic and landscape positions. Fire, grazing, and climatic history are also important, as is the sequence in which these factors occur. For example, fire that occurs in a site with a long history of grazing by ungulates may lead to different responses than fire in a site that has not been grazed. Moreover, a site that is burned and then grazed may show far different responses than a site that is grazed first and then burned. Thus, many community- and ecosystem-level responses to the main driving variables are well known and qualitatively predictable.

Quantitative predictions, on the other hand, are more challenging in tallgrass prairie because the system response to multiple forcing functions is characterized by legacies, time lags, complex interactions, and contingencies. Again, as McNaughton (1983b) noted, these contingencies are not capricious but probabilistic. For instance, a legacy of soil fertility develops in the absence of annual fire. This fertility leads to a shift in community structure from strong dominance by C_4 perennial grasses to greater abundances of C_3 forbs. After a time lag of 6 to 8 years following fire, forb biomass increases at the expense of grasses, and overall plant species diversity is highest at this time. The potential for soil fertility to enhance aboveground net primary production (ANPP) is expressed only when burning eliminates the litter layer, but this response is contingent upon sufficient soil moisture. When moisture, nutrients, and light are nonlimiting simultaneously, ANPP is maximized (Chapters 12, 13, this volume). This response to fire is predictable and well understood but was not evident without a 15-year data set. Maximum postfire ANPP is contingent upon the concurrent release of biotic processes from the constraints of multiple limiting resources related to current environmental conditions and legacies of previous land-use/climatic factors (Fig. 17.2).

The complex nonlinear dynamics and high species diversity that characterize tallgrass ecosystems also make grasslands ideal laboratories to test general ecological principles. Grasslands are amenable to experimental manipulations that can be replicated at a range of spatial and temporal scales. Mesic grasslands often respond rapidly to experimental treatments, providing answers to ecological questions more efficiently than similar experiments in forests or deserts. No better example of this exists than the numerous recent studies addressing the relationship between species-functional diversity and ecosystem processes and stability that have been carried out in grasslands (Rusch and Oesterheld 1997; Tilman et al. 1997). Both long- and short-term experiments at Cedar Creek Natural History Area (Tilman and Downing 1994; Tilman et al. 1996) have yielded evidence indicating that plant species diversity in experimental plots is related positively to ecosystem resistance and resilience. Dodd et al. (1994) reported

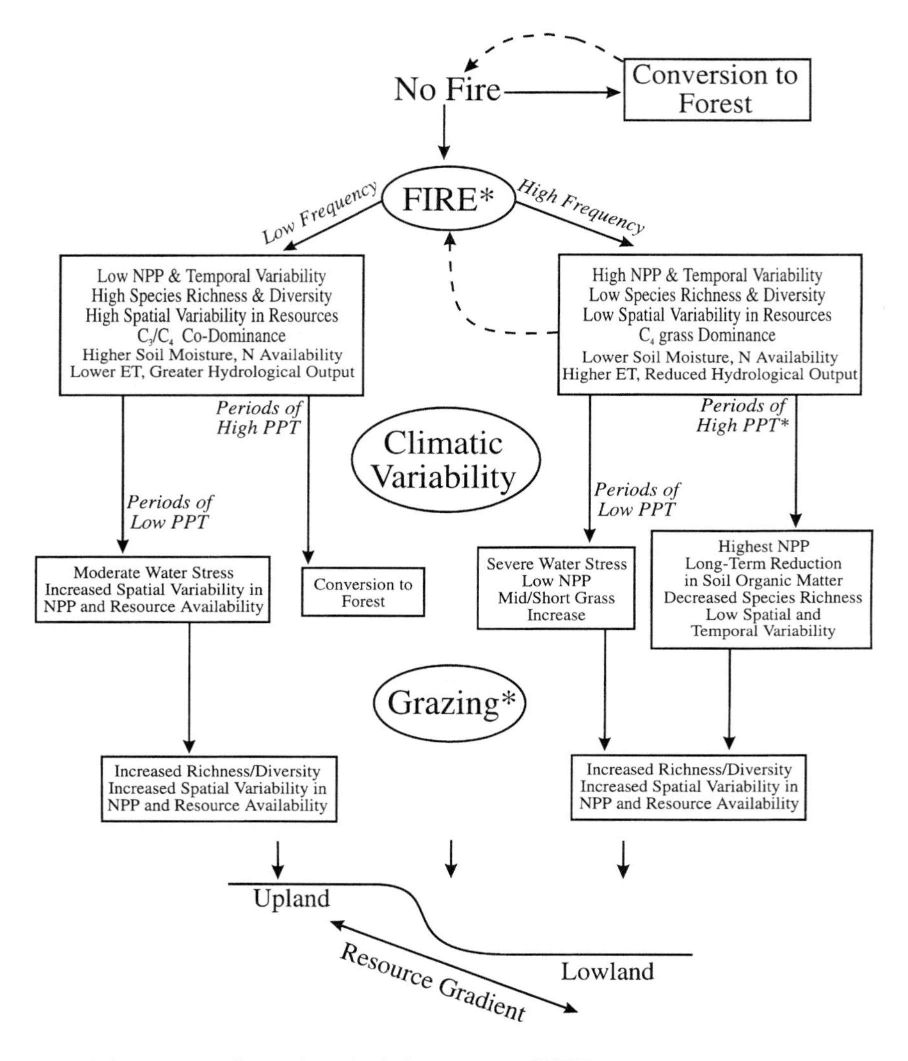

Figure 17.1. Flow diagram depicting the general relationships among fire, climatic variability, and grazing along topographic gradients at Konza Prairie. These factors all influence organismic through ecosystem responses independently and interactively. All proposed responses shown in boxes are being evaluated through long-term watershed-scale and small-scale experimental studies.

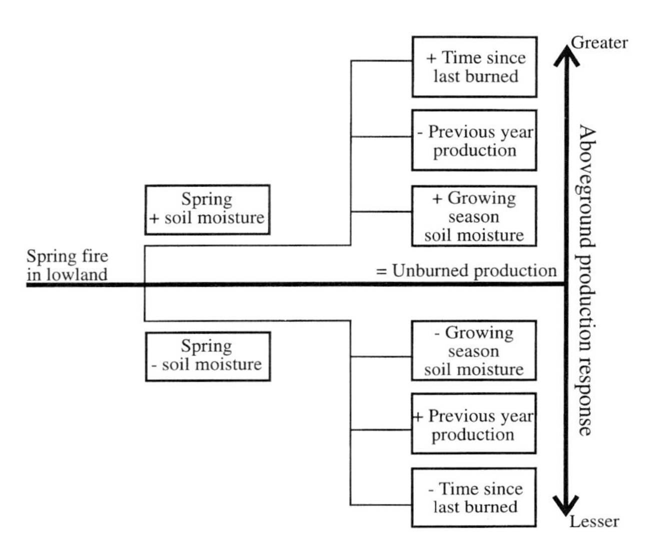

Figure 17.2. Aboveground primary production responses to fire in tallgrass prairie are influenced by a variety of factors (Chapter 12, this volume), and this "rule-based" model captures the array of factors that determine postfire responses. Maximum production after a fire (relative to unburned sites) occurs when spring and growing season soil moisture is high (indicated by +), and when in past year(s) fire was absent and production was low. The model represents a synthesis of 15 years of data and illustrates the value of long-term research.

that biomass stability was related positively to plant species diversity in the Park Grass Experiment. Frank and McNaughton (1991) reported that species diversity enhanced compositional stability in Yellowstone grasslands. On Konza Prairie, diversity enhanced biomass stability in lowland, but not upland, prairie (Collins and Benning 1996). Thus, research in grassland ecosystems has directly tested questions of general theoretical interest and helped to demonstrate the applied conservation value of biodiversity, although this topic remains controversial.

Despite the interest being generated by this research on diversity and stability, many questions remain (Johnson et al. 1996), and the overall generality of this relationship has yet to be determined. For example, Rodriguez and Gomez-Sal (1994) used the same analytical methods as Frank and McNaughton (1991) and found that high species diversity led to lower compositional stability in Spanish grasslands. In addition, the stability-diversity relationship often is cast in a scale-free context (Collins 1995), yet the data in support of this relationship often are derived from relatively small plots. Given that small-scale plant species diversity varies within an area, and that cumulative differences in richness are less dramatic at the scale of whole watersheds, the diversity-stability relationship has not been demonstrated to be applicable at larger spatial scales.

Another issue is whether or not the relationship between diversity and stability

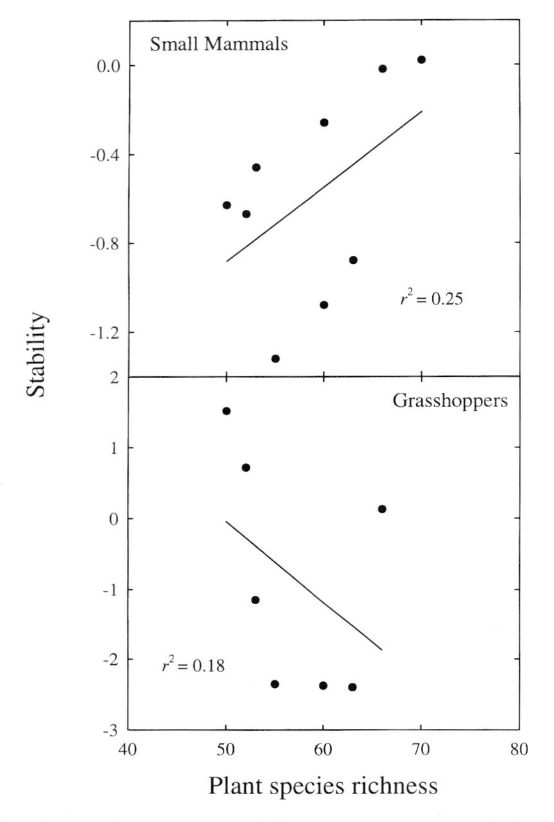

Figure 17.3. The relationship between stability of consumer populations and plant species richness during the drought of 1988. Stability was measured as the change in total abundance of small mammals or grasshoppers during a drought year compared with abundances before the drought. The x-axis is plant species richness in 1987, a year of above average precipitation. The y-axis is defined as $.5(\ln[\text{abundance}_{1988}/\text{abundance}_{1987}])$ following Tilman and Downing (1994).

is generalizable to other trophic levels. Data from Konza Prairie suggest that the relationship varies among consumer groups (Fig 17.3). Our long-term data show a positive relationship between plant species richness and stability of total abundance of small mammals but a negative relationship between plant species diversity and stability of total abundance of grasshoppers. These discrepancies do not necessarily imply that this line of inquiry is futile. Rather, the studies from Konza Prairie and elsewhere suggest that additional observational and experimental data are needed to determine the spatial and temporal domains of this controversial theoretical relationship. The ongoing experiments at Konza Prairie will provide valuable long-term data to test hypotheses derived from this long-standing debate in ecology.

Highlights of This Nonequilibrium System

One of the most interesting highlights associated with the lags and contingencies just described is that aboveground plant production does not accurately reflect relative nutrient availability in tallgrass prairie (Chapter 13, this volume). In most other systems, significant positive correlation occurs between production and nutrient availability (Grime 1979; Tilman 1982; Huston 1994), yet this does not appear to be the case at Konza Prairie. Nitrogen is considered to be the key soil resource that limits primary production and governs competitive interactions of vegetation in tallgrass prairie (Tilman 1988; Tilman and Wedin 1991a; Seastedt et al. 1991; Turner et al. 1997). Although ANPP is greatest following a fire in infrequently burned prairie, N availability is lower than in unburned prairie. In addition, production is often higher in annually burned prairie, where N availability is lowest, than in unburned prairie (Chapter 13, this volume). Although production can be enhanced by fertilizing with N, the greatest production response to fertilization occurs on annually burned sites. Greater soil N availability also has been documented at upland than at lowland sites, in spite of the fact that productivity is generally greater at lowlands. Thus, across the tallgrass prairie landscape, an inverse relationship exists between soil N availability and plant productivity.

The N-fire interaction impacts community heterogeneity as well. In most systems, disturbances are considered to be important sources of within-patch and between-patch heterogeneity (Pickett and White 1985; Kolasa and Pickett 1991). The opposite occurs with fire in tallgrass prairie. As fire frequency increases, spatial variability of soil nutrients, primary production, and compositional similarity decrease within burned patches (Chapters 9, 12, 13, this volume). Soils under frequent burning regimes tend to be warmer and drier in the spring and summer, and nutrients are low compared with those in soil in infrequently burned prairie. Under these conditions, dominance by C_4 perennial grasses increases dramatically, and species richness declines in response to competitive exclusion by the dominant grasses. As competitive exclusion occurs, the C_4 grasses account for more and more of the aboveground biomass, which reduces spatial variability in ANPP (Briggs and Knapp 1995; Chapter 12, this volume). Overall, between-patch heterogeneity also is decreased because the flora in burned patches is a nearly perfect subset of the vegetation found in unburned tallgrass prairie (Collins et al. 1995). Thus, tallgrass prairie does not conform to the widely held notion that disturbance increases within- and between-patch heterogeneity and species diversity.

Although the impacts of annual burning are well known and predictable, fire is only one of several commonly occurring natural disturbances in tallgrass ecosystems. One of the most pervasive disturbances in tallgrass prairie is herbivory, and ungulates disproportionately contribute to this activity. Ungulates typically remove 25 to 50% of aboveground production each year (Milchunas and Lauenroth 1993). On Konza Prairie, grazing by *Bos bison* uncouples the relationship of plant species diversity and fire frequency that developed under

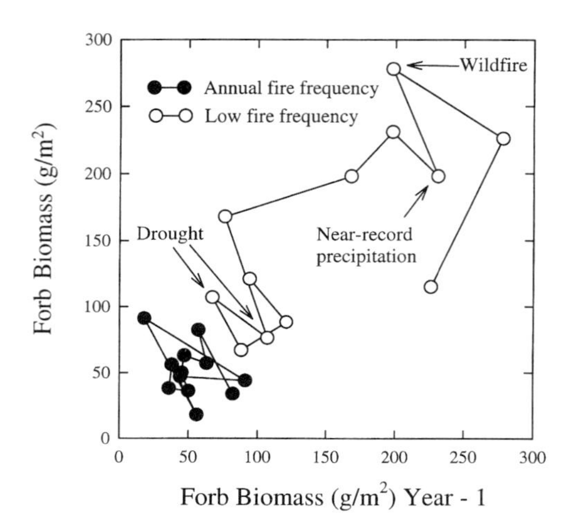

Figure 17.4. Temporal dynamics of forb biomass in adjacent annually burned and unburned watersheds on Konza Prairie. Biomass in any year is plotted relative to biomass in the previous year. In burned watersheds, forb biomass is low and variability is low. In unburned watersheds, variability was initially lower during the first 6 years of study, but 2 years of drought lead to rapid changes in forb production.

ungrazed conditions. For example, cover of forbs decreases with increasing fire frequency in the absence of grazing, but the opposite occurs on annually burned sites when large herbivores are present (Chapter 9, this volume). Overall, grazing leads to increased resource heterogeneity, enhanced plant species diversity, and altered patterns of primary production. Thus, fire and grazing interact to affect plant community structure and ecosystem function in tallgrass prairie.

As noted earlier, frequent burning dramatically affects the distribution and abundance of forb species in tallgrass prairie vegetation. Indeed, the abundance of forb species is reduced strongly by burning. Despite this reduction in cover, forbs contribute disproportionately to plant species diversity in this tallgrass prairie (Chapter 5, this volume). In effect, forbs are keystone species for biodiversity. For example, on four annually burned sites, the proportion of forb ANPP to total ANPP ranged from 1 to 8%, whereas the proportion of forb species richness to total species richness ranged from 63 to 70%, nearly a ninefold difference in importance. Thus, forbs are major sources of plant biodiversity in tallgrass prairie, and management practices greatly affect their overall abundances (Biondini et al. 1989). Given that many vertebrate and invertebrate species depend on forbs for food, cover, and habitat structure, the maintenance of forb species richness in prairies has tremendous consequences for overall biodiversity in grasslands. Climate change and unusual climatic events also may influence forb abundance in tallgrass prairie. In several watersheds on Konza Prairie, we have noted dramatic increases in forb abundance after a 2-year drought (Fig. 17.4). If climate change for this region includes an increase in such "uncommon" events (see later discussion), the possibility exists that additional keystone species or processes will be influenced to a greater degree than would be predicted based on changes in average climatic conditions.

The unique response dynamics of this ecosystem are not restricted to the terrestrial components. In particular, the streams on Konza Prairie do not conform to one of the fundamental models of structure and function in riverine systems. The river continuum concept (Vannote et al. 1980) has received extensive analysis and generated considerable debate. In grasslands, the river continuum is inverted compared with forested systems. Many first- and second-order streams in prairies occur in areas without a tree canopy. Therefore, these small reaches are primarily autochthonous (Chapter 11, this volume), which is opposite to the condition predicted by the river continuum concept. In addition, the terrestrial component of the tallgrass prairie exhibits fairly tight nutrient cycling, such that transfer of limiting resources from terrestrial to aquatic habitats is constrained. The result is that the terrestrial system is comparatively nutrient-rich compared with the aquatic system draining these watersheds. Exceptionally large C and N pools exist in prairie soils, but the "leakage" of these nutrients into the prairie streams is highly restricted.

The limited nutrient flow into prairie streams has significant consequences for aquatic food webs. Given that prairie streams have little organic C input from terrestrial vegetation, and that light levels are typically high in grassland streams, nutrient availability probably constrains primary production. When more nutrient-rich groundwater enters streams, it is stripped of nutrients as it flows downstream (Tate 1990). The net result is oligotrophic streamflow from intact prairie watersheds, a defining characteristic of tallgrass prairie (Chapter 1, this volume).

In addition to nutrient limitations, we believe that the streams in tallgrass prairie exhibit periodic extremes in discharge and temperatures that do not occur in most other systems. Even with 15 years of data, no "typical" year can be identified at Konza Prairie (Chapter 10, 11, this volume). Streams in more mesic regions have floods that occur during predictable times of year. Large desert streams are fed by snowmelt from high-elevation ecosystems or seasonal rainstorms. This profound variability in grassland lotic environments selects for organisms that have stress-resistant life stages, short generation times, rapid growth, rapid colonization potential, or combinations of these traits. These organisms are able to recolonize the systems within weeks after drying or flooding.

The extreme variability in streamflow and the lack of woody vegetation lining tallgrass prairie streams lead to unique in-stream decomposition patterns as well. Tallgrass prairie streams lack the abundant woody-debris dams found in forested streams (Chapter 11, this volume). As a result, the frequent and prolonged dry periods, coupled with the prevalence of scouring floods, allow for very little in-stream decomposition of leaf packs. Indeed, entire groups of stream detritivores (shredders) are missing from the aquatic fauna of tallgrass prairie streams (Chapter 11, this volume).

This high seasonal and annual variation in stream ecosystems leads us to the issue of climate change and its potential effects on terrestrial and aquatic habitats in grassland ecosystems. Several global-change models predict that average temperatures will increase in the Great Plains as atmospheric CO_2 concentrations increase during the next 50 to 100 years. Given that temperature extremes already exist in stream communities, an increase in growing season temperature might push aquatic systems beyond the tolerances of many species. However, these global-change models do not offer clear scenarios for changes in precipitation. Some models indicate an increase; others, a decrease. However, an increase in seasonal and annual variability in precipitation as climate changes is generally predicted (Houghton et al. 1990). The effects of this variability will be expressed differentially among aquatic and terrestrial systems in grasslands. As noted previously, temperature extremes already may threaten biodiversity in aquatic communities. However, increased variability in precipitation patterns might further exaggerate an already wildly variant flow regime in prairie streams.

In terrestrial systems, the predicted changes in temperature and precipitation patterns might be well within the evolved tolerances of many prairie plants (Chapter 16, this volume). However, this might not be true for many species of vertebrates and invertebrates. From the standpoint of biodiversity, Hayden (Chapter 2, this volume) suggests that grassland communities may serve as a "bellwether" for climate change. This suggestion is based on the potential limits to niche breadth in an assemblage of organisms that experiences multiple sources of stress from one year to the next. On the other hand, Seastedt et al. (Chapter 16, this volume) conclude that ecosystem processes in tallgrass prairie appear to be well buffered from the impacts of climate change. The overall implication of this would be a breakdown in linkages between community structure and ecosystem function under some climate change scenarios. Currently, grasslands appear to be the primary ecosystem in which such linkages have been demonstrated experimentally (Johnson et al. 1996). The breakdown of these linkages might have dramatic consequences for system stability and buffering capacity (Tilman et al. 1996).

Future Research Foci in This Nonequilibrium System

One of the positive end products of long-term ecological research at Konza Prairie has been the development of a detailed understanding of some of the biotic and abiotic interactions that regulate the structure, function, and dynamics of this tallgrass ecosystem. A second outcome of this long-term research program is the realization that there is much that we do not know about tallgrass ecosystems, especially as these ecosystems face stresses that are potentially well outside the historical environmental boundaries in which they developed. The impacts of human-generated global change, for example, have unknown consequences for tallgrass prairie ecosystems. Increases in atmospheric CO_2, intraseasonal and interannual variability in precipitation and temperature, changes in natural disturbance regimes, habitat fragmentation, and invasive alien species all might have significant negative repercussions for structure and function in tallgrass prairie ecosystems. Future foci for research at Konza Prairie will concentrate on two of the most immediate and pressing issues related to global change: the effect of climate variability and the role of restoration ecology in the conservation of biodiversity.

Climate Change

The amount and timing of precipitation inputs in grassland ecosystems are critical forcing functions that make these ecosystems particularly vulnerable to potential changes in climate. For the Central Plains, global circulation models predict decreased summer precipitation; increased summer temperatures (i.e., Karl et al. 1991); and, perhaps most important, increased variability in both the amounts and timing of rainfall events (Easterling 1990; Houghton et al. 1990). This is particularly significant because climatic variability is at least as important as mean climate values in determining the structure and function of grassland ecosystems (Gibson and Hulbert 1987; Briggs and Knapp 1995). Additionally, the Intergovernmental Panel on Climate Change has suggested that potential changes in the temporal patterns of precipitation will have more impact on the responses and feedback systems of North American grasslands than changes in the total amount of annual precipitation (Houghton et al. 1990). Thus, changes in patterns of water availability resulting from predicted climate changes are likely to greatly impact tallgrass prairie ecosystems.

To date, not enough is known about system response to increased variability to allow us to predict the magnitude of changes or how they will affect grassland ecosystems over long time scales. To address this problem, two experiments at Konza Prairie have been initiated to determine how key above- and belowground patterns and processes are altered in response to changes in the amount of precipitation and the variability of precipitation inputs. One of these experiments utilizes field-scale Rainfall Manipulation Plots in which the timing and amounts of rainfall events are experimentally manipulated, independently and in tandem. This will allow us to assess the effects of altered precipitation regimes on individual plant ecophysiological responses, plant community composition, and ecosystem-level processes. When results from this experiment are combined with those from an ongoing irrigation study (Knapp et al. 1994c), in which variability in precipitation is maintained at low and nonlimiting levels, we will be able to evaluate more comprehensively the role of precipitation variability in tallgrass prairie.

Fragmentation and Restoration

Tallgrass prairie is one of the most fragmented ecosystems in North America (Samson and Knopf 1994). Tilman et al. (1994) suggested that ecosystem fragmentation may greatly enhance the "extinction debt"; i.e., as habitats decrease in size, a concomitant decrease in species diversity will occur. However, a counterintuitive result of their analysis was that this extinction debt would have a greater impact on the widespread dominant species in communities rather than on the more sparsely distributed, rare species. This occurs because dominant species are limited in their ability to disperse and colonize additional habitats in fragmented landscapes. The dominant plant species have a significant control on community structure and function in grassland vegetation and form the template for food and habitat structure for many species in higher trophic levels (Vinton and Collins 1996). Thus, habitat fragmentation followed by the loss of dominant species may have important consequences for the preservation of tallgrass prairie ecosystems.

The tremendous productivity that characterizes tallgrass prairie environments is also the Achilles' heal of this ecosystem. Much of the tallgrass region in the United States has been converted to agricultural production and other forms of land use, such that the Great Plains landscape currently contains a large number of small, scattered prairie remnants (Steinauer and Collins 1996). Recently, Leach and Givnish (1996) suggested that plant species diversity had declined over a 40-year period in isolated prairie fragments in Wisconsin. This decline in diversity was attributed to the loss of large-scale natural disturbances such as fire. Thus, the potential for the extinction debt, as hypothesized by Tilman et al. (1996), to occur in grassland fragments looms large.

Although fragmentation is a widespread environmental threat to grassland ecosystems (Holt et al. 1995), attempts to restore degraded pastures and ecosystems may counteract some of the negative consequences of this problem. Numerous attempts have been made to restore characteristics of native ecosystems (Jordan et al. 1987), but species diversity tends to recover slowly in restored ecosystems (Howe 1994a). Our earlier research on fire and grazing at Konza Prairie has demonstrated clearly that species diversity is correlated positively with environmental heterogeneity. One of the variables that often is lacking in restoration efforts is spatial variation in resource availability. We hypothesize that the generally homogeneous and high nutrient conditions found at the start of restoration efforts (in former agricultural lands) promote dominance and low species diversity in restored grassland vegetation. To test this hypothesis, we are initiating a long-term prairie restoration experiment in which we will vary soil depth and nutrient availability (Fig 17.5) to determine their effects, singly and in combination, on the maintenance of species diversity in grasslands. We feel this experiment is important because it will address not only a specific need for restoration efforts in grasslands but also the more general phenomenon of the role of resource heterogeneity in the maintenance of species diversity in communities (e.g., Kolasa and Pickett 1991).

Fire in Tallgrass Prairie

Past research at Konza has focused primarily on the effect of fire on population, community, and ecosystem processes. Most of this research has been conducted on sites that have been burned in April of each year. One of the missing pieces of the fire puzzle at Konza is seasonality (Howe 1994b). To rectify this concern, the management plan for Konza Prairie has been modified to include new replicated burning treatments that will occur in early spring, late winter, and summer (Chapter 1, this volume).

An additional issue with the long-term burning studies on Konza Prairie is that several of the annually burned watersheds on which much of the earlier research was focused were incorporated into the management plan well before the LTER Program was started. In other words, we were not able to collect data on

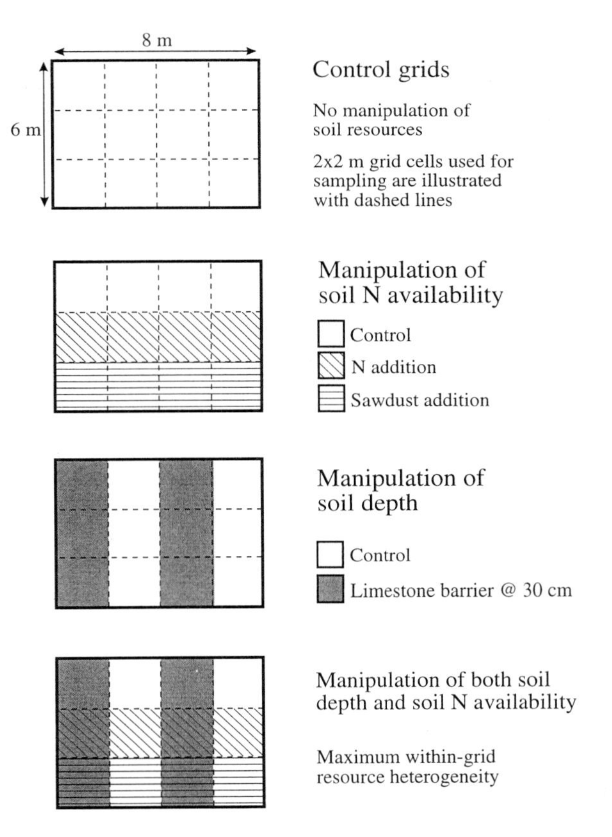

Figure 17.5. Schematic diagram of an experiment designed to determine the role of resource heterogeneity in the maintenance of species diversity during restoration of tallgrass prairie ecosystems. In a former agricultural field, soil depth will be varied by placing limestone barriers 30 cm below the soil surface, and nutrient availability will be altered by the addition of sawdust or N.

community and ecosystem processes when these burning treatments were initiated. Given the important legacy effects in grassland ecosystems, much of the current structure and function in annually burned grassland may reflect unknown starting conditions. For example, the CENTURY model predicts that N pools in prairies will decrease with annual burning (Ojima et al. 1990), but this prediction has not been substantiated at Konza Prairie (Chapter 13, this volume). One reason could be that a large pool of N existed in these prairie soils prior to annual burning, such that the predicted decline thus far has been buffered.

To address this issue, we will reverse the burning regime on four of the watersheds from which we already have long-term measurements of community composition, annual net primary production, and soil resource availability. Starting in 1999, we will stop burning two sites that have been burned annually since the early 1970s, and we will begin annual burning on two management units that have been burned infrequently (scheduled as 20-year fire intervals) over this time frame. Because we already have long-term measurements from these management units, reversing these burning treatments will allow us to more clearly document the effects of fire on community structure and ecosystem functioning in tallgrass prairie. In addition, this approach will permit us to tease apart the role of historical conditions, including community composition and pools of available resources, from the proximate effects of either annual burning or long-term fire suppression.

Large Ungulate Herbivory

Grazer population densities (stocking rates), temporal patterns of grazing (grazing systems), and the type of grazing animal (*B. bison* or domestic cattle) are primary determinants of grazing effects on grassland ecosystem structure and function. Much research has been conducted on effects of grazing systems and stocking rates on vegetation responses such as standing crop biomass, composition of forage and nonforage components, defoliation patterns, and carbohydrate reserves of forage plants, as well as on various animal performance responses. With the exception of a few previous studies on soil hydrologic properties and seedling emergence (e.g. Warren et al. 1986; Weigel et al. 1990), the effects of grazing systems and stocking rates on multiple components of biodiversity (e.g., species richness, morphological diversity, genetic diversity, spatial heterogeneity); resistance and resilience stability; and other long-term ecological consequences have not been studied. This is increasingly important as the demand for dwindling grassland resources intensifies (Samson and Knopf 1994).

Recently initiated research on Konza Prairie will investigate the long-term impacts of stocking rate, grazing systems, and type of grazing animal on a number of community-level attributes including biodiversity, spatial heterogeneity, invasibility, and resistance and resilience stability in grasslands. Twelve 90-ha management units will be used to compare experimentally the effects of cattle, *B. bison*, or no grazing on grassland biodiversity. Another eight units will be used to assess cattle and *B. bison* grazing treatments with different grazing periods and intensities. The goal of this research is to significantly increase understanding of ecological effects of different grazing management strategies and their impact on grassland structure and function. These studies will directly address effects of range management practices on both conservation of natural biodiversity and the resistance stability of grassland ecosystems to environmental change and stress. In addition, these long-term experiments will provide empirical tests of the recent theoretical and highly controversial predictions concerning relationships between grazing intensity and floristic diversity, and between diversity and community stability in grasslands.

Summary

Over the past 15 years, the Konza Prairie Long-Term Ecological Research Program has endeavored to compile a comprehensive data set that can enhance our understanding of population, community, and ecosystem structure and function in tallgrass prairie. By all forms of assessment, such as publications in scientific journals, contributions to education at all levels, and knowledge transfer to management interests, this program has been a success. The structure of the Konza Prairie research program fits well with the overall goals of LTER (Callahan 1984) to provide a comprehensive, site-based, long-term ecological research program designed to address questions that cannot be answered under typical short-term funding scenarios. Through this endeavor, the Konza Prairie LTER Program has developed a basic understanding of the role of fire in grassland ecosystems. In addition, research at Konza Prairie has addressed many fundamental conceptual issues in ecology. Most of these could be addressed only with data from long-term studies. At the same time, as knowledge of the tallgrass ecosystem accrues, new and more challenging questions arise. We hope that this volume has conveyed to the reader our collective understanding of this ecosystem; our excitement and enthusiasm for tallgrass prairies; and the many challenges to understanding that remain hidden in this complex, dynamic, nonequilibrium system.

Acknowledgments This is contribution no. 97-204-B from the Kansas Agricultural Experiment Station Kansas State University, Manhattan.

References

Aber, J. S. 1988. Upland chert gravels of east-central Kansas. Kansas Geological Survey, Guidebook Series 6:17–19.

Aber, J. S. 1991. The glaciation of northeast Kansas. Boreas 20:297–314.

Abrams, M. D. 1986. Historical development of gallery forests in northeast Kansas. Vegetatio 65:29–37.

Abrams, M. D. 1988. Effects of burning regime on buried seed banks and canopy coverage in Kansas tallgrass prairie. Southwestern Naturalist 33:65–70.

Abrams, M. D., A. K. Knapp, and L. C. Hulbert. 1986. A ten-year record of aboveground biomass in a Kansas tallgrass prairie: effects of fire and topographic position. American Journal of Botany 73:1509–1515.

Adem, J., and W. L. Donn. 1981. Progress in monthly climate forecasting with a physical model. Bulletin of the American Meteorological Society 62:1666–1675.

Albertson, F. W., and J. E. Weaver. 1944. Effects of drought, dust, and intensity of grazing on cover and yield of short-grass prairies. Ecological Monographs 14:1–29.

Allen, L. J., L. H. Harbers, R. R. Schalles, C. E. Owensby, and E. F. Smith. 1976. Range burning and fertilizing related to nutritive value of bluestem grass. Journal of Range Management 29:306–308.

Allen, M. F. 1991. The ecology of mycorrhizae. Cambridge University Press, Cambridge, UK.

Anderson, J. P. E., and K. H. Domsch. 1989. Ratios of microbial biomass carbon to total carbon in arable soils. Soil Biology and Biochemistry 21:471–479.

Anderson, K. L. 1965. Time of burning as it affects soil moisture in an ordinary upland bluestem prairie in the Flint Hills. Journal of Range Management 18:311–316.

Anderson, K. L., E. F. Smith, and C. E. Owensby. 1970. Burning bluestem range. Journal of Range Management 23:81–92.

Anderson, L. E., H. A. Crum, and W. R. Buck. 1990. List of the mosses of North America north of Mexico. The Bryologist 93:448–499.

Anderson, N. J., and J. R. Sedell. 1979. Detritus processing by macroinvertebrates in stream ecosystems. Annual Review of Entomology 24:351–377.

Anderson, R. C. 1982. An evolutionary model summarizing the roles of fire, climate, and grazing animals in the origin and maintenance of grasslands: an end paper. Pages 297–308 in J. R. Estes, R. J. Tyrl, and J. N. Brunken, editors, Grasses and grasslands: systematics and ecology. University of Oklahoma Press, Norman, Oklahoma, USA.

Anderson, R. C. 1990. The historic role of fire in the North American grassland. Pages 8–18 in S. L. Collins and L. L. Wallace, editors, Fire in North American tallgrass prairies. University of Oklahoma Press, Norman, Oklahoma, USA.

Anderson, R. C., and L. E. Brown. 1986. Stability and instability in plant communities following fire. American Journal of Botany 73:364–368.

Anderson, R. C., T. Leahy, and S. S. Dhillion. 1989. Numbers and biomass of selected insect groups on burned and unburned sand prairie. American Midland Naturalist 122:151–162.

Anthes, R. 1984. Enhancement of convective precipitation by mesoscale variations in vegetation covering in semi-arid regions. Journal of Climate and Meteorology 23:541–554.

Archer, S., D. S. Schimel, and E. H. Holland. 1995. Mechanisms of shrubland expansion: land use, climate, or CO_2. Climatic Change 29:91–99.

Axelrod, D. I. 1985. Rise of the grassland biome, central North America. Botanical Review 51:163–201.

Bare, J. E., and R. L. McGregor. 1970. An introduction to the phytogeography of Kansas. University of Kansas Science Bulletin 48:869–949.

Barker, W. T. 1969. The flora of the Kansas Flint Hills. University of Kansas Science Bulletin 48:525–584.

Barnes, C. P. 1959. The climatic environment of grassland. Pages 243–249 in H. B. Sprague, editor, Grasslands. American Association for the Advancement of Science, Washington, DC, USA.

Bartlett, C. A. 1988. Computer modeling of water yield from Kings Creek watershed. M.S. thesis. Kansas State University, Manhattan, Kansas, USA.

Bazzaz, F. A., and J. A. D. Parrish. 1982. Organization of grassland communities. Pages 233–254 in J. R. Estes, R. G. Tyrl, and J. N. Brunken, editors, Grasses and grasslands: systematics and ecology. University of Oklahoma Press, Norman, Oklahoma, USA.

Beck, H. V. 1949. The Quaternary geology of Riley County, Kansas. M.S. thesis. Kansas State University, Manhattan, Kansas, USA.

Bee, J. W., G. E. Glass, R. S. Hoffmann, and R. R. Patterson. 1981. Mammals in Kansas. University of Kansas, Museum of Natural History Public Education Series 7:1–300.

Belovsky, G. E., and A. Joern. 1995. The dominance of different regulating factors on rangeland grasshoppers. Pages 359–386 in N. Cappuccino and P. W. Price, editors, Population dynamics: new approaches and synthesis. Academic Press, San Diego, California, USA.

Belovsky, G. E., and J. B. Slade. 1993. The role of vertebrate and invertebrate predators in a grasshopper community. Oikos 68:193–201.

Benedict, R. A., P. W. Freeman, and H. H. Genoways. 1996. Prairie legacies—mammals. Pages 149–166 in F. B. Samson and F. L. Knopf, editors, Prairie conservation: preserving North America's most endangered ecosystem. Island Press, Washington, DC, USA.

Benning, T. L. 1993. Fire frequency and topoedaphic controls of net primary productivity in the tallgrass prairie. Ph.D. dissertation. University of Colorado, Boulder, Colorado, USA.

Benning, T. L., and T. R. Seastedt. 1995. Landscape-level interactions between topoedaphic features and nitrogen limitation in tallgrass prairie. Landscape Ecology 10:337–348.

Bentivenga, S. P., and B. A. D. Hetrick. 1991. Relationship between mycorrhizal activity, burning, and plant productivity in tallgrass prairie. Canadian Journal of Botany 69:2597–2602.

Bentivenga, S. P., and B. A. D. Hetrick. 1992a. The effect of prairie management practices on mycorrhizal symbiosis. Mycologia 84:522–527.

Bentivenga, S. P., and B. A. D. Hetrick. 1992b. Seasonal and temperature effects on mycorrhizal activity and dependence of cool- and warm-season tallgrass prairie grasses. Canadian Journal of Botany 70:1596–1602.

Bertin, R. I. 1989. Pollination biology. Pages 23–86 in W. G. Abrahamson, editor, Plant-animal interactions. McGraw-Hill Book Company, New York, New York, USA.

Biondini, M. E., W. K. Lauenroth, and O. E. Sala. 1991. Correcting estimates of net primary production: are we overestimating plant production in rangelands? Journal of Range Management 44:194–198.

Biondini, M E., and L. Manske. 1996. Grazing frequency and ecosystem processes in a northern mixed prairie, USA. Ecological Applications 6:239–256.

Biondini, M. E., A. A. Steuter, and C. E. Grygiel. 1989. Seasonal fire effects on the diversity patterns, spatial distribution, and community structure of forbs in the northern mixed prairie. Vegetatio 85:21–31.

Birney, E. C., W. E. Grant, and D. D. Baird. 1976. Importance of vegetative cover to cycles of *Microtus* populations. Ecology 57:1043–1051.

Bixler, S. H., and D. W. Kaufman. 1995a. Local distribution of prairie voles (*Microtus ochrogaster*) on Konza Prairie: effect of topographic position. Transactions of the Kansas Academy of Science 98:61–67.

Bixler, S. H., and D. W. Kaufman. 1995b. Prairie voles occur at low density in ungrazed tallgrass prairie in eastern Kansas. Prairie Naturalist 27:33–40.

Black, A. P., and R. F. Christman. 1963. Characteristics of colored surface waters. Journal of the American Water Works Association 55:753–770.

Blackburn, W. H. 1984. Impacts of grazing intensity and specialized grazing systems on watershed characteristics and responses. Pages 927–983 in National Research Council/National Academy of Sciences, editors, Developing strategies for rangeland management. Westview Press, Boulder, Colorado, USA.

Blad, B. L., and D. S. Schimel. 1992. An overview of surface radiance and biology studies in FIFE. Journal of Geophysical Research 97:18,829–18,835.

Blair, J. M. 1997. Fire, N availability, and plant response in grasslands: A test of the transient maxima hypothesis. Ecology 78:2359–2368.

Blankespoor, G. W. 1970. The significance of nest and nest site microclimate for the dickcissel, *Spiza americana*. Ph.D. dissertation. Kansas State University, Manhattan, Kansas, USA.

Bloom, A. F., F. S. Chapin, and H. A. Mooney. 1985. Resource limitation in plants: an economic analogy. Annual Review of Ecology and Systematics 16:363–392.

Boardman, N. K. 1977. Comparative photosynthetic rates of sun and shade plants. Annual Review of Plant Physiology 28:355–377.

Boddey, R. M., and J. Dobereiner. 1994. Biological nitrogen fixation associated with graminaceous plants. Pages 119–135 in Y. Okon, editor, *Azospirillum*/plant associations. CRC Press, Boca Raton, Florida, USA.

Bond, W. J. 1993. Keystone species. Pages 237–253 in E. D. Schulze and H. A. Mooney, editors, Biodiversity and ecosystem function. Springer-Verlag, New York, New York, USA.

Borchert, J. R. 1950. The climate of the central North American grassland. Annals of the Association of American Geographers 40:1–39.

Box, E. O., B. N. Holben, and V. Kalb. 1989. Accuracy of the AVHRR vegetation index as a predictor of biomass, primary productivity, and net CO_2 flux. Vegetatio 80:71–89.

Brady, N. C. 1990. The nature and properties of soils. 10th edition. Mcmillan, New York, New York, USA.

Bragg, T. B. 1982. Seasonal variations in fuel and fuel consumption by fires in a bluestem prairie. Ecology 63:7–11.

Bragg, T. B. 1995. The physical environment of Great Plains grasslands. Pages 49–81 in A. Joern and K. H. Keeler, editors, The changing prairie. Oxford University Press, Oxford, UK.

Bragg, T. B., and L. C. Hulbert. 1976. Woody plant invasion of unburned Kansas bluestem prairie. Journal of Range Management 29:19–23.

Brejda, J. J., R. J. Kremer, and J. R. Brown. 1994. Indications of associative nitrogen fixation in eastern gamagrass. Journal of Range Management 47:192–196.

Breymeyer, A. I., and G. M. Van Dyne, editors. 1980. Grasslands, systems analysis, and man. Cambridge University Press, Cambridge, UK.

Briggs, J. M., and S. L. Collins. 1994. Using remote sensing to determine heterogeneity in tallgrass prairie. Pages 113–119 in Proceedings of the Seventh Annual Erdas Users' Group Meeting. Atlanta, Georgia, USA.

Briggs, J. M., J. T. Fahnestock, L. E. Fischer, and A. K. Knapp. 1994. Aboveground biomass production in tallgrass prairie: effect of time since fire. Pages 165–170 in R. G. Wickett, P. D. Lewis, A. Woodliffe, and P. Pratt, editors, Proceedings of the Thirteenth North American Prairie Conference. Preney Print and Litho, Windsor, Ontario, Canada.

Briggs, J. M., and D. J. Gibson. 1992. Effects of fire on tree spatial patterns in a tallgrass prairie landscape. Bulletin of the Torrey Botanical Club 119:300–307.

Briggs, J. M., and A. K. Knapp. 1991. Estimating aboveground biomass in tallgrass prairie with the harvest method: determining proper sample size using jackknifing and Monte Carlo simulations. Southwestern Naturalist 36:1–6.

Briggs, J. M., and A. K. Knapp. 1995. Interannual variability in primary production in tallgrass prairie: climate, soil moisture, topographic position, and fire as determinants of aboveground biomass. American Journal of Botany 82:1024–1030.

Briggs, J. M., and M. D. Nellis. 1989. Landsat thematic mapper digital data for predicting aboveground biomass in a tallgrass prairie ecosystem. Pages 53–55 in T. B. Bragg and J. Stubbendieck, editors, Proceedings of the Eleventh North American Prairie Conference, University of Nebraska. University of Nebraska Press, Lincoln, Nebraska, USA.

Briggs, J. M., and M. D. Nellis. 1991. Seasonal variation of heterogeneity in the tallgrass prairie: a quantitative measure using remote sensing. Photogrammeric Engineering and Remote Sensing 57:407–411.

Briggs, J. M., D. R. Rieck, C. L. Turner, G. M. Henebry, D. G. Goodin, and M. D. Nellis. 1997. Spatial and temporal patterns of vegetation within the Flint Hills. Transactions of the Kansas Academy of Science 100:10–20.

Briggs, J. M., T. R. Seastedt, and D. J. Gibson. 1989. Comparative analysis of temporal and spatial variability in above-ground production in a deciduous forest and prairie. Holarctic Ecology 12:130–136.

Brillhart, D. E., and D. W. Kaufman. 1994. Temporal variation in coyote prey in tallgrass prairie of eastern Kansas. Prairie Naturalist 26:93–105.

Brillhart, D. E., and D. W. Kaufman. 1995. Spatial and seasonal variation in prey use by coyotes in north-central Kansas. Southwestern Naturalist 40:160–166.

Brillhart, D. E., G. A. Kaufman, and D. W. Kaufman. 1995. Small-mammal use of experimental patches of tallgrass prairie: influence of topographic position and fire history. Pages 59–65 in D. C. Hartnett, editor, Proceedings of the Fourteenth North American Prairie Conference, Kansas State University. Kansas State University, Manhattan, Kansas, USA.

Briske, D. D., and J. L. Butler. 1989. Density-dependent regulation of ramet populations within the bunchgrass *Schizachyrium scoparium*: interclonal versus intraclonal interference. Journal of Ecology 77:963–974.

Brown, J. H., and E. J. Heske. 1990. Control of a desert grassland transition by a keystone rodent guild. Science 250:1705–1708.

Brown, J. H., and J. Roughgarden. 1990. Ecology for a changing earth. Bulletin of the Ecological Society of America 71:173–188.

Bryson, R. A. 1966. Air masses, streamlines, and the boreal forest. Geographical Bulletin 8:228–269.

Bryson, R. A., and D. A. Baerreis. 1968. Climatic change and the Mill Creek culture of Iowa. Journal of the Iowa Archaeological Society 15:1–358.

Bryson, R. A., D. A. Baerreis, and W. M. Wendlend. 1970. The character of late-glacial and post-glacial climatic changes. Pages 53–74 in W. Dort and J. K. Jones, editors, Pleistocene and recent environments of the central Great Plains. University Press of Kansas, Lawrence, Kansas, USA.

Bryson, R. A., and F. K. Hare. 1974. Climates of North America. Elsevier Press, New York, New York, USA.

Budyko, M. I. 1974. Climate and life. International Geophysics Series 18. Academic Press, New York, New York, USA.

Burke, I. C., C. M. Yonker, W. J. Parton, C. V. Cole, K. Flach, and D. S. Schimel. 1989. Texture, climate and cultivation effects on soil organic matter content in U.S. grassland soils. Soil Science Society of America 53:800–805.

Byrne, F. E., M. R. Mudge, H. V. Beck, and R. H. Burton. 1949. Preliminary report and map on the geologic construction material resources in Riley County, Kansas. United States Geological Survey Open-File Report 49-128. Washington, DC, USA.

Cable, D. R. 1975. Influence of precipitation on perennial grass production in the semidesert Southwest. Ecology 56:981–986.

Callahan, J. T. 1984. Long-term ecological research. BioScience 34:363–367.

Carpenter, J. R. 1926. The grassland biome. Ecological Monographs 10:645–665.

Carpenter, S. R., and J. F. Kitchell. 1993. The trophic cascade in lakes. Cambridge University Press, Cambridge, UK.

Cerling, T. E., J. Quade, Y. Wang, and J. R. Bowman. 1989. Carbon isotopes in soils and palaeosoils as ecology and palaeoecology indicators. Nature 431:138–139.

Chapin, F. S., III. 1991. Integrated responses of plants to stress. BioScience 41:29–36.

Chapin, F. S., III, and R. A. Kedrowski. 1983. Seasonal changes in nitrogen and phosphorus fractions and autumn retranslocation in evergreen and deciduous taiga trees. Ecology 64:376–391.

Chapin, F. S., III, and S. J. McNaughton. 1989. Lack of compensatory growth under phosphorus deficiency in grazing-adapted grasses from the Serengeti Plains. Oecologia 79:551–557.

Chesson, P. L., and T. J. Case. 1986. Overview: non-equilibrium community theories:

chance, variability, history and coexistence. Pages 229–239 in J. Diamond and T. J. Case, editors, Community ecology. Harper and Row, New York, New York, USA.

Chiariello, N., J. C. Hickman, and H. A. Mooney. 1982. Endomycorrhizal role for interspecific transfer of phosphorus in a community of annual plants. Science 217:941–943.

Churchill, S. P. 1985. A synopsis of the Kansas mosses with keys and distribution maps. University of Kansas Science Bulletin 53:1–64.

Clark, B. K. 1989. Influence of plant litter and habitat structure on small mammal assemblages: experimental manipulations and field observations. Ph.D. dissertation. Kansas State University, Manhattan, Kansas, USA

Clark, B. K., and D. W. Kaufman. 1990. Short-term responses of small mammals to experimental fire in tallgrass prairie. Canadian Journal of Zoology 68:2450–2454.

Clark, B. K., and D. W. Kaufman. 1991. Effects of plant litter on foraging and nesting behavior of prairie rodents. Journal of Mammalogy 72:502–512.

Clark, B. K., D. W. Kaufman, E. J. Finck, and G. A. Kaufman. 1989. Small mammals in tallgrass prairie: patterns associated with grazing and burning. Prairie Naturalist 21:177–184.

Clark, B. K., D. W. Kaufman, G. A. Kaufman, and E. J. Finck. 1987. Use of tallgrass prairie by *Peromyscus leucopus*. Journal of Mammalogy 68:158–160.

Clark, B. K., D. W. Kaufman, G. A. Kaufman, and S. K. Gurtz. 1995. Population ecology of Elliot's short-tailed shrew and least shrew in ungrazed tallgrass prairie manipulated by experimental fire. Pages 87–92 in D. C. Hartnett, editor, Proceedings of the Fourteenth North American Prairie Conference, Kansas State University. Kansas State University, Manhattan, Kansas, USA.

Clark, B. K., D. W. Kaufman, G. A. Kaufman, S. K. Gurtz, and S. H. Bixler. 1992. Population ecology of thirteen-lined ground squirrels in ungrazed tallgrass prairie manipulated by fire. Pages 51–54 in D. A. Smith and C. A. Jacobs, editors, Proceedings of the Twelfth North American Prairie Conference, University of Northern Iowa. University of Northern Iowa, Cedar Falls, Iowa, USA.

Clark, F. E. 1977. Internal cycling of ^{15}nitrogen in shortgrass prairie. Ecology 58:1322–1333.

Clark, F. E., and E. A. Paul. 1970. The microflora of grassland. Advances in Agronomy 22:375–435.

Clark, J. S. 1988. Effect of climate change on fire regimes in northwestern Minnesota. Nature 334:233–235.

Clement, R. W. 1987. Floods in Kansas and techniques for estimating their magnitude and frequency on unregulated streams. United States Geological Survey Water-Resources Investigations Report 87-4008. Washington, DC, USA.

Clements, F. E. 1936. Nature and structure of the climax. Journal of Ecology 24:552–584.

Clements, F. E., and V. E. Shelford. 1939. Bio-ecology. Wiley, New York, New York, USA.

Cline, I. M. 1894. Summer hot winds on the Great Plains. American Meteorological Journal 51:135–140.

Coblentz, B. E. 1990. Exotic organisms: a dilemma for conservation biology. Conservation Biology 4:261–265.

Cody, M. L. 1966. The consistency of intra- and inter-continental grassland bird communities. American Naturalist 100:371–376.

Cody, M. L. 1968. On the methods of resource division in grassland bird communities. American Naturalist 102:107–147.

Coffin, D. P., and W. K. Lauenroth. 1988. The effects of disturbance size and frequency on a shortgrass plant community. Ecology 69:1609–1617.

Coffin, D. P., and W. K. Lauenroth. 1989. Spatial and temporal variation in the seed bank of a semiarid grassland. American Journal of Botany 76:53–58.

COHMAP Members. 1988. Climatic change of the last 18,000 years: observations and model simulations. Science 241:1043–1052.

Coleman, D. C., R. Andrews, J. E. Ellis, and J. S. Singh. 1976. Energy flow and partitioning in selected man-managed and natural ecosystems. Agro-Ecosystems 3:45–54.

Collins, H. H. 1959. Complete field guide to American wildlife. Harper and Brothers, New York, New York, USA.

Collins, J. T. 1993. Amphibians and reptiles in Kansas. University of Kansas, Museum of Natural History Public Education Series 13:1–397.

Collins, S. L. 1987. Interaction of disturbances in tallgrass prairie: a field experiment. Ecology 68:1243–1250.

Collins, S. L. 1990. Introduction: fire as a natural disturbance in tallgrass prairie ecosystems. Pages 3–7 in S. L. Collins and L. L. Wallace, editors, Fire in North American tallgrass prairies. University of Oklahoma Press, Norman, Oklahoma, USA.

Collins, S. L. 1992. Fire frequency and community heterogeneity in tallgrass prairie vegetation. Ecology 73:2001–2006.

Collins, S. L. 1995. The measurement of stability in grasslands. Trends in Ecology and Evolution 10:95–96.

Collins, S. L., and D. E. Adams. 1983. Succession in grasslands: thirty-two years of change in a central Oklahoma tallgrass prairie. Vegetatio 51:181–190.

Collins, S. L., and S. C. Barber. 1985. Effects of disturbance on diversity in mixed-grass prairie. Vegetatio 64:87–94.

Collins, S. L., and T. L. Benning. 1996. Spatial and temporal patterns in functional diversity. Pages 253–280 in K. Gaston, editor, Biodiversity: a biology of numbers and difference. Blackwell Science, Oxford, UK.

Collins, S. L., and D. J. Gibson. 1990. Effect of fire on community structure in tallgrass and mixed-grass prairie. Pages 81–98 in S. L. Collins and L. L. Wallace, editors, Fire in North American tallgrass prairies. University of Oklahoma Press, Norman, Oklahoma, USA.

Collins, S. L., and S. M. Glenn. 1991. Importance of spatial and temporal dynamics in species regional abundance and distribution. Ecology 72:654–664.

Collins, S. L., and S. M. Glenn. 1997. Effects of organismal and distance scaling on analysis of species distribution and abundance. Ecological Applications 7:543–551.

Collins, S. L., S. M. Glenn, and D. J. Gibson. 1995. Experimental analysis of intermediate disturbance and initial floristic composition: decoupling cause and effect. Ecology 76:486–492.

Collins, S. L., and L. L. Wallace. 1990. Fire in North American tallgrass prairies. University of Oklahoma Press, Norman, Oklahoma, USA.

Connell, J. H. 1978. Diversity in tropical rain forests and coral reefs. Science 199:1302–1310.

Conrad, H. S. 1939. Plant associations on land. American Midland Naturalist 21:1–27.

Copeland, J. H., R. A. Pielke, and T. G. F. Kittel. 1996. Potential climatic impacts of vegetation change: a regional modeling study. Journal of Geophysical Research 101:7409–7418.

Coupland, R. T. 1992. Ecosystems of the world 8A: Natural grasslands. Introduction and Western Hemisphere. Elsevier Press, Amsterdam, Netherlands.

Cowles, H. C. 1928. Persistence of prairies. Ecology 9:380–382.

Cross, F. B., and R. E. Moss. 1987. Historic changes in fish communities and aquatic habitats in plains streams of Kansas. Pages 155–165 in W. J. Matthews and D. C. Heins, editors, Community and evolutionary ecology of North American stream fishes. University of Oklahoma Press, Norman, Oklahoma, USA.

Cummins, K. W. 1974. Structure and function of stream ecosystems. BioScience 24:631–641.

Cummins, K. W., and M. J. Klug. 1979. Feeding ecology of stream invertebrates. Annual Review of Ecology and Systematics 10:147–172.

Cummins, K. W., and R. W. Merritt. 1984. Ecology and distribution of aquatic insects. Pages 59–65 in R. W. Merritt and K. W. Cummins, editors, An introduction to the aquatic insects of North America. 2nd edition. Kendall/Hunt, Dubuque, Iowa, USA.

Dahlman, R. C., and C. L. Kucera. 1965. Root productivity and turnover in native prairie. Ecology 46:84–89.

Dahlman, R. C., and C. L. Kucera. 1969. Carbon-14 cycling in the root and soil components of a prairie ecosystem. Pages 652–660 in D. J. Nelson and F. C. Evans, editors, Proceedings of the Second National Symposium on Radioecology. Division Technical Information, USAECTID-4500 (CONF-670503). Oak Ridge, Tennessee, USA.

Damhoureyeh, S. A. 1996. Effects of bison and cattle grazing on growth, reproduction, and abundances of five tallgrass prairie forbs. M.S. thesis. Kansas State University, Manhattan, Kansas, USA.

D'Angelo, D. J., J. R. Webster, S. V. Gregory, and J. L. Meyer. 1993. Transient storage in Appalachian and Cascade mountain streams as related to hydraulic characteristics. Journal of the North American Benthological Society 2:223–235.

Daubenmire, R. 1968. Ecology of fire in grasslands. Advances in Ecological Research 5:209–266.

Day, T. A., and J. K. Detling. 1990. Grassland patch dynamics and herbivore grazing preference following urine deposition. Ecology 63:180–188.

DeAngelis, D. L., J. C. Waterhouse, W. M. Post, and R. V. O'Neill. 1985. Ecological modeling and disturbance evaluation. Ecological Modeling 29:399–419.

Delcourt, P. A., and H. R. Delcourt. 1993. Paleoclimates, paleovegetation, and paleofloras during the late Quaternary. Pages 71–94 in Flora of North America Editorial Committee, editors, Flora of North America north of Mexico. Volume 1, Introduction. Oxford University Press, New York, New York, USA.

DeLucia, E. H., S. A. Heckathorn, and T. A. Day. 1992. Effects of soil temperature on growth, biomass allocation and resource acquisition of *Andropogon gerardii*. New Phytologist 120:543–549.

Detling, J. K. 1988. Grasslands and savannas: Regulation of energy flow and nutrient cycling by herbivores. Pages 131–148 in L. R. Pomeroy and J. J. Alberts, editors, Concepts of ecosystems ecology. Ecological Studies 67. Springer-Verlag, New York, New York, USA.

Dodd, M. E., J. Silvertown, K. McConway, J. Potts, and M. Crawley. 1994. Stability in the communities in the Park Grass Experiment: the relationship between species richness, soil pH, and biomass variability. Proceedings of the Royal Society of London B 346:185–193.

Dodds, W. K., M. K. Banks, C. S. Clennan, C. W. Rice, D. Sotomayor, E. Strauss, and W. Yu. 1996a. Biological properties of soil and subsurface sediments under grassland and cultivation. Soil Biology and Biochemistry 28:837–846.

Dodds, W. K., J. M. Blair, G. M. Henebry, J. K. Koelliker, R. Ramundo, and C. M. Tate. 1996b. Nitrogen transport from tallgrass prairie watersheds. Journal of Environmental Quality. 25:973–981.

Dodds, W. K., R. E. Hutson, A. C. Eichem, M. A. Evans, D. A. Gudder, K. M. Fritz, and L. J. Gray. 1996c. The relationship of flood, drying, flow and light to primary production and producer biomass in a prairie stream. Hydrobiologia 333:151–159.

Dodds, W. K., C. A. Randel, and C. C. Edler. 1996d. Microcosms for aquatic research: application to colonization of various sized particles by ground-water microorganisms. Ground Water 34:756–759.

Doran, J. W., and T. B. Parkin. 1994. Defining and assessing soil quality. Pages 3–21 in J. W. Doran, D. C. Coleman, D. F. Bezdicek, and B. A. Stewart, editors, Defining soil quality for a sustainable environment, Special Publication No. 35. Soil Science Society of America, Madison, Wisconsin, USA.

Dort, W., Jr. 1987. Salient aspects of the terminal zone of continental glaciation in Kansas. Pages 55–66 in W. C. Johnson, editor, Quaternary environments of Kansas. Kansas Geological Survey Guidebook Series 5. Lawrence, Kansas, USA.

Duell, A. B. 1990. Effects of burning on infiltration, overland flow, and sediment loss on tallgrass prairie. M.S. thesis. Kansas State University, Manhattan, Kansas, USA.

Dyksterhuis, E. J. 1958. Ecological principles in range evaluation. Botanical Review 24:253–272.

Easterling, W. E. 1990. Climate trends and prospects. Pages 32–55 in R. N. Sampson and D. Hair, editors, Natural resources for the 21st century. Island Press, Washington, DC, USA.

Editorial Committee of the Flora of North America. 1993. Flora of North America. Volume 2, Ferns and gymnosperms. Oxford University Press, New York, New York, USA.

Edler, C., and W. K. Dodds. 1992. Characterization of a groundwater community dominated by *Caecidotea tridentata* (Isopoda). Pages 91–99 in J. A. Stanford and J. J. Simons, editors, Proceedings of the First International Conference on Groundwater Ecology. United States Environmental Protection Agency, American Water Research Association, Bethesda, Maryland, USA.

Edler, C., and W. K. Dodds. 1996. The ecology of a subterranean isopod, *Caecidotea tridentata*. Freshwater Biology 35:249–259.

Eichem, A. C., W. K. Dodds, C. M. Tate, and C. Edler. 1993. Microbial decomposition of elm and oak leaves in a karst aquifer. Applied and Environmental Microbiology 59:3592–3596.

Eisele, K. A., D. S. Schimel, L. A. Kapuska, and W. J. Parton. 1990. Effects of available P and N:P ratios on non-symbiotic dinitrogen fixation in tallgrass prairie soils. Oecologia 79:471–474.

Elliott, E. T., H. W. Hunt, and D. E. Walter. 1988. Detrital foodweb interactions in North American grassland ecosystems. Agriculture, Ecosystems and Environment. 24:41–56.

Engle, D. M., and P. M. Bultsma. 1984. Burning of northern mixed prairie during drought. Journal of Range Management 37:398–401.

Erwin, W. J., and R. H. Stasiak. 1979. Vertebrate mortality during the burning of a reestablished prairie in Nebraska. American Midland Naturalist 101:247–249.

Evans, E. W. 1983. The influence of neighboring hosts on colonization of prairie milkweeds by a seed-feeding bug. Ecology 64:648–653.

Evans, E. W. 1984. Fire as a natural disturbance to grasshopper assemblages of tallgrass prairie. Oikos 43:9–16.

Evans, E. W. 1988a. Community dynamics of prairie grasshoppers subjected to periodic fire: predictable trajectories or random walks in time? Oikos 52:283–292.

Evans, E. W. 1988b. Grasshopper (Insecta: Orthoptera: Acrididae) assemblages of tallgrass prairie: influences of fire frequency, topography, and vegetation. Canadian Journal of Zoology 66:1495–1501.

Evans, E. W. 1989. Interspecific interactions among phytophagous insects of tallgrass prairie: an experimental test. Ecology 70:435–444.

Evans, E. W. 1990. Dynamics of an aggregation of blister beetles (Coleoptera: Meloidae) attacking a prairie legume. Journal of the Kansas Entomological Society 63:616–625.

Evans, E. W. 1992. Absence of interspecific competition among tallgrass prairie grasshoppers during a drought. Ecology 73:1038–1044.

Evans, E. W., J. M. Briggs, E. J. Fink, D. J. Gibson, S. W. James, D. W. Kaufman, and T. R. Seastedt. 1989a. Is fire a disturbance in grasslands? Pages 159–161 in T. B. Bragg and J. Stubbendieck, editors, Proceedings of the Eleventh North American Prairie Conference, University of Nebraska. University of Nebraska Press, Lincoln, Nebraska, USA.

Evans, E. W., R. A. Rogers, and D. J. Opfermann. 1983. Sampling grasshoppers (Orthroptera: Acrididae) on burned and unburned tallgrass prairie: night trapping vs. sweeping. Environmental Entomology 12:1449–1454.

Evans, E. W., and T. R. Seastedt. 1995. The relations of phytophagous invertebrates and range plants. Pages 580–634 in D. J. Bedunah and R. E. Sosebee, editors, Wildland plants: physiological ecology and development morphology. Society for Range Management, Denver, Colorado, USA.

Evans, E. W., C. C. Smith, and R. P. Gendron. 1989b. Timing of reproduction in a prairie legume: seasonal impacts of insects consuming flowers and seeds. Oecologia 78: 220–230.

Fahnestock, J. T., and A. K. Knapp. 1993. Water relations and growth of tallgrass prairie forbs in response to selective herbivory by bison. International Journal of Plant Science 154:432–440.

Fahnestock, J. T., and A. K. Knapp. 1994. Responses of forbs and grasses to selective grazing by bison: interactions between herbivory and water stress. Vegetatio 115: 123–131.

Fajer, E. D., M. D. Bowers, and F. A. Bazzaz. 1989. The effects of enriched carbon dioxide atmospheres on plant-insect herbivore interactions. Science 243:1198–2000.

Fay, P. A., and D. C. Hartnett. 1991. Constraints on growth and allocation patterns of *Silphium integrifolium* (Asteraceae) caused by a cynpid gall wasp. Oecologia 88:243–250.

Fay, P. A., D. C. Hartnett, and A. K. Knapp. 1996. Plant tolerance of gall-insect attack and gall-insect performance. Ecology 77:521–534.

Fay, P. A., and R. W. Samenus. 1993. Gall wasp (Hymenoptera: Cynipidae) mortality in a spring tallgrass prairie fire. Environmental Entomology 22:1333–1337.

Field, C. B., C. P. Lund, N. R. Chiariello, and B. E. Mortimer. 1997. CO_2 effects on the water budget of grassland microcosm communities. Global Change Biology 67:251–306.

Finck, E. J. 1986. Birds wintering on the Konza Prairie Research Natural Area. Pages 91–94 in G. K. Clambey and R. H. Pemble, editors, Proceedings of the Ninth North American Prairie Conference, North Dakota State University. North Dakota State University, Fargo, North Dakota, USA.

Finck, E. J., D. W. Kaufman, G. A. Kaufman, S. K. Gurtz, B. K. Clark, L. J. McLellan, and B. S. Clark. 1986. Mammals of the Konza Prairie Research Natural Area, Kansas. Prairie Naturalist 18:153–166.

Fisher, S. G. 1986. Structure and dynamics of desert streams. Pages 119–139 in G. Whitford, editor, Pattern and process in desert ecosystems. University of New Mexico Press, Albuquerque, New Mexico, USA.

Fisher, S. G., L. J. Gray, N. B. Grimm, and D. E. Busch. 1982. Temporal succession in a desert stream ecosystem following flash flooding. Ecological Monographs 52:93–110.

Fischer-Walter, L. E., D. C. Hartnett, B. A. D. Hetrick, and A. P. Schwab. 1996. Interspecific nutrient transfer in a tallgrass prairie plant community. American Journal of Botany 83:180–184.

Fisk, M. 1995. Nitrogen dynamics in an alpine landscape. Ph.D. dissertation. University of Colorado, Boulder, Colorado, USA.

Forman, R. T. T., and M. Godron. 1986. Landscape ecology. Wiley, New York, New York, USA.

Forman, S. L., E. A. Bettis, T. J. Kemmis, and B. B. Miller. 1992. Chronological evidence for multiple periods of loess deposition during the late Pleistocene in the Missouri and Mississippi River valley, United States: implications for the activity of the Laurentide ice sheet. Palaeogeography, Palaeoclimatology, Palaeoecology 89:71–83.

Fowler, A. C., R. L. Knight, T. L. George, and L. C. McEwan. 1991. Effects of avian predation on grasshopper populations in North Dakota grasslands. Ecology 72:1775–1781.

Frank, D. A., and R. D. Evans. 1997. Effects of native grazers on grassland N cycling in Yellowstone National Park. Ecology 78:2238–2248.

Frank, D. A., and R. S. Inouye. 1994. Temporal variation in actual evapotranspiration of terrestrial ecosystems: patterns and ecological implications. Journal of Biogeography 21:401–411.

Frank, D. A., R. S. Inouye, N. Huntly, G. W. Minshall, and J. E. Anderson. 1994. The biogeochemistry of a north-temperate grassland with native ungulates: nitrogen dynamics in Yellowstone National Park. Biogeochemistry 26:163–188.

Frank, D. A., and S. J. McNaughton. 1991. Stability increases with diversity in plant communities: empirical evidence from the 1988 Yellowstone drought. Oikos 62:360–362.

Frank, D. A., and S. J. McNaughton. 1992. The ecology of plants, large mammalian herbivores, and drought in Yellowstone National Park. Ecology 73:2043–2058.

Frank, D. A., and Y. Zhang. 1997. Ammonia volatilization from a seasonally and spatially grazed grassland: Yellowstone National Park. Biogeochemistry 36:189–203.

Franklin, J. F. 1989. Importance and justification of long-term studies in ecology. Pages 3–19 in G. E. Likens, editor, Long-term studies in ecology. Springer-Verlag, New York, New York, USA.

Fredlund, G. G., and P. J. Jaumann. 1987. Late Quaternary palynological and paleobotanical records from the central Great Plains. Pages 167–178 in W. C. Johnson, editor, Quaternary environments of Kansas. Kansas Geological Survey Guidebook Series 5. Lawrence, Kansas, USA.

Freeman, C. C. 1977. An annotated list of the vascular flora of Konza Prairie Research Natural Area. NSF undergraduate research paper. Division of Biology, Kansas State University, Manhattan, Kansas, USA.

Freeman, C. C., and D. J. Gibson. 1987. Additions to the vascular flora of Konza Prairie Research Natural Area, Kansas. Transactions of the Kansas Academy of Science 90:81–84.

Freeman, C. C., and L. C. Hulbert. 1985. An annotated list of the vascular flora of Konza Prairie Research Natural Area, Kansas. Transactions of the Kansas Academy of Science 88:84–115.

French, N. R., editor. 1979. Perspectives in grassland ecology: results and applications of the US/IBP Grassland Biome Study. Springer-Verlag, New York, New York, USA.

French, N. R., R. K. Steinhorst, and D. M. Swift. 1979. Grassland biomass trophic pyramids. Pages 59–87 in N. R. French, editor. Perspectives in grassland ecology: results and applications of the US/IBP Grassland Biome Study. Springer-Verlag, New York, New York, USA.

Fretwell, S. D. 1972. Populations in a seasonal environment. Princeton University Press, Princeton, New Jersey, USA.

Fretwell, S. D. 1977. The regulation of plant communities by the food chains exploiting them. Perspectives in Biology and Medicine 20:169–185.

Fretwell, S. D. 1987. Food chain dynamics: the central theory to ecology? Oikos 50:291–301.

Fritz, K. M., and W. K. Dodds. 1996. The effects of drying and flood upon the macroinvertebrate community of a tallgrass prairie stream. Bulletin of the North American Benthological Society 13:212.

Frye, J. C. 1955. The erosional history of the Flint Hills. Transactions of the Kansas Academy of Science 58:79–86.

Frye, J. C., and A. B. Leonard. 1952. Pleistocene geology of Kansas. Kansas Geological Survey Bulletin 99, Lawrence, Kansas, USA.

Garcia, F. O. 1992. Carbon and nitrogen dynamics and microbial ecology in tallgrass prairie. Ph.D. dissertation. Kansas State University, Manhattan, Kansas, USA.

Garcia, F. O., and C. W. Rice. 1994. Microbial biomass dynamics in tallgrass prairie. Soil Science Society of America Journal 58:816–823.

Gentry, A. H. 1986. Endemism in tropical versus temperate plant communities. Pages 153–181 in M. E. Soulé, editor, Conservation biology: the science of scarcity and diversity. Sinauer, Sunderland, Massachusetts, USA.

Ghiorse, W. C., and J. T. Wilson. 1988. Microbial ecology of the terrestrial subsurface. Advances in Applied Microbiology 33:107–172.

Gibson, D. J. 1988. Regeneration and fluctuations of tallgrass prairie vegetation in response to burning frequency. Bulletin of the Torrey Botanical Club 115:1–12.

Gibson, D. J. 1989. Effects of animal disturbance on tallgrass prairie vegetation. American Midland Naturalist 121:144–154.

Gibson, D. J., C. C. Freeman, and L. C. Hulbert. 1990. Effects of small mammal and invertebrate herbivory on plant species richness and abundance in tallgrass prairie. Oecologia 84:169–175.

Gibson, D. J., and B. A. D. Hetrick. 1988. Topographic and fire effects on the composition and abundance of VA-mycorrhizal fungi in tallgrass prairie. Mycologia 80:433–441.

Gibson, D. J., and L. C. Hulbert. 1987. Effects of fire, topography and year-to-year climatic variation on species composition in tallgrass prairie. Vegetatio 72:175–185.

Gibson, D. J., T. R. Seastedt, and J. M. Briggs. 1993. Management practices in tallgrass prairie: large- and small-scale experimental effects on species composition. Journal of Applied Ecology 30:247–255.

Gigon, A., and A. Leutert. 1996. The dynamic keyhole-key model of coexistence to explain diversity of plants in limestone and other grasslands. Journal of Vegetation Science 7:29–40.

Gilliam, F. S. 1987. The chemistry of wet deposition for a tallgrass prairie ecosystem: inputs and interactions with plant canopies. Biogeochemistry 4:203–217.

Glaze, S. L., M. D. Ransom, and W. A. Wehmueller. 1994. Sodium and gypsum accumulation in polygenetic soils of north central Kansas, U.S.A. Transactions of the Fifteenth World Congress of Soil Science. Acapulco, Mexico. 6B:101–102.

Gleason, H. A. 1939. The individualistic concept of the plant association. American Midland Naturalist 21:92–110.

Gleeson, S. K., and D. Tilman. 1992. Plant allocation and the multiple limitation hypothesis. American Naturalist 139:1322–1343.

Glenn, S. M., and S. L. Collins. 1990. Patch structure in tallgrass prairie: dynamics of satellite species. Oikos 57:229–236.

Glenn, S. M., and S. L. Collins. 1992. Effects of scale of disturbance on rates of immigration and extinction of species in prairies. Oikos 63:273–280.

Glenn, S. M., S. L. Collins, and D. J. Gibson. 1992. Disturbances in tallgrass prairie: local versus regional effects on community heterogeneity. Landscape Ecology 7:243–252.

Glenn-Lewin, D. C., L. A. Johnson, T. W. Jurik, A. Akey, M. Leoschke, and T. Rosburg. 1990. Fire in central North American grasslands: vegetative reproduction, seed germination, and seedling establishment. Pages 28–45 in S. L. Collins and L. L. Wallace, editors, Fire in North American tallgrass prairies. University of Oklahoma Press, Norman, Oklahoma, USA.

Goldsmith, F. B., C. M. Harrison, and A. J. Morton. 1986. Description and analysis of vegetation. Pages 437–524 in P. D. Moore and S. B. Chapman, editors, Methods in plant ecology. Blackwell Scientific Publications, Oxford, UK.

Graham, A. 1993. History of the vegetation: Cretaceous (Maastrichtian)-Tertiary. Pages 57–70 in Flora of North America Editorial Committee, editors, Flora of North America north of Mexico. Volume 1, Introduction. Oxford University Press, New York, New York, USA.

Grahmner, K. 1990. Respiration of soil and vegetation from a tallgrass prairie rangeland. M.S. thesis. University of Nebraska, Lincoln, Nebraska, USA.

Grant, W. E., E. C. Birney, N. R. French, and D. M. Swift. 1982. Structure and productivity of grassland small mammal communities related to grazing-induced changes in vegetative cover. Journal of Mammalogy 63:248–260.

Gray, L. J. 1984. Benthic macroinvertebrates of Rattlesnake Creek: similarities in community structure with desert streams. Transactions of the Kansas Academy of Science 86:9.

Gray, L. J. 1989. Emergence, production, and export of aquatic insects from a tallgrass prairie stream. Southwestern Naturalist 34:313–318.

Gray, L. J. 1993. Response of insectivorous birds to emerging aquatic insects in riparian habitats of a tallgrass prairie stream. American Midland Naturalist 129:288–300.

Gray, L. J. 1997. Organic matter dynamics in Kings Creek. Journal of the North American Benthological Society 16:50–54.

Gray, L. J., and K. W. Johnson. 1988. Trophic structure of benthic macroinvertebrates in Kings Creek. Transactions of the Kansas Academy of Science 91:178–184.

Great Plains Flora Association. 1977. Atlas of the flora of the Great Plains. Iowa State University Press, Ames, Iowa, USA.

Great Plains Flora Association. 1986. Flora of the Great Plains. University Press of Kansas, Lawrence, Kansas, USA.

Grime, J. P. 1973. Control of species density in herbaceous vegetation. Journal of Environmental Management 1:151–167.

Grime, J. P. 1979. Plant strategies and vegetation processes. Wiley, New York, New York, USA.

Groffman, P. M., C. W. Rice, and J. M. Tiedje. 1993. Denitrification in a tallgrass prairie landscape. Ecology 74:855–862.

Groffman, P. M., and C. L. Turner. 1995. Plant productivity and nitrogen gas fluxes in a tallgrass prairie landscape. Landscape Ecology 10:255–266.

Grubb, T. C., Jr., and L. Greenwald. 1982. Sparrows and a brushpile: foraging responses to different combinations of predation risk and energy cost. Animal Behaviour 30:637–640.

Grzybowski, J. A. 1982. Population structure in grassland bird communities during winter. Condor 84:637–640.

Gurtz, M. E., G. R. Marzolf, K. T. Killingbeck, D. L. Smith, and J. V. McArthur. 1982. Organic matter loading and processing in a pristine stream draining a tallgrass prairie/riparian forest watershed. Contribution No. 230 of the Kansas Water Resources Research Institute, Manhattan, Kansas, USA. 78 pages.

Gurtz, M. E., G. R. Marzolf, K. T. Killingbeck, D. L. Smith, and J. V. McArthur. 1988. Hydrologic and riparian influences on the import and storage of coarse particulate organic matter in a prairie stream. Canadian Journal of Fisheries and Aquatic Sciences 45:655–665.

Gurtz, M. E., and C. M. Tate. 1988. Hydrologic influences on leaf decomposition in a channel and adjacent bank of a gallery forest stream. American Midland Naturalist 120:11–21.

Hadley, E. B., and B. J. Kieckhefer. 1963. Productivity of two prairie grasses in relation to fire frequency. Ecology 44:389–395.

Hairston, N. G., F. E. Smith, and L. B. Slobodkin. 1960. Community structure, population control, and competition. American Naturalist 94:421–425.

Hake, D. R., J. Powell, J. K. McPherson, P. L. Claypool, and G. L. Dunn. 1984. Water stress of tallgrass prairie plants in central Oklahoma. Journal of Range Management 37:147–151.

Halda-Alija, L. 1996. Diversity and distribution of denitrifying bacteria in soil and subsurface sediments under grassland and cultivated soil. Ph.D. dissertation, Kansas State University, Mahanttan, Kansas, USA.

Ham, J. M., and A. K. Knapp. 1998. Fluxes of CO_2, water vapor, and energy from a prairie ecosystem during the seasonal transition from carbon sink to carbon source. Agricultural and Forest Meteorology 89:1–14.

Ham, J. M., C. E. Owensby, P. I. Coyne, and D. J. Bremer. 1995. Fluxes of CO_2 and water vapor from a prairie ecosystem exposed to ambient and elevated CO_2. Agricultural and Forest Meteorology 77:73–93.

Hanski, I. 1982. Dynamics of regional distribution: the core and satellite species hypothesis. Oikos 38:210–221.

Hare, F. K. 1951. Some climatological problems of the Arctic and sub-Arctic. Pages 952–964 in T. F. Malone, editor, Compendium of meteorology. American Meteorology Society, Boston, Massachusetts, USA.

Harper, P. P. 1978. Variations in the production of emerging insects from a Quebec stream. Verhandlungen der Internationale Vereinigung fur Theoretische und Angewandt Limnologie 20:1317–1323.

Hartnett, D. C. 1989. Density- and growth stage–dependent responses to defoliation in two rhizomatous grasses. Oecologia 80:414–420.

Hartnett, D. C. 1990. Size-dependent allocation to seed and vegetative reproduction in four clonal composites. Oecologia 84:254–259.

Hartnett, D. C. 1991. Effects of fire in tallgrass prairie on growth and reproduction of prairie coneflower (*Ratibida columnifera*: Asteraceae). American Journal of Botany 78:429–435.

Hartnett, D. C. 1993. Regulation of clonal growth and dynamics of *Panicum virgatum* in tallgrass prairie: effects of neighbor removal and nutrient addition. American Journal of Botany 80:1114–1120.

Hartnett, D. C., B. A. D. Hetrick, G. W. T. Wilson, and D. J. Gibson. 1993. VA-mycorrhizal influence on intra- and interspecific neighbor interactions among co-occurring prairie grasses. Journal of Ecology 81:787–795.

Hartnett, D. C., K. R. Hickman, and L. E. Fischer-Walter. 1996. Effects of bison grazing, fire, and topography on floristic diversity in tallgrass prairie. Journal of Range Management. 49:413–420.

Hartnett, D. C., and K. H. Keeler. 1995. Population processes. Pages 82–99 in A. Joern and K. H. Keeler, editors, The changing prairie: North American grasslands. Oxford University Press, Oxford, UK.

Hartnett, D. C., R. J. Samenus, L. E. Fischer, and B. A. D. Hetrick. 1994. Plant demographic responses to mycorrhizal symbiosis in tallgrass prairie. Oecologia 99:21–26.

Hartnett, D. C., A. A. Steuter, and K. R. Hickman. 1997. Comparative ecology of native versus introduced ungulates. Pages 72–101 in F. Knopf and F. Samson, editors. Ecology and conservation of Great Plains vertebrates. Springer-Verlag, New York, New York, USA.

Hayden, B. P. 1981. Secular variation in Atlantic Coast extratropical cyclones. Monthly Weather Review 109:159–167.

Hayden, B. P. 1982. Season-to-season cyclone frequency prediction. Monthly Weather Review 110:239–253.

Hayden, B. P. 1984. A systematic error in the Northern Hemisphere sea-level pressure data set. Monthly Weather Review 112:2354–2357.

Hayden, B. P. 1994. Global biosphere requirements for general circulation models. Pages 263–276 in W. K. Michener, J. W. Brunt, and S. G. Stafford, editors, Environmental information management and analysis. Taylor and Francis, London, UK.

Hayes, D. C. 1985. Seasonal nitrogen translocation in big bluestem during drought conditions. Journal of Range Management 38:406–410.

Hayes, D. C., and T. R. Seastedt. 1987. Root dynamics of tallgrass prairie in wet and dry years. Canadian Journal of Botany 65:787–791.

Hayes, D. C., and T. R. Seastedt. 1989. Nitrogen dynamics of soil water in burned and unburned tallgrass prairie. Soil Biology and Biochemistry 21:1003–1007.

Heat-Moon, W. L. 1991. PrairyErth: a deep map. Houghton Mifflin, Boston, Massachusetts, USA.

Heaton, T. H. E. 1986. Isotopic studies of nitrogen pollution in the hydrosphere and atmosphere: a review. Chemical Geology 59:87–102.

Heckathorn, S. A., and E. H. DeLucia. 1991. Effect of leaf rolling on gas exchange and leaf temperature of *Andropogon gerardii* and *Spartina pectinata*. Botanical Gazette 152:263–268.

Heckathorn, S. A., and E. H. DeLucia. 1994. Drought-induced nitrogen retranslocation in perennial C_4 grasses of tallgrass prairie. Ecology 75:1877–1886.

Heckathorn, S. A., and E. H. DeLucia. 1995. Ammonia volatilization during drought in perennial C_4 grasses of tallgrass prairie. Oecologia 101:361–365.

Heckathorn, S. A., and E. H. Delucia. 1996. Retranslocation of shoot nitrogen to rhizomes and roots in prairie grasses may limit loss of N to grazing and fire during drought. Functional Ecology 10:396–400.

Heinrich, M. L., and D. W. Kaufman. 1985. Herpetofauna of the Konza Prairie Research Natural Area, Kansas. Prairie Naturalist 17:101–112.

Henebry, G. M. 1993. Detecting change in grasslands using measures of spatial dependence with Landsat TM data. Remote Sensing of the Environment 46:223–234.

Henebry, G. M., and H. Su. 1993. Using landscape trajectories to assess the effects of radiometric rectification. International Journal of Remote Sensing 14:2417–2423.

Henry, A. J. 1930. The calendar year as a time unit in drought statistics. Monthly Weather Review 59:150–154.

Herbel, C. H., and K. L. Anderson. 1959. Response of true prairie vegetation on major Flint Hills range sites to grazing treatment. Ecological Monographs 29:171–186.

Hetrick, B. A. D., and J. Bloom. 1983. Vesicular-arbuscular mycorrhizal fungi associated with native tall grass prairie and cultivated winter wheat. Canadian Journal of Botany 61:2140–2146.

Hetrick, B. A. D., D. C. Hartnett, G. W. T. Wilson, and D. J. Gibson. 1994. Effects of mycorrhizae, phosphorous availability, and plant density on yield relationships among competing tallgrass prairie grasses. Canadian Journal of Botany 72:168–176.

Hetrick, B. A. D., D. G. Kitt, and G. W. T. Wilson. 1988. Mycorrhizal dependence and growth habit of warm-season and cold-season tallgrass prairie plants. Canadian Journal of Botany 66:1376–1380.

Hetrick, B. A. D., G. W. T. Wilson, and D. C. Hartnett. 1989. Relationship between mycorrhizal dependence and competitive ability of two tallgrass prairie grasses. Canadian Journal of Botany 67:2608–2615.

Hetrick, B. A. D., G. W. T. Wilson, and C. E. Owensby. 1990. The influence of mycorrhizae on big bluestem rhizome regrowth and clipping tolerance. Journal of Range Management 43:286–290.

Hetrick, B. A. D., G. W. T. Wilson, and T. C. Todd. 1992. Relationships of mycorrhizal symbiosis, rooting strategy, and phenology among tallgrass prairie forbs. Canadian Journal of Botany 70:1521–1528.

Hobbs, N. T., D. S. Schimel, C. E. Owensby, and D. J. Ojima. 1991. Fire and grazing in tallgrass prairie: contingent effects on nitrogen budgets. Ecology 72:1374–1382.

Hobbs, R. J., and H. A. Mooney. 1995. Spatial and temporal variability in California annual grassland: results from a long-term study. Journal of Vegetation Science 6:43–56.

Holdridge, L. R. 1947. Determination of world plant formations from simple climatic data. Science 105:267–368.

Holland, E. A., and J. K. Detling. 1990. Plant response to herbivory and belowground nitrogen cycling. Ecology 71:1040–1049.

Holland, E. A., W. J. Parton, J. K. Detling, and D. L. Coppock. 1992. Physiological responses of plant populations to herbivory and their consequences for ecosystem nutrient flow. American Naturalist 140:685–706.

Holt, R. D., D. M. Debinski, J. E. Diffendorfer, M. S. Gaines, E. A. Martinko, G. R. Robinson, and G. C. Ward. 1995. Perspectives from an experimental study of fragmentation in an agroecosystem. Pages 147–176 in D. Glen, editor, Arable ecosystems for the twenty-first century. Wiley, New York, New York, USA.

Hooker, K. L., and G. R. Marzolf. 1987. Differential decomposition of leaves in grassland and gallery forest reaches of Kings Creek. Transactions of the Kansas Academy of Science 90:17–24.

Horak, G. J. 1985. Kansas prairie chickens. Kansas Fish and Game Commission, Wildlife Bulletin 3:1–65.

Horn, H. S. 1974. The ecology of secondary succession. Annual Review of Ecology and Systematics 5:25–37.

Horton, R. E. 1945. Erosional development of streams and their drainage basins. Geological Society of America Bulletin 56:275–370.

Houghton, J. T., G. J. Jenkins, and J. J. Ephraums, editors. 1990. Climate change: The IPCC Scientific Assessment. World Meteorological Organization, Cambridge University Press, Cambridge, UK.
Howe, H. F. 1994a. Managing species diversity in tallgrass prairies: assumptions and implications. Conservation Biology 8:691–704.
Howe, H. F. 1994b. Response of early- and late-flowering plants to fire season in experimental prairies. Ecological Applications 4:121–133.
Hubbard, K. G. 1994. Spatial variability of daily weather variables in the high plains of the USA. Agricultural and Forest Meteorology 68:29–41.
Hulbert, L. C. 1969. Fire and litter effects in undisturbed bluestem prairie in Kansas. Ecology 50:874–877.
Hulbert, L. C. 1973. Management of Konza Prairie to approximate pre-white-man fire influences. Pages 14–16 in L. C. Hulbert, editor, Proceedings of the Third Midwest Prairie Conference, Kansas State University. Kansas State University, Manhattan, Kansas, USA.
Hulbert, L. C. 1976. Tentative list of vascular plants on Konza Prairie Research Natural Area. Division of Biology, Kansas State University, Manhattan, Kansas, USA.
Hulbert, L. C. 1985. History and use of Konza Prairie Research Natural Area. Prairie Scout 5:63–93.
Hulbert, L. C. 1986. Fire effects on tallgrass prairie. Pages 138–142 in G. K. Clambey and R. H. Pemble, editors, Proceedings of the Ninth North American Prairie Conference, North Dakota State University. North Dakota State University, Fargo, North Dakota, USA.
Hulbert, L. C., 1988. Causes of fire effects in tallgrass prairie. Ecology 69:46–58.
Hulbert, L. C., and J. K. Wilson. 1983. Fire interval effects on flowering of grasses in Kansas bluestem prairie. Pages 255–257 in C. L. Kucera, editor, Proceedings of the Seventh North American Prairie Conference, Southwest Missouri State University. Southwest Missouri State University, Springfield, Missouri, USA.
Humbert, C. E. 1990. Application of AGNPS model to watersheds in Northeast Kansas. M.S. thesis. Kansas State University, Manhattan, Kansas, USA.
Hunt, H. W. 1977. A simulation model for decomposition in grasslands. Ecology 58:469–484.
Hunt, H. W., D. C. Coleman, E. R. Ingham, E. T. Elliott, J. C. Moore, S. L. Rose, C. P. P. Reid, and C. R. Morley. 1987. The detrital food web in a shortgrass prairie. Biology and Fertility of Soils 3:57–68.
Hunter, M. D., and P. W. Price. 1992. Playing chutes and ladders: heterogeneity and the relative roles of bottom-up and top-down forces in natural communities. Ecology 73:724–732.
Huntly, N. 1991. Herbivores and the dynamics of communities and ecosystems. Annual Review of Ecology and Systematics 22:477–503.
Huston, M. 1979. A general hypothesis of species diversity. American Naturalist 133:81–101.
Huston, M. A. 1994. Biological diversity. Cambridge University Press, New York, New York, USA.
Huston, M. A., and D. L. DeAngelis. 1994. Competition and coexistence: the effects of resource transport and supply rates. American Naturalist 144: 954–977.
Illies, J. 1975. A new attempt to estimate production in running waters. Verhandlungen der Internationale Vereinigung fur Theoretische und Angewandt Limnologie 19:1705–1711.
Ingham, E. R., J. A. Trofymow, R. N. Ames, H. W. Hunt, C. R. Morely, J. C. Moore, and

D. C. Coleman. 1986a. Trophic interactions and nitrogen cycling in a semi-arid grassland soil: I. Seasonal dynamics of the natural populations, their interactions, and effects on nitrogen cycling. Journal of Applied Ecology 23:597–614.

Ingham, E. R., J. A. Trofymow, R. N. Ames, H. W. Hunt, C. R. Morely, J. C. Moore, and D. C. Coleman. 1986b. Trophic interactions and nitrogen cycling in a semi-arid grassland soil: II. System responses to removal of different groups of soil microbes or fauna. Journal of Applied Ecology 23:615–630.

Ingham, R. E., and J. K. Detling. 1984. Plant-herbivore interactions in a North American mixed-grass prairie. III. Soil nematode populations and root biomass on *Cynomys ludovicianus* colonies and adjacent uncolonized areas. Oecologia 63:307–313.

Ingham, R. E., and J. K. Detling. 1990. Effects of root-feeding nematodes on aboveground net primary production in a North American grassland. Plant Soil 121:279–281.

Insam, H. 1990. Are the soil microbial biomass and basal respiration governed by the climatic regime? Soil Biology and Biochemistry 22:525–532.

Insam, H., D. Parkinson, and K. H. Domsch. 1989. Influence of macroclimate on soil microbial biomass. Soil Biology and Biochemistry 21:211–221.

Isern, T. D. 1983. Farming in the Flint Hills: a photographic essay. Kansas History 6:221–236.

Jackson, J. K., and S. G. Fisher. 1986. Secondary production, emergence, and export of aquatic insects in a Sonoran Desert stream. Ecology 67:629–638.

James, F. C. 1971. Ordinations of habitat relationships among breeding birds. Wilson Bulletin 83:215–236.

James, S. W. 1982. Effects of fire and soil type on earthworm populations in a tallgrass prairie. Pedobiologia 24:37–40.

James, S. W. 1984. New records of earthworms from Kansas (Oligochaeta: Acanthodrilidae, Lumbricidae, Megascolecidae). Prairie Naturalist 16:91–95.

James, S. W. 1988. The post-fire environment and earthworm populations in tallgrass prairie. Ecology 69:476–483.

James, S. W. 1991. Soil, nitrogen, phosphorus, and organic matter processing by earthworms in tallgrass prairie. Ecology 72:2101–2109.

James, S. W. 1992a. Localized dynamics of earthworm populations in relation to bison dung in North American tallgrass prairie. Soil Biology and Biochemistry 24:1471–1476.

James, S. W. 1992b. Seasonal and experimental variation in population structure of earthworms in tallgrass prairie. Soil Biology and Biochemistry 24:1445–1449.

James, S. W., and M. R. Cunningham. 1989. Feeding ecology of some earthworms in Kansas tallgrass prairie. American Midland Naturalist 121:78–83.

James, S. W., and T. R. Seastedt. 1986. Nitrogen mineralization by native and introduced earthworms: effects on big bluestem growth. Ecology 67:1094–1097.

Jantz, D. R., R. F. Harner, H. T. Rowland, and D. A. Gier. 1975. Soil survey of Riley County and part of Geary County, Kansas. Soil Conservation Service, United States Department of Agriculture, Washington, DC, USA.

Jaramillo, V. J., and J. K. Detling. 1992a. Small-scale grazing in a semi-arid North American grassland. I. Tillering, N uptake, and retranslocation in simulated urine patches. Journal of Applied Ecology 29:1–8.

Jaramillo, V. J., and J. K. Detling. 1992b. Small-scale grazing in a semi-arid North American grassland. II. Cattle grazing of simulated urine patches. Journal of Applied Ecology 19:9–13.

Jekanoski, R. D., and D. W. Kaufman. 1993. Experimental observations of the cutting and climbing of vegetation by hispid cotton rats. Prairie Naturalist 25:249–254.

Jekanoski, R. D., and D. W. Kaufman. 1995. Use of simulated herbaceous canopy by foraging rodents. American Midland Naturalist 133:304–311.

Jenkinson, D. S. 1988. Determination of microbial biomass carbon and nitrogen in soil. Pages 368–386 in J. R. Wilson, editor, Advances in nitrogen cycling in agricultural ecosystems. CAB International, Wallingford, UK.

Jenkinson, D. S., and J. N. Ladd. 1981. Microbial biomass in soil: measurement and turnover. Pages 415–471 in E. A. Paul and J. N. Ladd, editors, Soil biochemistry. Volume 5. Marcel Dekker, New York, New York, USA.

Jenny, H. 1930. A study on the influence of climate upon the nitrogen and organic matter content of the soil. Research Bulletin 152. Missouri Agricultural Experiment Station, Columbia, Missouri, USA.

Jenny, H. 1941. Factors of soil formation. McGraw-Hill, New York, New York, USA.

Jenny, H. 1980. The soil resource. Ecological Studies 37. Springer-Verlag, New York, New York, USA.

Jewell, M. E. 1927. Aquatic biology of the prairie. Ecology 8:289–298.

Jewett, J. M. 1941. The geology of Riley and Geary Counties, Kansas. Kansas Geological Survey Bulletin 39. Lawrence, Kansas, USA.

Joern, A. 1992. Variable impact of avian predation on grasshopper assemblies in sandhills grassland. Oikos 64:458–463.

Johnson, K. H., K. A. Vogt, H. J. Clark, O. J. Schmitz, and D. J. Vogt. 1996. Biodiversity and the productivity and stability of ecosystems. Trends in Ecology and Evolution 11:372–377.

Johnson, S. R., and A. K. Knapp. 1993. The effect of fire on gas exchange and aboveground production in annually vs. biennially burned *Spartina pectinata* wetlands. Wetlands 13:299–303.

Johnson, W. C., and C. W. Martin. 1987. Holocene alluvial-stratigraphic studies from Kansas and adjoining states of the east-central Plains. Kansas Geological Survey, Guidebook Series 5:109–122.

Jones, C. G., J. H. Lawton, and M. Shachak. 1994. Organisms as ecological engineers. Oikos 69:373–386.

Jones, J. K., Jr., D. M. Armstrong, R. S. Hoffmann, and C. Jones. 1983. Mammals of the northern Great Plains. University of Nebraska Press, Lincoln, Nebraska, USA.

Jordan, W. R., III, M. E. Gilpin, and J. D. Aber. 1987. Restoration ecology. Cambridge University Press, Cambridge, UK.

Junge, C. E. 1958. The distribution of ammonium and nitrate in rain water over the United States. Transactions of the American Geophysical Union 39:241–248.

Kahl, R. B., T. S. Baskett, J. A. Ellis, and J. N. Burroughs. 1985. Characteristics of summer habitats of selected nongame birds in Missouri. University of Missouri, Agricultural Experiment Station Research Bulletin 1056:1–155.

Kansas Agricultural Statistics. 1992. Bluestem pasture report. Kansas State Board of Agriculture, Topeka, Kansas, USA.

Kantrud, H. A. 1981. Grazing intensity effects on the breeding avifauna of North Dakota native grassland. Canadian Field-Naturalist 95:404–417.

Kareiva, P., and M. Anderson. 1988. Spatial aspects of species interactions: the wedding of models and experiments. Pages 35–50 in A. Hastings, editor, Community ecology. Springer-Verlag, New York, New York, USA.

Karl, T. R., R. R. Heim, and R. G. Quayle. 1991. The greenhouse effect in central North America: if not now, when? Science 251:1058–1061.

Kaufman, D. W., and S. H. Bixler. 1995. Prairie voles impact plants in tallgrass prairie. Pages 117–121 in D. C. Hartnett, editor, Proceedings of the Fourteenth North American Prairie Conference, Kansas State University. Kansas State University, Manhattan, Kansas, USA.

Kaufman, D. W., E. J. Finck, and G. A. Kaufman. 1990. Small mammals and grassland fires. Pages 46–80 in S. L. Collins and L. L. Wallace, editors, Fire in North American tallgrass prairies. University of Oklahoma Press, Norman, Oklahoma, USA.

Kaufman, D. W., S. K. Gurtz, and G. A. Kaufman. 1988a. Movements of deer mice in response to prairie fire. Prairie Naturalist 20:225–229.

Kaufman, D. W., and G. A. Kaufman. 1990a. House mice (*Mus musculus*) in natural and disturbed habitats in Kansas. Journal of Mammalogy 71:428–432.

Kaufman, D. W., and G. A. Kaufman. 1990b. Influence of plant litter on patch use by foraging *Peromyscus maniculatus* and *Reithrodontomys megalotis*. American Midland Naturalist 124:195–198.

Kaufman, D. W., G. A. Kaufman, and E. J. Finck. 1983. Effects of fire on rodents in tallgrass prairie of the Flint Hills region of eastern Kansas. Prairie Naturalist 15:49–56.

Kaufman, D. W., G. A. Kaufman, and E. J. Finck. 1989. Rodents and shrews in ungrazed tallgrass prairie manipulated by fire. Pages 173–177 in T. B. Bragg and J. Stubbendieck, editors, Proceedings of the Eleventh North American Prairie Conference, University of Nebraska. University of Nebraska Press, Lincoln, Nebraska, USA.

Kaufman, D. W., G. A. Kaufman, and E. J. Finck. 1993a. Small mammals of wooded habitats of the Konza Prairie Research Natural Area, Kansas. Prairie Naturalist 25:27–32.

Kaufman, D. W., G. A. Kaufman, and E. J. Finck. 1995a. Temporal variation in abundance of *Peromyscus leucopus* in wooded habitats of eastern Kansas. American Midland Naturalist 133:7–17.

Kaufman, G. A., D. E. Brillhart, and D. W. Kaufman. 1993b. Are deer mice a common prey of coyotes? Prairie Naturalist 25:295–304.

Kaufman, G. A., and D. W. Kaufman. 1997. Ecology of small mammals in prairie landscapes. Pages 207–243 in F. L. Knopf and F. B. Samson, editors, Ecology and conservation of Great Plains vertebrates. Springer-Verlag, New York, New York, USA.

Kaufman, G. A., D. W. Kaufman, D. E. Brillhart, and E. J. Finck. 1995b. Effect of topography on the distribution of small mammals on the Konza Prairie Research Natural Area, Kansas. Pages 97–102 in D. C. Hartnett, editor, Proceedings of the Fourteenth North American Prairie Conference, Kansas State University. Kansas State University, Manhattan, Kansas, USA.

Kaufman, G. A., D. W. Kaufman, and E. J. Finck. 1988b. Influence of fire and topography on habitat selection by *Peromyscus maniculatus* and *Reithrodontomys megalotis* in ungrazed tallgrass prairie. Journal of Mammalogy 69:342–352.

Kavenaugh, J. 1988. Community composition and emergence phenology of Chironomidae (Diptera) in a prairie stream with different flow regimes. M.S. thesis, University of Kansas, Lawrence, Kansas, USA.

Kazmaier, R. 1993a. Checklist of the amphibians and reptiles of the Konza Prairie Research Natural Area. Konza Prairie Office, Mahattan, Kansas, USA.

Kazmaier, R. 1993b. Checklist of the plants of the Konza Prairie Research Natural Area. Konza Prairie Office, Manhattan, Kansas, USA.

Keeler, K. H. 1991. Survivorship and recruitment in a long-lived prairie perennial, *Ipomoea leptophylla* (Convolvulaceae). American Midland Naturalist 126:44–60.

Keeling, C. D., T. P. Whorf, M. Wahlen, and J. van der Plicht. 1995. Interannual extremes in the rate of rise of atmospheric carbon dioxide since 1980. Nature 375:666–670.

Kelting, R. W. 1954. Effects of moderate grazing on the composition and plant production of a native tall-grass prairie in central Oklahoma. Ecology 35:200–207.

Kemp, P. R., D. G. Waldecker, C. E. Owensby, J. F. Reynolds, and R. A. Virginia. 1994. Effects of elevated CO_2 and nitrogen fertilization pretreatments on decomposition of tallgrass prairie leaf litter. Plant Soil 165:115–127.

Kemp, W. P., S. J. Harvey, and K. M. O'Neill. 1990. Patterns of vegetation and grasshopper community composition. Oecologia 83:299–308.

Killingbeck, K. T., D. L. Smith, and G. R. Marzolf. 1982. Chemical changes in tree leaves during decomposition in a tallgrass prairie stream. Ecology 63:585–589.

Klatt, B. J., and L. L. Getz. 1987. Vegetation characteristics of *Microtus ochrogaster* and *M. pennsylvanicus* habitats in east-central Illinois. Journal of Mammalogy 68:569–577.

Knapp, A. K. 1984a. Effect of fire in tallgrass prairie on seed production of *Vernonia baldwinii* Torr. (Compositae). Southwestern Naturalist 29:242–243.

Knapp, A. K. 1984b. Post-burn differences in solar radiation, leaf temperature, and water stress influencing production in a lowland tallgrass prairie. American Journal of Botany 71:220–227.

Knapp, A. K. 1984c. Water relations and growth of three grasses during wet and drought years in a tallgrass prairie. Oecologia 65:35–43.

Knapp, A. K. 1985a. Early season production and microclimate associated with topography in a C_4-dominated grassland. Oecologia Plantarum 6:337–346.

Knapp, A. K. 1985b. Effect of fire and drought on the ecophysiology of *Andropogon gerardii* and *Panicum virgatum* in a tallgrass prairie. Ecology 66:1309–1320.

Knapp, A. K. 1993. Gas exchange dynamics in C_3 and C_4 grasses: consequences of differences in stomatal conductance. Ecology 74:113–123.

Knapp, A. K., M. D. Abrams, and L. C. Hulbert. 1985. An evaluation of beta attenuation for estimating aboveground biomass in a tallgrass prairie. Journal of Range Management 38:556–558.

Knapp, A. K., M. Cocke, E. P. Hamerlynck, and C. E. Owensby. 1994a. Effect of elevated CO_2 on stomatal density and distribution in a C_4 grass and a C_3 forb under field conditions. Annals of Botany 74:595–599.

Knapp, A. K., S. L. Conard, and J. M. Blair. 1998. Determinants of soil CO_2 flux from a sub-humid grassland: effects of fire and history. Ecological Applications. In press.

Knapp, A. K., J. T. Fahnestock, S. J. Hamburg, L. B. Statland, T. R. Seastedt, and D. S. Schimel. 1993a. Landscape patterns in soil-plant water relations and primary production in tallgrass prairie. Ecology 74:549–560.

Knapp, A. K., J. T. Fahnestock, and C. E. Owensby. 1994b. Elevated CO_2 alters dynamic stomatal responses to sunlight in a C_4 grass. Plant, Cell and Environment 17:189–195.

Knapp, A. K., and F. S. Gilliam. 1985. Response of *Andropogon gerardii* (Poaceae) to fire-induced high vs. low irradiance environments in tallgrass prairie: leaf structure and photosynthetic pigments. American Journal of Botany 72:1668–1671.

Knapp, A. K., E. P. Hamerlynck, J. M. Ham, and C. E. Owensby. 1996. Responses in stomatal conductance to elevated CO_2 in 12 grassland species that differ in growth form. Vegetatio 125:31–41.

Knapp, A. K., E. P. Hamerlynck, and C. E. Owensby. 1993b. Photosynthetic and water relations responses to elevated CO_2 in the C_4 grass *Andropogon gerardii*. International Journal of Plant Sciences 154:459–466.

Knapp, A. K., and L. C. Hulbert. 1986. Production, density, and height of flower stalks of three grasses in annually burned and unburned eastern Kansas tallgrass prairie: a four-year record. Southwestern Naturalist 31:235–241.

Knapp, A. K., J. K. Koelliker, J. T. Fahnestock, and J. M. Briggs. 1994c. Water relations and biomass responses to irrigation across a topographic gradient in tallgrass prairie. Pages 215–220 in R. G. Wickett, P. D. Lewis, A. Woodliffe, and P. Pratt, editors, Proceedings of the Thirteenth North American Prairie Conference. Preney Print and Litho, Windsor, Ontario, Canada.

Knapp, A. K., and T. R. Seastedt. 1986. Detritus accumulation limits productivity in tallgrass prairie. BioScience 36:662–668.

Knight, C. L., J. M. Briggs, and M. D. Nellis. 1994. Expansion of gallery forest on Konza Prairie Research Natural Area, Kansas, USA. Landscape Ecology 9:117–125.

Knopf, F. L. 1994. Avian assemblages on altered grasslands. Studies in Avian Biology 15:247–257.

Knox, J. C. 1983. Responses of river systems to Holocene climates. Pages 26–41 in H. E. Wright, Jr., editor, Late Quaternary environments of the United States: the Holocene. University of Minnesota Press, Minneapolis, Minnesota, USA.

Knutson, H., and J. B. Campbell. 1976. Relationships of grasshoppers (Acrididae) to burning, grazing, and range sites of native tallgrass prairie in Kansas. Proceedings of the Sixth Tall Timbers Conference on Ecological Animal Control by Habitat Management. Tall Timbers Research Station, Tallahassee, Florida, USA.

Koelliker, J. K., and A. B. Duell. 1990. Effect of burning on surface hydrology of tallgrass prairie. American Society of Agricultural Engineers 1990 International Winter Meeting, Chicago, Illinois, USA.

Koelliker, J. K., M. E. Gurtz, and G. R. Marzolf. 1985. Watershed research at Konza-Tallgrass Prairie. Pages 862–867 in Hydraulics and hydrology in the small computer age. Proceedings of the American Society of Civil Engineering Hydraulics Division, Specialty Conference. American Society of Civil Engineering, New York, New York, USA.

Koelling, M. R., and C. L. Kucera. 1965. Dry matter losses and mineral leaching in bluestem standing crop and litter. Ecology 66:1309–1320.

Kolasa, J., and S. T. A. Pickett, editors. 1991. Ecological heterogeneity. Springer-Verlag, New York, New York, USA.

Kollmorgen, W. M., and D. S. Simonett. 1965. Grazing operations in the Flint Hills-bluestem pastures of Chase County, Kansas. Annals of the Association of American Geography 55:260–290.

Komarek, E. V. 1968. Lightning and lightning fires as an ecological force. Pages 169–197 in Proceedings of the Eighth Annual Tall Timbers Forest Ecology Conference. Tall Timbers Research Station, Tallahassee, Florida, USA.

Korom, S. F. 1992. Natural denitrification in the saturated zone: a review. Water Resources Research 28:1657–1668.

Krebs, C. J. 1991. The experimental paradigm and long-term population studies. Ibis 133 (Supplement 1):3–8.

Krebs, J. S., and R. G. Barry. 1970. The Arctic front and the tundra-taiga boundary in Eurasia. Geographical Review 60:548–554.

Kreitler, C. W. 1975. Determining the source of nitrogen in ground water by nitrogen isotope studies. Bureau of Economic Geology, Report of Investigations No. 83. University of Texas, Austin, Texas, USA.

Kucera, C. L. 1981. Grasslands and fire. Pages 90–111 in H. A. Mooney, T. M. Bonnicksen, N. L. Christensen, J. E. Lotan, and W. A. Reiners, editors, Fire regimes and ecosystem properties, United States Forest Service, General Technical Report WO-26. Washington, DC, USA.

Kucera, C. L., and R. C. Dahlman. 1968. Root-rhizome relationships in fire-treated stand of big bluestem, *Andropogon gerardii* Vitman. American Midland Naturalist 80:268–271.

Kucera, C. L., R. C. Dahlman, and M. Koelling. 1967. Total net productivity and turnover on an energy basis for tallgrass prairie. Ecology 48:536–541.

Kucera, C. L. and J. H. Ehrenreich. 1962. Some effects of annual burning on central Missouri prairie. Ecology 43:334–346.

Kucera, C. L., and D. R. Kirkham. 1971. Soil respiration studies in tallgrass prairie in Missouri. Ecology 52:912–915.

Kucera, C. L., and M. Koelling. 1964. The influence of fire on composition of central Missouri prairie. American Midland Naturalist 72:142–147.

Küchler, A. W. 1974a. A new vegetation map of Kansas. Ecology 55:568–604 and map supplement.

Küchler, A. W. 1974b. Potential natural vegetation: revised edition map. American Geographical Society, United States Geological Survey, Washington, DC, USA.

Kutzbach, J. E. 1987. Model simulations of the climatic patterns during the deglaciation of North America. Pages 425–446 in W. F. Ruddiman and H. E. Wright, Jr., editors, North America and adjacent oceans during the last deglaciation. Volume K-3 of The geology of North America. Geological Society of America, Boulder, Colorado, USA.

Lal, R., J. Kimble, E. Levine, and B. A. Stewart. 1995. Soil and global change. CRC Lewis Publishing, Boca Raton, Florida, USA.

Lanyon, W. E. 1957. The comparative biology of the meadowlarks (*Sturnella*) in Wisconsin. Publication of the Nuttall Ornithological Club 1:1–67.

Lauenroth, W. K., and O. E. Sala. 1992. Long-term forage production of North American shortgrass steppe. Ecological Applications 2:397–403.

Lauver, C. L. 1989. Preliminary classification of the natural communities of Kansas. Kansas Biological Survey Report No. 50:1–21. Lawrence, Kansas, USA.

Laycock, W. A., and Conrad, P. W. 1967. Effect of grazing on soil compaction as measured by bulk density on a high cattle range. Journal of Range Management 20:136–140.

Leach, M. K., and T. J. Givnish. 1996. Ecological determinants of species loss in remnant prairies. Science 273:1555–1558.

Le Houerou, H. N., R. L. Bingham, and W. Skerbek. 1988. Relationship between the variability of primary production and the variability of annual precipitation in world arid lands. Journal of Arid Environments 15:1–18.

Leopold, A. 1949. A Sand County almanac. Oxford University Press, Oxford, UK.

Lieth, H., and R. H. Whittaker, editors. 1975. Primary productivity of the biosphere. Springer-Verlag, New York, New York, USA.

Likens, G. E. 1983. A priority for ecological research. Bulletin of the Ecological Society of America 64:234–243.

Little, E. L. 1971. Atlas of United States trees. Volume 1. Conifers and important hardwoods. United States Department of Agriculture, Miscellaneous Publication Number 1146, Washington, DC, USA.

Little, E. L. 1977. Atlas of United States trees. Vol. 4. Minor eastern hardwoods. United States Department of Agriculture, Miscellaneous Publication No. 1342. Washington, DC, USA.

Lodge, D. J., W. H. McDowell, and C. P. McSwiney. 1994. The importance of nutrient pulses in tropical forests. Trends in Evolution and Ecology 9:384–387.

Lomolino, M. V., J. C. Creighton, G. D. Schnell, and C. L. Certain. 1995. Ecology and conservation of the endangered American burying beetle (*Nicrophorus americanus*). Conservation Biology 9:605–614.

Looman, J. 1983. Grassland as natural or semi-natural vegetation. Pages 173–184 in W. Holzner, M. J. A. Werger, and I. Ikusima, editors, Man's impact on vegetation. Junk Publishing, The Hague, Netherlands.

Louda, S. M., M. A. Potvin, and S. K. Collinge. 1990. Predispersal seed predation, post-dispersal seed predation, and competition in the recruitment of seedlings in a native thistle in sandhills prairie. American Midland Naturalist 124:105–113.

Lovett, G. M. 1994. Atmospheric deposition of nutrients and pollutants in North America: an ecological perspective. Ecological Applications 4:627–628.

Lugo, A. F. 1992. The search for carbon sinks in the tropics. Water, Air, Soil Pollution 64:3–9.

Lussenhop, J. 1976. Soil arthropod response to prairie burning. Ecology 57:88–98.

Lussenhop, J. 1981. Microbial and microarthropod detrital processing in a prairie soil. Ecology 62:964–972.

Lynch, J. A., V. C. Bowersox, and C. Simmons. 1995. Precipitation chemistry trends in the United States: 1980–1993. National Atmospheric Deposition Program, Natural Resource Ecology Laboratory, Summary Report, Fort Collins, Colorado, USA.

Macpherson, G. L. 1992a. Ground-water chemistry under tallgrass prairie, central Kansas, USA. Pages 809–812 in Y. K. Kharaka and A. S. Maest, editors, Water-rock interactions. Proceedings of the Seventh International Symposium on Water-Rock Interaction, Park City, Utah, USA. A. A. Balkema Publishers, Rotterdam, Netherlands.

Macpherson, G. L. 1992b. Variations in shallow ground-water chemistry at Konza Prairie LTER Site, Kansas, USA, 1991. EOS, Transactions, American Geophysical Union 73:126–127.

Macpherson, G. L. 1993. Source(s), fate, and residence time of nitrate in ground-water: A comparison of carbonate and alluvial aquifers. Contributions of the Kansas Water Resourses Research Institute, Manhattan, Kansas, USA. 305:1–39.

Macpherson, G. L. 1994. Nitrate at the Konza Prairie NSF-EPSCoR site, Kansas, USA. EOS Transactions, American Geophysical Union 75:262.

Macpherson, G. L. 1996. Hydrogeology of thin-bedded limestones: the Konza Prairie Long-Term Ecological Research Site, Northeastern Kansas. Journal of Hydrology 186:191–228.

Macpherson, G. L., and M. K. Schulmeister. 1994. Source(s), fate, and residence time of nitrate in ground-water: a comparison of carbonate and alluvial aquifers (Year 2). Contributions of the Kansas Water Resources Research Institute, Manhattan, Kansas, USA. 312:1–81.

Madden, J. L. 1973. An emerging agricultural economy: Kansas, 1860–1880. Kansas Historical Quarterly 39:101–114.

Malin, J. C. 1942. An introduction to the history of the bluestem-pasture region of Kansas. Kansas Historical Quarterly 11:3–28.

Mandel, R. D. 1994. Holocene landscape evolution in the Pawnee River valley, southwestern Kansas. Kansas Geological Survey Bulletin 236. Lawrence, Kansas, USA.

Marple, L. 1975. An annotated checklist of plants on the Konza Prairie Research Natural Area. NSF undergraduate research paper. Division of Biology, Kansas State University, Manhattan, Kansas, USA.

Martin, C. E., F. S. Harris, and F. J. Norman. 1991. Ecophysiological responses of C_3 forbs and C_4 grasses to drought and rain on a tallgrass prairie in northeastern Kansas. Botanical Gazette 152:257–262.

Matthews, W. J. 1988. North American prairie streams as systems for ecological study. Journal of the North American Benthological Society 7:387–409.

McArthur, J. V., M. E. Gurtz, C. M. Tate, and F. S. Gillian. 1985. The interaction of biological and hydrological phenomena that mediate the qualities of water draining native tallgrass prairie on the Konza Prairie Research Natural Area. Pages 478–482 in

Perspectives on nonpoint source pollution. United States Environmental Protection Agency 440/5–85–001, Washington, DC, USA.
McArthur, J. V., and G. R. Marzolf. 1986. Interactions of the bacterial assemblages of a prairie stream with dissolved organic carbon from riparian vegetation. Hydrobiologia 134:193–199.
McArthur, J. V., G. R. Marzolf, and J. E. Urban. 1985. Response of bacteria isolated from a pristine prairie stream to concentration and source of soluble organic carbon. Applied and Environmental Microbiology 49:238–241.
McGill, W. B., K. R. Cannon, J. A. Robertson, and F. D. Cook. 1986. Dynamics of soil microbial biomass and water-soluble organic C in Breton L after 50 years of cropping two rotations. Canadian Journal of Soil Science 66:1–19.
McGill, W. B., H. B. W. Hunt, R. G. Woodmansee, and J. O. Reuss. 1981. PHOENIX, a model of the dynamics of carbon and nitrogen in grassland soils. Ecological Bulletin (Stockholm) 33:49–115.
McGregor, R. L. 1955. Taxonomy and ecology of Kansas Hepaticae. University of Kansas Science Bulletin 37:55–141.
McGregor, R. L. 1968. A C-14 date for the Muscotah Marsh. Transactions of the Kansas Academy of Science 71:85–86.
McLaren, B. E., and R. O. Peterson. 1994. Wolves, moose and tree rings on Isle Royale. Science 266:1555–1558.
McMillan, B. R., and D. W. Kaufman. 1994. Differences in use of interspersed woodland and grassland by small mammals in northeastern Kansas. Prairie Naturalist 26:107–116.
McMillan, B. R., D. W. Kaufman, G. A. Kaufman, and R. S. Matlack. 1997. Mammals of the Konza Prairie Research Natural Area: new observations and an updated species list. Prairie Naturalist. In press.
McMurphy, W. E., and K. L. Anderson. 1965. Burning Flint Hills range. Journal of Range Management 18:265–269.
McNaughton, S. J. 1979. Grazing as an optimization process: grass-ungulate relationships in the Serengeti. American Naturalist 113:691–703.
McNaughton, S. J. 1983a. Compensatory plant growth as a response to herbivory. Oikos 40:329–336.
McNaughton, S. J. 1983b. Serengeti grassland ecology: the role of composite environmental factors and contingency in community organization. Ecological Monographs 53:291–320.
McNaughton, S. J. 1985. Ecology of a grazing ecosystem: the Serengeti. Ecological Monographs 55:259–294.
McNaughton, S. J., M. B. Coughenour, and L. L. Wallace. 1982. Interactive processes in grassland ecosystems. Pages 167–193 in J. R. Estes, R. J. Tyrl, and J. N. Brunken, editors, Grasses and grasslands: systematics and ecology. University of Oklahoma Press, Norman, Oklahoma, USA.
Melillo, J. M., T. V. Callaghan, F. I. Woodward, E. Salata, and S. K. Sinha. 1990. Effects on ecosystems. Pages 286–308 in J. T. Houghton, G. J. Jenkins, and J. J. Ephraums, editors, Climate change: the IPPC assessment. Cambridge University Press, New York, New York, USA.
Mengel, R. M. 1970. The North American central plains as an isolating agent in bird speciation. Pages 279–340 in W. Dort, Jr., and J. K. Jones, Jr., editors, Pleistocene and recent environments of the central Great Plains. University of Kansas, Department of Geology Publication 3, Lawrence, Kansas, USA.
Merrill, G. L. 1991a. Bryophytes of Konza Prairie Research Natural Area, Kansas. Bryologist 94:383–391.

Merrill, G. L. 1991b. New records for Kansas mosses. Transactions of the Kansas Academy of Science 92:70–78.

Merrill, G. L. 1991c. New records for Kansas mosses, II. Transactions of the Kansas Academy of Science 94:22–29.

Merrill, G. L. 1991d. New records for Kansas mosses, III. Evansia 8:25–31.

Merrill, G. L., and S. L. Timme. 1991. Bryophytes of the Kansas Ecological Reserves. Pages 46–51 in W. D. Kettle and D. O. Whittemore, editors, Ecology and hydrogeology of the Kansas Ecological Reserves and the Baker University Wetlands. Kansas Geological Survey, Open-file Report 91–31. Lawrence, Kansas, USA.

Meyer, J., T. Crocker, D. D'Angelo, W. Dodds, S. Findlay, M. Oswood, D. Repert, and D. Toetz. 1993. Stream research in the long-term ecological research network. Long-term Ecological Research Network Office, Long-term Ecological Research Publication No. 15. Seattle, Washington, USA.

Milchunas, D. G., and W. K. Lauenroth. 1993. Quantitative effects of grazing on vegetation and soils over a global range of environments. Ecological Monographs 63:327–366.

Milchunas D. G., O. E. Sala, and W. K. Lauenroth. 1988. A generalized model of the effects of grazing by large herbivores on grassland community structure. American Naturalist 132:87–106.

Miller, K. B., and R. R. West. 1993. Reevaluation of Wolfcampian cyclotherms in northeastern Kansas: significance of subaerial exposure and flooding surfaces. Pages 1–26 in Current research on Kansas geology. Kansas Geological Survey Bulletin 235. Lawrence, Kansas, USA.

Mitchell, V. L. 1969. The regionalization of climate in montane areas. Ph.D. dissertation. University of Wisconsin, Madison, Wisconsin, USA.

Moen, J., and S. L. Collins. 1996. Trophic interactions and plant species richness along a productivity gradient. Oikos 76:603–607.

Morris, D. R., D. A. Zuberer, and R. W. Weaver. 1985. Nitrogen fixation by intact grass-soil cores using ^{15}N and acetylene reduction. Soil Biology and Biochemistry 17:87–91.

Mosse, B., D. P. Stribley, and F. LeTacon. 1981. Ecology of mycorrhizae and mycorrhizal fungi. Advances in Microbial Ecology 5:137–210.

Mudge, M. R. 1955. Early Pleistocene geomorphic history of Wabaunsee, southeastern Riley, and southern Pottawatomie Counties, Kansas. Transactions of the Kansas Academy of Science 58:271–281.

Munger, J. W., and S. J. Eisenreich. 1983. Continental-scale variations in precipitation chemistry. Environmental Science and Technology 17:32A-42A.

Myneni, R. B., C. D. Keeling, C. J. Tucker, G. Asrar, and R. R. Nemani. 1997. Increased growth in the northern high latitudes from 1981 to 1991. Nature 386:698–702.

Nagel, H. G., T. Nightengale, and N. Dankert. 1991. Regal fritillary butterfly population estimation and natural history on Rowe Sanctuary, Nebraska. Prairie Naturalist 23:145–152.

Naiman, R. J., and J. R. Sedell. 1980. Relationships between metabolic parameters and stream order in Oregon. Canadian Journal of Fisheries and Aquatic Sciences 37:834–847.

Nellis, M. D., and J. M. Briggs. 1987. Micro-based Landsat TM data processing for tallgrass prairie monitoring in the Konza Prairie Research Natural Area, Kansas. Papers and Proceedings of the Applied Geography Conference 10:76–80.

Nellis, M. D., and J. M. Briggs. 1989. The impact of spatial scale on Konza Landscape Classification using textural analysis. Landscape Ecology 2:93–100.

Nie, D., H. He, M. B. Kirkham, and E. T. Kanemasu. 1992. Photosynthesis of a C_3 and a C_4 grass under elevated CO_2. Photosynthetica 26:189–198.

Norton, D. C., and D. P. Schmitt. 1978. Community analyses of plant-parasitic nematodes in the Kasow Prairie, Iowa. Journal of Nematology 10:171–176.

Noss, R. F., and A. Y. Cooperrider. 1994. Saving nature's legacy: protecting and restoring biodiversity. Island Press, Washington, DC, USA.

Odum. E. P. 1969. The strategy of ecosystem development. Science 164:262–270.

Ojima, D. S. 1987. The short-term and long-term effects of burning on tallgrass prairie ecosystem properties and dynamics. Ph.D. dissertation. Colorado State University, Fort Collins, Colorado, USA.

Ojima, D. S., B. O. M. Dirks, E. P. Glenn, C. E. Owensby, and J. O. Scurlock. 1993a. Assessment of C budget for grasslands and drylands of the world. Water, Air and Soil Pollution 70:95–109.

Ojima, D. S., W. J. Parton, D. S. Schimel, and C. E. Owensby. 1990. Simulated impacts of annual burning on prairie ecosystems. Pages 118–132 in S. L. Collins and L. L. Wallace, editors, Fire in North American tallgrass prairies. University of Oklahoma Press, Norman, Oklahoma, USA.

Ojima, D. S., W. J. Parton, D. S. Schimel, J. M. O. Scurlock, and T. G. F. Kittel. 1993b. Modeling the effects of climatic and CO_2 changes on grassland storage of soil C. Water, Air and Soil Pollution 70:643–657.

Ojima, D. S., D. S. Schimel, W. J. Parton, and C. E. Owensby. 1994. Long- and short-term effects of fire on nitrogen cycling in tallgrass prairie. Biogeochemistry 24:67–84.

Oksanen, L., S. D. Fretwell, J. Arruda, and P. Niemela. 1981. Exploitation ecosystems in gradients of primary productivity. American Naturalist 118:240–261.

Old, S. M. 1969. Microclimate, fire, and plant production in an Illinois prairie. Ecological Monographs 39:355–384.

O'Lear, H. 1996. Effects of altered soil water on litter decomposition and microarthropod composition. M.S. thesis. Kansas State University, Manhattan, Kansas, USA.

O'Lear, H. A., T. R. Seastedt, J. M. Briggs, J. M. Blair, and R. A. Ramundo. 1996. Fire and topographic effects on decomposition rates and nitrogen dynamics of buried wood in tallgrass prairie. Soil Biology and Biochemistry 28:323–329.

Olff, H., and J. P. Bakker. 1991. Long-term dynamics of standing crop and species composition after cessation of fertilizer application to mown grassland. Journal of Applied Ecology 28:1040–1052.

Orchard, V. A., and F. J. Cook. 1983. Relationships between soil respiration and soil moisture. Soil Biology Biochemistry 15:447–453.

Orr, C. C. 1965. Nematodes in native prairie soil of Kansas and the plants with which they are associated. Ph.D. dissertation. Kansas State University, Manhattan, Kansas, USA.

Ostfeld, R. S. 1988. Fluctuations and constancy in populations of small rodents. American Naturalist 131:445–452.

Overland, J. E., and R. W. Prisendorfer. 1982. A significance test for principal components applied to a cyclone climatology. Monthly Weather Review 110:1–4.

Owensby, C. E., and K. L. Anderson. 1967. Yield responses to time of burning in the Kansas Flint Hills. Journal of Range Management 20:12–16.

Owensby, C. E., and K. L. Anderson. 1969. Effect of clipping date on loamy upland bluestem range. Journal of Range Management 22:351–354.

Owensby, C. E., L. M. Auen, and P. I. Coyne. 1993a. Biomass production in a nitrogen fertilized tallgrass prairie ecosystem exposed to ambient and elevated levels of CO_2. Plant and Soil 165:105–113.

Owensby, C. E., P. I. Coyne, and L. M. Auen. 1993b. Nitrogen and phosphorus dynamics of a tallgrass prairie ecosystem exposed to elevated carbon dioxide. Plant, Cell and Environment 16:843–850.

Owensby, C. E., P. I. Coyne, J. M. Ham, L. A. Auen, and A. K. Knapp. 1993c. Biomass production in a tallgrass prairie ecosystem exposed to ambient and elevated CO_2. Ecological Applications 3:644–653.

Owensby, C. E., J. M. Ham, A. K. Knapp, D. Bremer, and L. M. Auen. 1997. Water vapor fluxes and their impact under elevated CO_2 in a C_4 tallgrass prairie. Global Change Biology 3:189–195.

Owensby, C. E., J. M. Ham, A. K. Knapp, C. W. Rice, P. I. Coyne, and L. M. Auen. 1996. Ecosystem level responses of tallgrass prairie to elevated CO_2. Pages 175–193 in H. A. Mooney and G. W. Koch, editors, Carbon dioxide and terrestrial ecosystems. Academic Press, New York, New York, USA.

Owensby, C. E., R. M. Hyde, and K. L. Anderson. 1970. Effects of clipping and supplemental nitrogen and water on loamy upland bluestem range. Journal of Range Management 23:341–346.

Packham, R. F. 1964. Studies of organic color in natural water. Proceedings of the Society of Water Treatment Examination 13:316–334.

Paine, R. T. 1969. A note on trophic complexity and community stability. American Naturalist 103:91–93.

Parker, L. W., P. F. Santos, J. Phillips, and W. G. Whitford. 1984. Carbon and nitrogen dynamics during decomposition of litter and roots of a Chihuahuan desert annual, *Lepidium lasiocarpum*. Ecological Monographs 54:339–360.

Parrish, J. A. D., and F. A. Bazzaz. 1976. Niche separation in roots of successional plants. Ecology 57:1281–1288.

Parrish, J. A. D., and F. A. Bazzaz. 1979. Difference in pollination niche relationships in early and late successional plant communities. Ecology 60:597–610.

Parton, W. J., D. S. Ojima, C. V. Cole, and D. S. Schimel. 1994. A general model for soil organic matter dynamics: sensitivity to litter chemistry, texture, and management. Pages 147–167 in Soil Science Society of America, Special Publication 39. Madison, Wisconsin, USA.

Parton, W. J., and P. G. Risser. 1979. Simulated impact of management practices upon the tallgrass prairie. Pages 135–155 in N. R. French, editor, Perspectives in grassland ecology: results and applications of the US/IBP Grassland Biome Study. Springer-Verlag, New York, New York, USA.

Parton, W. J., D. S. Schimel, C. V. Cole, and D. S. Ojima. 1987. Analysis of factors controlling soil organic matter levels in Great Plains grasslands. Soil Science Society of America Journal 51:1173–1179.

Parton, W. J., M. D. Scurolci, D. S. Ojima, T. G. Gilmanor, R. J. Scholos, D. S. Schimel, T. Kirehner, J. C. Menaut, T. Seastedt, L. G. Moya, A. Kamnalrat, and J. I. Kinyamario. 1993. Observation and modeling of biomass and soil organic matter dynamics for the grassland biome worldwide. Global Biogeochemical Cycles 7:785–809.

Pastor, J., B. Dewey, R. J. Naiman, P. F. McInnes, and Y. Cohen. 1993. Moose browsing and soil fertility in the boreal forests of Isle Royale National Park. Ecology 74:467–480.

Pastor J., and W. M. Post. 1993. Linear regressions do not predict the transient responses of eastern North American forests to CO_2-induced climate change. Climate Change 23:111–119.

Pastor J., M. A. Stillwell, and D. Tilman. 1987. Little bluestem litter dynamics in Minnesota old fields. Oecologia 72:327–330.

Paul, E. A., F. E. Clark, and V. O. Biederbeck. 1979. Microorganisms. Pages 87–96 in R.

T. Coupland, editor, Grassland ecosystems of the world: analysis of grasslands and their uses. Cambridge University Press, Cambridge, UK.

Paul, E. A., W. R. Horwath, D. Harris, R. Follett, S. Leavitt, B. A. Kimball, and K. Pregitzer. 1995. Establishing the pool sizes and fluxes of CO_2 emissions from soil organic matter turnover. Pages 297–308 in R. Lal, J. Kimble, E. Levine, and B. A. Stewart, editors, Soils and global change. CRC Lewis Publishers. Boca Raton, Florida, USA.

Peet, M., R. Anderson, and M. S. Adams. 1975. Effect of fire on big bluestem production. American Midland Naturalist 94:15–26.

Peet, R. K., D. C. Glenn-Lewin, and J. Walker-Wolf. 1983. Prediction of man's impact on plant species diversity. Pages 41–54 in W. Holzner, M. J. A. Werger, and I. Ikusima, editors, Man's impact on vegetation. Junk Publishing, The Hague, Netherlands.

Penfound, W. T. 1964. The relation of grazing to plant succession in the tall grass prairie. Journal of Range Management 17:256–260.

Peterson, N. J. 1983. The effects of fire, litter, and ash on flowering in *Andropogon gerardii*. Pages 21–24 in R. Brewer, editor, Proceedings of the Eighth North American Prairie Conference, Western Michigan University. Western Michigan University, Kalamazoo, Michigan, USA.

Peterson, R. C., and K. W. Cummins. 1974. Leaf processing in a woodland stream. Freshwater Biology 4:343–368.

Peterson, S. K., G. A. Kaufman, and D. W. Kaufman. 1985. Habitat selection by small mammals of the tall-grass prairie: experimental patch choice. Prairie Naturalist 17:65–70.

Pfeiffer, K. E., and D. C. Hartnett. 1995. Bison selectivity and grazing responses of little bluestem in tallgrass prairie. Journal of Range Management 48:26–31.

Phillips, P. 1936. The distribution of rodents in overgrazed and normal grasslands in Oklahoma. Ecology 17:673–679.

Pickett, S. T. A. 1980. Non-equilibrium coexistence in plants. Bulletin of the Torrey Botanical Club 107:238–248.

Pickett, S. T. A., and P. S. White, editors. 1985. The ecology of natural disturbance and patch dynamics. Academic Press, New York, New York, USA.

Pielke, R. A., and R. Avissar. 1990. Influence of landscape structure on local and regional climate. Landscape Ecology 4:133–155.

Pimm, S. L. 1991. The balance of nature? University of Chicago Press, Chicago, USA.

Platt, W. J. 1975. The colonization and formation of equilibrium plant species associations on badger disturbances in a tall-grass prairie. Ecological Monographs 45:285–305.

Platt, W. J., G. R. Hill, and S. Clark. 1974. Seed production in a prairie legume (*Astragalus canadensis* L.): interactions between pollination, pre-dispersal seed predation, and plant density. Oecologia 17:55–63.

Platt, W. J., and I. M. Weis. 1977. Resource partitioning and competition within a guild of fugitive prairie plants. American Naturalist 111:479–513.

Polis, G. A., and D. R. Strong. 1996. Food web complexity and community dynamics. American Naturalist 147:813–846.

Polunin, N. 1960. Introduction to plant geography and some related sciences. McGraw-Hill, New York, New York, USA.

Pomes, M. L. 1995. A study of aquatic humic substances and hydrogeology in a prairie watershed: Use of fulvic and humic acids as tracers of ground-water flow. Ph.D. dissertation. University of Kansas, Lawrence, Kansas, USA.

Pomes, M. L., and E. M. Thurman. 1991. Variation of dissolved-organic-carbon concentrations in soil waters from a tallgrass prairie and gallery-forest watershed, northeast Kansas. EOS, Transactions, American Geophysical Union 72:205–206.

Power, M. E. 1992. Top-down and bottom-up forces in food webs: do plants have primacy? Ecology 73:733–746.

Price, P. W., N. Cobb, T. P. Craig, G. W. Fernandes, J. K. Itami, S. Mopper, and R. W. Preszler. 1989. Insect herbivore population dynamics on trees and shrubs: new approaches relevant to latent and eruptive species and life table development. Pages 1–38 in E. A. Bernays, editor, Insect-plant interactions. Volume 2. CRC Press, Boca Raton, Florida, USA.

Pulliam, H. R., and G. S. Mills. 1977. The use of space by wintering sparrows. Ecology 58:1393–1399.

Quisenberry, S. C., and B. Purcell. 1993. New record for the American burying beetle (*Nicrophorus americanus*) in Bryan County, OK. Proceedings of the Oklahoma Academy of Science 73:77.

Rabinowitz, D. 1978. Abundance and diaspore weight in rare and common prairie grasses. Oecologia 37:213–219.

Rabinowitz, D., and J. K. Rapp. 1980. Seed rain in a North American tallgrass prairie. Journal of Applied Ecology 17:793–802.

Raison, R. J. 1979. Modifications of the soil environment by vegetation fires, with particular reference to nitrogen transformations: a review. Plant and Soil 51:73–108.

Ramundo, R. A., and T. R. Seastedt. 1990. Site-specific underestimation of wetfall NH_4^+ using NADP data. Atmospheric Environment 24A:3093–3095.

Ramundo, R. A., C. M. Tate, and T. R. Seastedt. 1992. Effects of tallgrass prairie vegetation on the concentration and seasonality of nitrate-nitrogen in soil water and streams. Pages 9–12 in D. D. Smith and C. A. Jacobs, editors, Proceedings of the Twelfth Midwest Prairie Conference, University of Northern Iowa. University of Northern Iowa, Iowa, USA.

Rapp, J. K., and D. Rabinowitz. 1985. Colonization and establishment of Missouri prairie plants on artificial soil disturbances. I. Dynamics of forb and graminoid seedlings and shoots. American Journal of Botany 72:1618–1628.

Ratcliffe, B. C., and M. L. Jameson. 1992. New Nebraska occurrences of the endangered American burying beetle (Coleoptera: Silphidae). Coleopterists' Bulletin 46:421–425.

Raunkiaer, C. 1934. The life forms of plants and statistical plant geography. Clarendon Press, Oxford, UK.

Read, D. J. 1984. The structure and function of the vegetative mycelium of mycorrhizal roots. Pages 215–240 in D. H. Jennings and A. D. M. Rayner, editors, The ecology and physiology of fungal mycelium. Cambridge University Press, Cambridge, UK.

Redmann, R. E. 1978. Plant and soil water potentials following fire in a northern mixed grassland. Journal of Range Management 31:443–445.

Redmann, R. E. 1985. Adaptation of grasses to water stress: leaf rolling and stomate distribution. Annals of the Missouri Botanical Garden 72:833–845.

Reed, P. G. 1988. National list of plant species that occur in wetlands: central plains (region 5). U.S. Fish and Wildlife Service Biological Report No. 88. Washington, DC, USA.

Reice, S. R. 1994. Nonequilibrium determinants of biological community structure. American Scientist 82:424–435.

Reichman, O. J., J. Benedix, and T. R. Seastedt. 1993. Distinct animal-generated edge effects in a tallgrass prairie community. Ecology 74:1281–1285

Reynolds, T. D. 1980. Effects of different land management practices on small mammal populations. Journal of Mammalogy 61:558–561.

Reynolds, T. D., and C. H. Trost. 1980. The response of native vertebrate populations to created wheatgrass planting and grazing by sheep. Journal of Range Management 33:122–125.

Rice, C. W., and F. O. Garcia. 1994. Biologically active pools of carbon and nitrogen in tallgrass prairie soil. Pages 201–208 in J. W. Doran, D. C. Coleman, D. F. Bezdicek, and B. A. Stewart, editors, Defining soil quality for a sustainable environment. Special Publication No. 35. Soil Science Society of America, Madison, Wisconsin, USA.

Rice, C. W., F. O. Garcia, C. O. Hampton, and C. E. Owensby. 1994. Microbial response in tallgrass prairie to elevated CO_2. Plant and Soil 165:67–74.

Rice, E. L., and R. L. Parenti. 1978. Causes of decreases in productivity in undisturbed tallgrass prairie. American Journal of Botany 65:1091–1097.

Risser, P. G. 1988. Abiotic controls on primary productivity and nutrient cycles in North American grasslands. Pages 116–129 in L. R. Pomeroy and J. J. Alberts, editors, Concepts of ecosystem ecology. A comparative view. Springer-Verlag, New York, New York, USA.

Risser, P. G., C. E. Birney, H. D. Blocker, S. W. May, W. J. Parton, and J. A. Wiens. 1981. The true prairie ecosystem. US/IBP Synthesis Series 16. Hutchinson Ross Publishing, Stroudsburg, Pennsylvania, USA.

Risser, P. G., J. R. Karr, and R. T. T. Forman. 1984. Landscape ecology: directions and approaches. Illinois Natural History Survey Special Publication No. 2. Champaign, Illinois, USA.

Risser, P., and W. J., Parton. 1982. Ecological analysis of a tallgrass prairie: nitrogen cycle. Ecology 63:1342–1351.

Robbins, C. S., J. R. Sauer, R. S. Greenberg, and S. Droege. 1989. Population declines in North American birds that migrate to the neotropics. Proceedings of the National Academy of Science 86:7658–7662.

Rodriguez, M. A., and A. Gomez-Sal. 1994. Stability may decrease with diversity in grassland communities: empirical evidence from the 1986 Cantabrian Mountains (Spain) drought. Oikos 71:177–180.

Rose, J. K. 1936. Intercorrelation between climate variables in the corn belt. Monthly Weather Review 64:76–82.

Rosensweig, M. L. 1968. Net primary productivity of terrestrial communities: predictions from climatological data. American Naturalist 102:40–45.

Ross, K. L. 1995. Geomorphology of the N4D watershed, Konza Prairie Research Natural Area, Riley and Geary Counties, Kansas. M.S. thesis. Kansas State University, Manhattan, Kansas, USA.

Rousseau, D. D., and G. Kukla. 1994. Late Pleistocene climate record in the Eustis loess section, Nebraska, based on land snail assemblages and magnetic susceptibility. Quaternary Research 42:176–187.

Rusch, G. M., and M. Oesterheld. 1997. Relationship between productivity and species and functional group diversity in grazed and non-grazed Pampas grassland. Oikos 78:519–526.

Rustiati, E. L., and D. W. Kaufman. 1993. Effect of prairie-fire ash on food choice by deer mice and hispid cotton rats. Prairie Naturalist 25:305–308.

Ryder, R. A. 1980. Effects of grazing on bird habitats. Pages 51–66 in R. M. DeGraff, editor, Proceedings of the Workshop on Management of Western Forests and Grasslands for Nongame Birds. United States Forest Service, General Technical Report INT-86. Washington, DC, USA.

Sala, O. E., W. J. Parton, L. A. Joyce, and W. K. Lauenroth. 1988. Primary production of the central grassland region of the United States. Ecology 69:40–45.

Samson, F., and F. Knopf. 1994. Prairie conservation in North America. BioScience 44:418–421.

Samson, F. B., and F. L. Knopf. 1996. Prairie conservation: preserving North America's most endangered ecosystem. Island Press, Washington, DC, USA.

Sauer, C. O. 1950. Grassland climax, fire and man. Journal of Range Management 3:16–21.

Schartz, R. L., and J. L. Zimmerman. 1971. The time and energy budget of the male dickcissel (*Spiza americana*). Condor 73:65–76.

Schimel, D. S., T. G. F. Kittel, A. K. Knapp, T. R. Seastedt, W. J. Parton, and V. B. Brown. 1991. Physiological interactions along resource gradients in a tallgrass prairie. Ecology 72:672–684.

Schimel, D. S., W. J. Parton, F. J. Adamsen, R. G. Woodmansee, R. L. Sneft, and M. A. Stillwell. 1986. The role of cattle in the volatile loss of nitrogen from a shortgrass steppe. Biogeochemistry 2:39–52.

Schimel, D. S., W. J. Parton, T. G. F. Kittel, D. S. Ojima, and C. V. Cole. 1990. Grassland biogeochemistry: links to atmospheric processes. Climate Change 17:13–25.

Schimel, D. S., VEMAP Participants, and B. H. Braswell. 1997. Continental scale variability in ecosystem processes: models, data, and the role of disturbance. Ecological Monographs 67:251–271.

Schlesinger, W. H., J. F. Reynolds, G. L. Cunningham, L. R. Huenneke, W. H. Jarrell, R. A. Virginia, and W. G. Whitford. 1990. Biological feedbacks in global desertification. Science 247:1043–1048.

Schmitt, D. P., and D. C. Norton. 1972. Relationships of plant parasitic nematodes to sites in native Iowa prairies. Journal of Nematology 4:200–206.

Schmitz, O. J. 1993. Trophic exploitation in grassland food chains: simple models and a field experiment. Oecologia 93:327–335.

Scott, J. A., N. R. French, and J. W. Leetham. 1979. Grassland biomass trophic pyramids. Pages 89–105 in N. R. French, editor, Perspectives in grassland ecology. Springer-Verlag, New York, New York, USA.

Seastedt, T. R. 1984a. Belowground macroarthropods of annually burned and unburned tallgrass prairie. American Midland Naturalist 111:405–408.

Seastedt, T. R. 1984b. Microarthropods of burned and unburned tallgrass prairie. Journal of the Kansas Entomological Society 57:468–476.

Seastedt, T. R. 1985a. Canopy interception of nitrogen in bulk precipitation by annually burned and unburned tallgrass prairie. Oecologia 66:88–92.

Seastedt, T. R. 1985b. Maximization of primary and secondary productivity by grazers. American Naturalist 126:559–564.

Seastedt, T. R. 1988. Mass, nitrogen and phosphorus dynamics in foliage and root detritus of tallgrass prairie. Ecology 69:59–65.

Seastedt, T. R. 1995. Soil systems and nutrient cycles on the North American prairie. Pages 157–174 in A. Joern and K. H. Keeler, editors, The changing prairie. Oxford University Press, Oxford, UK.

Seastedt, T. R., and J. M. Briggs. 1991. Long-term ecological questions and considerations for taking long-term measurements: lessons from the LTER and FIFE programs on tallgrass prairie. Pages 153–172 in P. G. Risser, editor, Long-term ecological research: An international perspective. Wiley, New York, New York, USA.

Seastedt, T. R., J. M. Briggs, and D. J. Gibson. 1991. Controls of nitrogen limitation in tallgrass prairie. Oecologia 87:72–79.

Seastedt, T. R., C. C. Coxwell, D. S. Ojima, and W. J. Parton. 1994. Controls and plant and soil carbon in a semihumid temperate grassland. Ecological Applications 4:344–353.

Seastedt, T. R., and D. A. Crossley, Jr. 1984. The influence of arthropods on ecosystems. BioScience 34:157–161.

Seastedt, T. R., and D. C. Hayes. 1988. Factors influencing nitrogen concentrations in soil water in a North American tallgrass prairie. Soil Biology and Biochemistry 20:725–729.

Seastedt, T. R., D. C. Hayes, and N. J. Petersen. 1986. Effects of vegetation, burning, and mowing on soil macroarthropods of tallgrass prairie. Pages 99–102 in G. K. Clambey and R. H. Pemble, editors, Proceedings of the Ninth North American Prairie Conference. Tri-College Press, Fargo, North Dakota, USA.

Seastedt, T. R., S. W. James, and T. C. Todd. 1988a. Interactions among soil invertebrates, microbes, and plant growth in the tallgrass prairie. Agriculture, Ecosystems and Environment 24:219–228.

Seastedt, T. R., and A. K. Knapp. 1993. Consequences of non-equilibrium resource availability across multiple time scales: the transient maxima hypothesis. American Naturalist 141:621–633.

Seastedt, T. R., W. J. Parton, and D. S. Ojima. 1992. Mass loss and nitrogen dynamics of decaying litter of grasslands: the apparent low nitrogen immobilization potential of root detritus. Canadian Journal of Botany 70:384–391.

Seastedt, T. R., and R. A. Ramundo. 1990. The influence of fire on belowground processes of tallgrass prairies. Pages 99–117 in S. L. Collins and L. L. Wallace, editors, Fire in North American tallgrass prairies. University of Oklahoma Press, Norman, Oklahoma, USA.

Seastedt, T. R., R. A. Ramundo, and D. C. Hayes. 1988b. Maximization of densities of soil animals by foliage herbivory: graphical and conceptual models and empirical evidence. Oikos 51:243–248.

Seastedt, T. R., R. A. Ramundo, and D. C. Hayes. 1989. Silica, nitrogen, and phosphorus dynamics of tallgrass prairie. Pages 205–209 in T. Bragg and J. Stubbendieck, editors, Proceedings of the Eleventh North American Prairie Conference, University of Nebraska. University of Nebraska Printing, Lincoln, Nebraska, USA.

Seastedt, T. R., T. C. Todd, and S. W. James. 1987. Experimental manipulations of arthropod, nematode, and earthworm communities in a North American tallgrass prairie. Pedobiologia 30:9–17.

Sellers, P. J., F. G. Hall, G. Asrar, D. E. Strebel, and R. E. Murphy. 1988. The First ISLSCP Field Experiment (FIFE). Bulletin of the American Meteorological Society 69:22–27.

Senft, R. L., M. B. Coughenour, D. W. Bailey, L. R. Rittenhouse, O. E. Sala, and D. E. Swift. 1987. Large herbivore foraging and ecological hierarchies. BioScience 37:789–799.

Sharrow, S. H., and H. A. Wright. 1977. Effect of fire, ash, and litter on soil nitrate, temperature, moisture, and tobosagrass production in the rolling plains. Journal of Range Management 30:266–270.

Silvertown, J., M. E. Dodd, K. McConway, J. Potts, and M. Crawley. 1994. Rainfall, biomass variation, and community composition in the Park Grass Experiment. Ecology 75:2430–2437.

Sims, P. L., and J. S. Singh. 1978. The structure and function of ten western North American grasslands. III. Net primary production, turnover, and efficiencies of energy capture and water use. Journal of Ecology 66:573–597.

Smith, D. L. 1986. Leaf litter processing and associated invertebrate fauna in a tallgrass prairie stream. American Midland Naturalist 116:78–86.

Smith, G. N. 1991. Geomorphology and geomorphic history of the Konza Prairie Research Natural Area, Riley and Geary Counties, Kansas. M.S. thesis. Kansas State University, Manhattan, Kansas, USA.

Smith, R. M., P. C. Twiss, R. K. Krauss, and M. J. Brown. 1970. Dust deposition in relation to site, season, and climatic variables. Soil Science Society of America Proceedings 34:112–117.

Smith, S. D., and R. S. Nowak. 1990. Ecophysiology of plants in the intermountain lowlands. Pages 179–241 in C. B. Osmond, L. F. Pitelka, and G. M. Hidy, editors, Plant biology of the basin and range. Springer-Verlag, New York, New York, USA.

Smith, T. M., and H. H. Shugart. 1993. The transient response of terrestrial carbon storage to a perturbed climate. Nature 361:523–526.

Smolik, J. D. 1977. Effect of nematicide treatment on growth of range grasses in field and greenhouse studies. Pages 257–260 in J. K. Marshall, editor, Proceedings of Belowground Ecosystem Symposium, Colorado State University. Colorado State University Range Science Series No. 26. Fort Collins, Colorado, USA.

Smolik, J. D., and J. L. Dodd. 1983. Effect of water and nitrogen, and grazing on nematodes in a shortgrass prairie. Journal of Range Management 36:744–748.

Smolik, J. D., and J. L. Lewis. 1982. Effect of range condition on density and biomass of nematodes in a mixed prairie ecosystem. Journal of Range Management 35:657–663.

Soil Conservation Service. 1981. Land resource regions and major land resource areas of the United States. United States Department of Agriculture, Soil Conservation Service, Agricultural Handbook 296. United States Government Printing Office, Washington, DC, USA.

Soil Survey Labo atory Staff. 1996. Soil Survey Laboratory methods manual. United States Department of Agriculture, Natural Resources and Conservation Service, National Soil Survey Center, Soil Survey Investigations Report No. 42, Version 3. Lincoln, Nebraska, USA.

Soil Survey Staff. 1994. Keys to soil taxonomy. 6th edition. Pocahontas Press, Blacksburg, Virginia, USA.

Sotamayor, D. 1996. Microbial ecology and denitrification in tallgrass prairie and cultivated soils. Ph.D. dissertation. Kansas State University, Manhattan, Kansas, USA.

Sotomayor, D., and C. W. Rice. 1996. Denitrification beneath grassland and cultivated soils. Soil Science Society of America Journal 60:1822–1828.

Sparling, G. P. 1992. Ratio of microbial biomass to soil organic carbon as a sensitive indicator of changes in soil organic matter. Australian Journal of Soil Research 30: 195–207.

Spencer, J. S., J. K. Strickler, and W. J. Moyer. 1984. Kansas forest inventory. United States Department of Agriculture, North Central Forest Experiment Station, Saint Paul, Minnesota, USA.

Sprugel, D. G. 1991. Disturbance, equilibrium, and environmental variability: what is natural vegetation in a changing environment? Biological Conservation 58:1–18.

Stalter, R., D. W. Kincaid, and E. E. Lamont. 1991. Life forms of the flora at Hempstead Plains, New York, and a comparison with four other sites. Bulletin of the Torrey Botanical Club 118:191–194.

Stanton, N. L., M. Allen, and M. Campion. 1981. The effect of the pesticide carbofuran on soil organisms and root and shoot production in shortgrass prairie. Journal of Applied Ecology 18:417–431.

Steinauer, E. M. 1994. Effects of urine deposition on small-scale patch structure and vegetative patterns in tallgrass and sandhills prairies. Ph.D. dissertation. University of Oklahoma, Norman, Oklahoma, USA.

Steinauer, E. M., and S. L. Collins. 1995. Effects of urine deposition on small-scale patch structure in prairie vegetation. Ecology 76:1195–1205.

Steinauer, E. M., and S. L. Collins. 1996. Prairie ecology: the tallgrass prairie. Pages 39–52 in F. B. Samson and F. J. Knopf, editors, Prairie conservation: preserving North America's most endangered ecosystem. Island Press, Covelo, California, USA.

Stevenson, F. J. 1985. Geochemistry of soil humic substances. Pages 13–52 in G. R. Aiken, D. M. McKnight, and R. L. Wershaw, editors, Humic substances in soil, sediment, and water: geochemistry, isolation, and characterization. Wiley, New York, New York, USA.

Strahler, A. H. 1952. Hypsometric (area-altitude) analysis of erosional topography. Geological Society of America Bulletin 63:1117–1142.

Strahler, A. H. 1964. Quantitative geomorphology of drainage basin and channel networks. Pages 39–76 in V. T. Chow, editor, Handbook of applied hydrology. McGraw-Hill, New York, New York, USA.

Strahler, A. H., C. E. Woodcock, and J. A. Smith. 1986. On the nature of models in remote sensing. Remote Sensing of the Environment 20:121–139.

Strain, B. R., and J. D. Cure. 1985. Direct effects of increasing carbon dioxide on vegetation. United States Department of Energy, Carbon Dioxide Research Division Technical Report No. DOE/ER-0238. Washington, DC, USA.

Strauss, E. 1995. Protozoa-bacteria interactions in subsurface sediments and the subsequent effects on nitrification. M.S. thesis. Kansas State University, Manhattan, Kansas, USA.

Strong, D. R. 1992. Are trophic cascades all wet? Differentiation and donor-control in speciose systems. Ecology 73:747–754.

Strong D. R., J. H. Lawton, and T. R. S. Southwood. 1984. Insects on plants: community patterns and mechanisms. Harvard University Press, Cambridge, Massachusetts, USA.

Stuckey, R. L., and T. M. Barkley. 1993. Weeds. Pages 193–198 in Flora of North America Editorial Committee, editor, Flora of North America north of Mexico. Vol. 1, Introduction. Oxford University Press, New York, New York, USA.

Su, H., J. M. Briggs, A. K. Knapp, J. M. Blair, and J. R. Krummel. 1996. Detecting spatial and temporal patterns of aboveground production in a tallgrass prairie using remotely-sensed data. Proceedings of the 1996 International Geoscience and Remote Sensing Symposium 4:2361–2365.

Su, H., E. T. Kanemasu, M. D. Ranson, and S. Yang. 1990. Separability of soils in a tallgrass prairie using SPOT and DEM data. Remote Sensing of the Environment 33:157–163.

Sullivan, S. 1996. Towards a non-equlibrium ecology: perspectives from an arid land. Journal of Biogeography 23:1–5.

Svejcar, T. J. 1990. Response of *Andropogon gerardii* to fire in the tallgrass prairie. Pages 19–27 in S. L. Collins and L. L. Wallace, editors, Fire in North American tallgrass prairies. University of Oklahoma Press, Norman, Oklahoma, USA.

Svejcar, T. J., and J. A. Browning. 1988. Growth and gas exchange of *Andropogon gerardii* as influenced by burning. Journal of Range Management 41:239–244.

Swan, J. M. A. 1970. An example of some ordination problems by use of simulated vegetation data. Ecology 51:89–102.

Swengel, A. B. 1996. Effects of fire and hay management on abundance of prairie butterflies. Biological Conservation 76:73–85.

Swift, M. J., O. W. Heal, and J. M. Anderson. 1979. Decomposition in terrestrial ecosystems. Blackwell Scientific Publications, Oxford, UK.

Taitt, M. J., and C. J. Krebs. 1985. Population dynamics and cycles. Pages 567–620 in R. H. Tamarin, editor, Biology of New World *Microtus*. American Society of Mammalogists, Special Publication 8. Shippensburg, Pennsylvania, USA.

Tate, C. M. 1985. A study of temporal and spatial variation in nitrogen concentrations in a tallgrass prairie stream. Ph.D. dissertation. Kansas State University, Manhattan, Kansas, USA.

Tate, C. M. 1990. Patterns and controls of nitrogen in tallgrass prairie streams. Ecology 71:2007–2018.

Tate, C. M., and M. E. Gurtz. 1986. Comparison of mass loss, nutrients, and invertebrates associated with elm leaf litter decomposition in perennial and intermittent reaches of tallgrass prairie streams. Southwestern Naturalist 31:511–520.

Thompson, M. C., and C. Ely. 1989. Birds in Kansas. Volume 1. University of Kansas, Museum of Natural History Public Education Series 11:1–404.

Thompson, M. C., and C. Ely. 1992. Birds in Kansas, Volume 2. University of Kansas, Museum of Natural History Public Education Series 12:1–424.

Thorne, R. F. 1993. Phytogeography. Pages 132–153 in Flora of North America Editorial Committee, editor, Flora of North America north of Mexico. Vol. 1, Introduction. Oxford University Press, New York, New York, USA.

Tilman, D. 1982. Resource competition and community structure. Princeton University Press, Princeton, New Jersey, USA.

Tilman, D. 1987. Secondary succession and patterns of plant dominance along experimental nitrogen gradients. Ecological Monographs 57:189–214.

Tilman, D. 1988. Plant strategies and the dynamics and structure of plant communities. Princeton University Press, Princeton, New Jersey, USA.

Tilman, D., and J. A. Downing. 1994. Biodiversity and stability in grasslands. Nature 367:363–365.

Tilman D., J. Knops, D. Wedin, P. Reich, M. Ritchie, and E. Siemann. 1997. The influence of functional diversity and composition on ecosystem processes. Science 277:1300–1302.

Tilman, D., R. M. May, C. L. Lehman, and M. A. Nowak. 1994. Habitat destruction and the extinction debt. Nature 371:65–66.

Tilman, D., and D. Wedin. 1991a. Dynamics of nitrogen competition between successional grasses. Ecology 72:1038–1049.

Tilman, D., and D. Wedin. 1991b. Plant traits and resource reduction for five grasses growing on a nitrogen gradient. Ecology 72:685–700.

Tilman, D., D. Wedin, and J. Knops. 1996. Productivity and sustainability influenced by biodiversity in grassland ecosystems. Nature 379:718–720.

Tjepkema, J. D. 1975. Nitrogenase activity in the rhizosphere of *Panicum virgatum*. Soil Biology and Biochemistry 7:179–180.

Tjepkema, J. D., and R. H. Burris. 1976. Nitrogenase activity associated with some Wisconsin prairie grasses. Plant and Soil 45:81–94.

Todd, T. C. 1996. Effects of management practices on nematode community structure in tallgrass prairie. Applied Soil Ecology 3:235–246.

Todd, T. C., S. W. James, and T. R. Seastedt. 1992. Soil invertebrate and plant responses to mowing and carbofuran application in a North American tallgrass prairie. Plant Soil 144:117–124.

Towne, E. G. 1995. Influence of fire frequency and burning date on the proportion of reproductive tillers in big bluestem and Indian grass. Pages 75–78 in D. C. Hartnett, editor, Prairie biodiversity. Proceedings of the Fourteenth North American Prairie Conference, Kansas State University. Kansas State University, Manhattan, Kansas, USA.

Towne, E. G., and A. K. Knapp. 1996. Biomass and density responses in tallgrass prairie legumes to annual fire and topographic position. American Journal of Botany 83:175–179.

Towne, G., and C. Owensby. 1984. Long-term effects of annual burning at different dates in ungrazed Kansas tallgrass prairie. Journal of Range Management 37:392–397.

Transeau, E. N. 1935. The prairie peninsula. Ecology 16:423–437.

Trenberth, K. E., and D. A. Paolino. 1980. The Northern Hemisphere sea level pressure data set: trends, errors, and discontinuities. Monthly Weather Review 108:855–972.

Turner, C. L., J. M. Blair, R. J. Shartz, and J. C. Neel. 1997. Soil N and plant responses to fire, topography, and supplemental N in tallgrass prairie. Ecology 78:1832–1843.

Turner, C. L., and A. K. Knapp. 1996. Responses of a C_4 grass and three C_3 forbs to variation in nitrogen and light in tallgrass prairie. Ecology 77:1738–1749.

Turner, C. T., J. R. Kneisler, and A. K. Knapp. 1995. Comparative gas exchange and nitrogen responses of the dominant C_4 grass, *Andropogon gerardii*, and five C_3 forbs to fire and topographic position in tallgrass prairie during a wet year. International Journal of Plant Science 156:216–226.

Turner, C. L., T. R. Seastedt, and M. I. Dyer. 1993. Maximization of aboveground grassland production: the role of defoliation frequency, intensity, and history. Ecological Applications 3:175–186.

Turner, C. L., T. R. Seastedt, M. I. Dyer, T. G. F. Kittel, and D. S. Schimel. 1992. Effects of management and topography on the radiometric response of tallgrass prairie. Journal of Geophysical Research 97:18,855–18,866.

Turner, M. G. 1989. Landscape ecology: the effect of pattern on process. Annual Review of Ecology and Systematics 20:171–197.

Turner, M. G. 1990. Spatial and temporal analysis of landscape patterns. Landscape Ecology 4:21–30.

United States Geological Survey. 1996. Water resources data for Kansas, Water Year 1995. Lawrence, Kansas, USA.

Uno, G. E. 1989. Dynamics of plants in buffalo wallows. Pages 431–444 in J. H. Bock and Y. B. Linhart, editors, The evolutionary ecology of plants. Westview Press, Boulder, Colorado, USA.

Vannote, R. L., G. W. Minshall, K. W. Cummins, J. R. Sedell, and C. E. Cushing. 1980. The river continuum concept. Canadian Journal of Fisheries and Aquatic Sciences 37:130–137.

van Veen, J. A., J. N. Ladd, and M. J. Frissel. 1984. Modeling C and N turnover through the microbial biomass in soil. Plant and Soil 76:257–274.

van Veen, J. A., and E. A. Paul. 1981. Organic C dynamics in grassland soils. I. Background information and computer simulation. Canadian Journal of Soil Science 61:185–201.

Vinton, M. A., and S. L. Collins. 1996. Landscape heterogeneity gradients and habitat structure in native grasslands of the central Great Plains. Pages 3–19 in F. L. Knopf and F. B. Samson, editors, Ecology of Great Plains vertebrates and their habitats. Springer-Verlag, New York, New York, USA.

Vinton, M. A., and D. C. Hartnett. 1992. Effects of bison grazing on *Andropogon gerardii* and *Panicum virgatum* in burned and unburned tallgrass prairie. Oecologia 90:374–382.

Vinton, M. A., D. C. Hartnett, E. J. Finck, and J. M. Briggs. 1993. Interactive effects of fire, bison (*Bison bison*) grazing, and plant community composition in tallgrass prairie. American Midland Naturalist 129:10–18.

Vogl, R. J. 1974. Effects of fire on grasslands. Pages 139–194 in T. T. Kozlowski and C.

E. Ahlgren, editors, Fire and ecosystems. Academic Press, New York, New York, USA.
Voigt, J. W., and J. E. Weaver. 1951. Range condition classes of native midwestern pastures: an ecological analysis. Ecological Monographs 21:39–60.
Wallace, J. B. 1988. Aquatic invertebrate research. Pages 257–268 in W. T. Swank and D. A. Crossley, Jr., editors, Forest hydrology and ecology at Coweeta. Springer-Verlag, New York, New York, USA.
Wallace, L. L. 1990. Comparative photosynthetic responses of big bluestem to clipping versus grazing. Journal of Range Management 43:58–61.
Walter, D. E., and E. K. Ikonen. 1989. Species, guilds, and functional groups: taxonomy and behavior in nematophagous arthropods. Journal of Nematology 21:315–327.
Walter, H. 1985. Vegetation of the earth. 3rd edition. Springer-Verlag, New York, New York, USA.
Walter, H., and S. Breckel. 1985. Ecological systems of the geobiosphere. Springer-Verlag, Berlin, Germany.
Wardle, D. A., and D. Parkinson. 1990. Interactions between microclimatic variables and the soil microbial biomass. Biology and Fertility of Soils 9:273–280.
Warembourg, F. R., and E. A. Paul. 1977. Seasonal transfers of assimilated C-14 in grassland: plant production and turnover, soil and plant respiration. Soil Biology and Biochemistry 9:259–301.
Warren, S. D., C. J. Scifres, and P. D. Teel. 1987. Response of grassland arthropods to burning: a review. Agriculture, Ecosystems and Environment 19:105–130.
Warren, S. D., T. L. Thurow, W. H. Blackburn, and N. E. Garza. 1986. The influence of livestock trampling under intensive rotation grazing on soil hydrologic characteristics. Journal of Range Management 39:491–495.
Weakly, H. E. 1962. History of drought in Nebraska. Journal of Soil and Water Conservation 17:271–273.
Weaver, J. E. 1954. North American prairie. Johnsen Publishing, Lincoln, Nebraska, USA.
Weaver, J. E. 1958. Summary and interpretation of underground development in natural grassland communities. Ecological Monographs 28:55–78.
Weaver, J. E. 1968. Prairie plants and their environment. University of Nebraska Press, Lincoln, Nebraska, USA.
Weaver, J. E., and F. W. Albertson. 1944. Nature and degree of recovery of grasslands from the great drought of 1933 to 1940. Ecological Monographs 14:394–497.
Weaver, J. E., and W. W. Hansen. 1941. Native midwestern pastures: their origin, composition, and degeneration. University of Nebraska Conservation and Survey Division Bulletin 22:1–93.
Weaver, J. E., and N. W. Rowland. 1952. Effects of excessive natural mulch on development, yield, and structure of native grassland. Botanical Gazette 114:1–19.
Webb, W. L., W. K. Lauenroth, S. R. Szarek, and R. S. Kinerson. 1983. Primary production and abiotic controls in forests, grasslands, and desert ecosystems in the United States. Ecology 64:134–151.
Webb, W. L., S. Szarek, W. Lauenroth, R. Kinerson, and M. Smith. 1978. Primary productivity and water use on native forest, grassland, and desert ecosystems. Ecology 59:1239–1247.
Webster, J. R., M. E. Gurtz, J. J. Hains, J. L. Meyer, W. T. Swank, J. B. Waide, and J. B. Wallace. 1983. Stability of stream ecosystems. Pages 355–395 in J. R. Barnes and G. W. Minshall, editors, Stream ecology. Plenum Press, New York, New York, USA.
Wedin, D. A., and J. Pastor. 1993. Nitrogen mineralization dynamics in grass monocultures. Oecologia 96:186–192.

Wedin, D. A., and D. Tilman. 1990. Species effects on nitrogen cycling: a test with perennial grasses. Oecologia 84:433–441.

Wedin, D. A., and D. Tilman. 1993. Competition among grasses along a nitrogen gradient: initial conditions and mechanisms of competition. Ecological Monographs 63:199–229.

Wedin, D. A., and D. Tilman. 1996. Influence of nitrogen loading and species composition on the carbon balance of grasslands. Science 274:1720–1723.

Wehmueller, W. A. 1996. Genesis and morphology of soils on the Konza Prairie Research Natural Area, Riley and Geary Counties, Kansas. M.S. thesis. Kansas State University, Manhattan, Kansas, USA.

Wehmueller, W. A., M. D. Ransom, and W. D. Nettleton. 1994. Micromorphology of polygenetic soils in a small watershed, north central Kansas, U.S.A. Pages 247–255 in A. J. Ringrose-Voase and G. S. Humphreys, editors, Soil micromorphology: studies in management and genesis. Proceedings of the Ninth International Working Meeting on Soil, Developments in Soil Science 22. Elsevier Science, Amsterdam, Netherlands.

Weigel, J. R., C. M. Britton, and G. R. McPherson. 1990. Trampling effects from short-duration grazing on tobosagrass range. Journal of Range Management 43:92–95.

Wells, P. V. 1970a. Historical factors controlling vegetation patterns and floristic distributions in the central plains region of North America. Pages 211–221 in W. Dort, Jr., and J. K. Knox, Jr., editors, Pleistocene and recent environments of the central Great Plains. University of Kansas Department of Geology Special Publications No. 3. University of Kansas Press, Lawrence, Kansas, USA.

Wells, P. V. 1970b. Postglacial vegetational history of the Great Plains. Science 167:1574–1582.

Wells, P. V., and J. D. Stewart. 1987. Spruce charcoal, conifer macrofossils, and landsnail and small-vertebrate faunas in Wisconsinan sediments on the High Plains of Kansas. Pages 129–140 in W. C. Johnson, editor, Quaternary environments of Kansas. Kansas Geological Survey Guidebook Series 5. Lawrence, Kansas, USA.

Wendland, W. M. 1980. Holocene climatic reconstructions on the Prairie Peninsula. Pages 139–148 in D. C. Anderson and H. A. Semken, Jr., editors, The Cherokee excavations: Holocene ecology and human adaptations in northwestern Iowa. Academic Press, New York, New York, USA.

Wendland, W. M., and R. A. Bryson. 1981. Northern Hemisphere airstream regions. Monthly Weather Review 9:255–270.

Westoby, M., B. Walker, and I. Noy-Meir. 1989. Opportunistic management for rangelands not at equilibrium. Journal of Range Management 42:266–274.

Wetselaar, R., and G. D. Farquhar. 1980. Nitrogen losses from tops of plants. Pages 263–302 in N. C. Brady, editor, Advances in agronomy. Volume 33. Academic Press, New York, New York.

Whelan, R. J. 1995. The ecology of fire. Cambridge University Press, Cambridge, UK.

Whittaker, R. H. 1951. A criticism of the plant association and climatic climax concepts. Northwest Scientist 25:17–31.

Whittaker, R. H. 1953. A consideration of the climax theory: the climax as a population and pattern. Ecological Monographs 23:41–78.

Whittaker, R. H. 1975. Communities and ecosystems, 2nd edition. Macmillan, New York, New York, USA.

Whitehead, D. C. 1995. Grassland nitrogen. CAB International, Wallingford, UK.

Whitford, W. G., V. Meentemeyer, T. R. Seastedt, K. Cromack, Jr., D. A. Crossley, Jr., P. Santos, R. L. Todd, and J. B. Waide. 1981. Tests of the AET model of litter decomposition in deserts and a clear-cut forest. Ecology 62:275–277.

Whittemore, D. O. 1980. Geochemistry of natural waters in the Konza Prairie. Kansas Geological Survey Open-File Report 80-20. Lawrence, Kansas, USA.

Wiens, J. A. 1969. An approach to the study of ecological relationships among grassland birds. Ornithological Monographs 8:1–93.

Wiens, J. A. 1973. Pattern and process in grassland bird communities. Ecological Monographs 43:237–270.

Wiens, J. A. 1974. Climatic instability and the "ecological saturation" of bird communities in North American grasslands. Condor 76:385–400.

Wiens, J. A. 1997. Metapopulation dynamics and landscape ecology. Pages 43–62 in I. A. Hanski and M. E. Gilpin, editors, Metapopulation biology. Academic Press, London, UK.

Wiens, J. A., and M. I. Dyer. 1975. Rangeland avifaunas: their composition, energetics, and role in the ecosystem. Pages 146–182 in Proceedings of the Symposium on Management of Forest and Range Habitats for Nongame Birds. United States Forest Service, General Technical Report WO-1. Washington, DC, USA.

Wiens, J. A., and J. T. Rotenberry. 1979. Diet niche relationships among North American grassland and shrubsteppe birds. Oecologia 42:253–292.

Williams, M. H. 1978. The ordination of incidence data. Journal of Ecology 66:911–920.

Williams, R. E., B. W. Allred, R. M. De Nio, and H. E. Paulsen, Jr. 1968. Conservation, development and use of the world's rangelands. Journal of Range Management 21:355–360.

Wilson, S. D., and J. M. Shay. 1990. Competition, fire, and nutrients in a mixed-grass prairie. Ecology 71:1959–1967.

Wilson, S. D., and D. Tilman. 1991. Components of plant competition along an experimental gradient of nitrogen availability. Ecology 72:1050–1065.

Woodmansee, R. G. 1978. Additions and losses of nitrogen in grassland ecosystems. BioScience 28:448–453.

Wright, H. A., and A. W. Bailey. 1982. Fire ecology. Wiley, New York, New York, USA.

Wright, H. E., Jr. 1970. Vegetational history of the central plains. Pages 157–172 in W. Dort, Jr., and J. K. Knox, Jr., editors, Pleistocene and recent environments of the central Great Plains. University of Kansas Department of Geology Special Publications No. 3. University of Kansas Press, Lawrence, Kansas, USA.

Wu, J., and O. L. Loucks. 1995. From balance of nature to hierarchical patch dynamics: a paradigm shift in ecology. Quarterly Review of Biology 70:439–466.

Yeates, G. W., T. Bongers, R. G. M. de Goede, D. W. Freckman, and S. S. Georgieva. 1993. Feeding habits in soil nematode families and genera: an outline for soil ecologists. Journal of Nematology 25:315–331.

Zak, D. R., D. Tilman, R. P. Parmenter, C. W. Rice, F. M. Fisher, J. Vose, D. Milchunas, and C. W. Martin. 1994. Plant production and soil microorganisms in late-successional ecosystems: a continental-scale study. Ecology 75:2333–2347.

Zeller, D. E., editor. 1968. The stratigraphic successions in Kansas. Kansas Geological Survey Bulletin 189:1–81.

Zimmerman, J. L. 1971. The territory and its density-dependent effect in *Spiza americana*. Auk 88:591–612.

Zimmerman, J. L. 1983. Cowbird parasitism of dickcissels in different habitats and at different nest densities. Wilson Bulletin 95:7–22.

Zimmerman, J. L. 1984. Nest predation and its relationship to habitat and nest density in dickcissels. Condor 86:68–72.

Zimmerman, J. L. 1985. The birds of Konza Prairie Research Natural Area, Kansas. Prairie Naturalist 17:185–192.

Zimmerman, J. L. 1988. Breeding season habitat selection by the Henslow's sparrow (*Ammodramus henslowii*) in Kansas. Wilson Bulletin 100:85–94.

Zimmerman, J. L. 1992. Density-independent factors affecting the avian diversity of the tallgrass prairie community. Wilson Bulletin 104:85–94

Zimmerman, J. L. 1993. The birds of Konza: the avian ecology of the tallgrass prairie. University Press of Kansas, Lawrence, Kansas, USA.

Zimmerman, J. L. 1997. Avian community responses to fire, grazing, and drought in the tallgrass prairie. Pages 167–180 in F. L. Knopf and F. B. Samson, editors, Ecology and conservation of Great Plains vertebrates. Springer-Verlag, New York, New York, USA.

Zimmerman, J. L., and E. J. Finck. 1989. Philopatry and correlates of territorial fidelity in male dickcissels. North American Bird Bander 14:83–85.

Zimmerman, U. D., and C. L. Kucera. 1977. Effects of composition changes on productivity and biomass relationships in tallgrass prairie. American Midland Naturalist 97:465–469.

Index